D0078922

In Praise of *Designing SOCs with Configured Cores...*

Designing SOCs with Configured Cores is an essential reference for system-on-chip designers. This well-written book gives a practical introduction to three basic techniques of modern SOC design: use of optimized standard CPU and DSP processors cores, application-specific configuration of processor cores, and system-level design of SOCs using configured cores as the key building block. Readers will find it is often the first book they reach for in defining and designing their next chip.

Chris Rowen, President and CEO, Tensilica, Inc.

We're poised on the brink of a revolution in computing. Instead of fixed-architecture processors suited only to general-purpose computing or special-purpose digital signal processing tasks, we're moving to system-on-chip (SoC) devices containing multiple processor cores, each configured to perform specific tasks with extreme performance while consuming ultra-low power. The surf is up—and this book tells us how to ride the wave!

Clive "Max" Maxfield, President, TechBites Interactive and author of *The Design Warrior's Guide to FPGAs*.

Steve Leibson's book is a gentle introduction to the art of digital electronics design capturing an important moment in the creation of the complete system on a chip. Generously dotted with block diagrams and snippets of code using the vehicle furnished by Xtensa technology, the book takes the reader through the evolution that led to multiple cores and configurable engines.

Max Baron, Senior Analyst at The Microprocessor Report.

About the Author

Steve Leibson is Technology Evangelist, Strategic Marketing Manager at Tensilica, Inc. He has decades of experience as a hardware and software design engineer, engineering manager, and design consultant. His experience includes the development of large digital systems as a design engineer while working at Hewlett-Packard's Desktop Computer Division and at EDA workstation pioneer Cadnetix, and covering such topics as an award-winning technology Journalist and Editor-in-chief for EDN magazine and the Microprocessor Report. Since joining Tensilica, he has published numerous technical articles relating to chip-level system design in trade and engineering-society magazines published in countries around the world including the US, UK, France, Germany, Italy, Japan, South Korea, and China.

Designing SOCs with Configured Cores

The Morgan Kaufmann Series in Systems on Silicon
Series Editor: Wayne Wolf, Princeton University

The rapid growth of silicon technology and the demands of applications are increasingly forcing electronics designers to take a systems-oriented approach to design. This has lead to new challenges in design methodology, design automation, manufacture and test. The main challenges are to enhance designer productivity and to achieve correctness on the first pass. *The Morgan Kaufmann Series in Systems on Silicon* presents high-quality, peer-reviewed books authored by leading experts in the field who are uniquely qualified to address these issues.

The Designer's Guide to VHDL, Second Edition
Peter J. Ashenden

The System Designer's Guide to VHDL-AMS
Peter J. Ashenden, Gregory D. Peterson, and Darrell A Teegarden

Readings in Hardware/Software Co-Design
Edited by Giovanni De Micheli, Rolf Ernst, and Wayne Wolf

Modeling Embedded Systems and SoCs
Axel Jantsch

ASIC and FPGA Verification: A Guide to Component Modeling
Richard Munden

Multiprocessor Systems-on-Chips
Edited by Ahmed Amine Jerraya and Wayne Wolf

Comprehensive Functional Verification
Bruce Wile, John Goss, and Wolfgang Roesner

Customizable Embedded Processors: Design Technologies and Applications
Edited by Paolo Ienne and Rainer Leupers

Networks on Chips: Technology and Tools
Edited by Giovanni De Micheli and Luca Benini

Designing SOCs with Configured Cores: Unleashing the Tensilica Xtensa and Diamond Cores
Steve Leibson

VLSI Test Principles and Architectures: Design for Testability
Edited by Laung-Terng Wang, Cheng-Wen Wu, and Xiaoqing Wen

Contact Information

Charles B. Glaser
Senior Acquisitions Editor
Elsevier
(Morgan Kaufmann; Academic Press; Newnes)
(781) 313-4732
c.glaser@elsevier.com
http://www.books.elsevier.com

Wayne Wolf
Professor
Electrical Engineering, Princeton University
(609) 258-1424
wolf@princeton.edu
http://www.ee.princeton.edu/~wolf/

Designing SOCs with Configured Cores

Unleashing the Tensilica Xtensa and Diamond Cores

Steve Leibson

AMSTERDAM • BOSTON • HEIDELBERG • LONDON •
NEW YORK • OXFORD • PARIS • SAN DIEGO •
SAN FRANCISCO • SINGAPORE • SYDNEY • TOKYO
Morgan Kaufmann Publishers is an imprint of Elsevier

Acquisitions Editor	Charles Glaser
Publishing Services Manager	George Morrison
Production Editor	Dawnmarie Simpson
Production Assistant	Melinda Ritchie
Cover Design	Paul Hodgson
Cover Illustration	Stephen A. Emery Jr.
Composition	Charon Tec
Copyeditor	Charon Tec
Proofreader	Charon Tec
Indexer	Charon Tec
Interior printer	The Maple-Vail Book Manufacturing Group
Cover printer	Phoenix Color Corporation

Morgan Kaufmann Publishers is an imprint of Elsevier.
500 Sansome Street, Suite 400, San Francisco, CA 94111

This book is printed on acid-free paper.

Library of Congress Cataloging-in-Publication Data
Leibson, Steve.
Designing SOCs with configured cores : unleashing the Tensilica Xtensa and diamond cores / Steve Leibson.
p.cm.
Includes bibliographical references and index.
ISBN-13: 978-0-12-372498-4 (hardcover: alk. paper)
ISBN-10: 0-12-372498-8 (hardcover: alk. paper)
1. Embedded computer systems—Design and construction.
2. Systems on a chip—Design and construction. I. Title
TK7895.E42L45 2006
621.3815—dc22 2006011723

ISBN 13: 978-0-12-372498-4
ISBN 10: 0-12-372498-8

For information on all Morgan Kaufmann publications,
visit our Web site at www.mkp.com or www.books.elsevier.com

Printed in the United States of America
06 07 08 09 10 5 4 3 2 1

Contents

Foreword

This book is about System-on-Chip design using configurable processors. Steve has taken up some extremely important points that are stemming from the fact that to keep on improving the performance of systems implemented on silicon we have to consider multi-processor SoC (MPSOC), not just as an alternative but as The Alternative. The ITRS roadmap points to tens or hundreds of processing engines in a single SoC design. Multiple processors imply inter-processor communication and a diversity of specialized processors. Communication and computation.

I was happy to see that Steve has absorbed some ideas about Networks-on-Chip from all the invited talks he had to sit through with me in the International Symposium on System-on-Chip last fall. Beyond that, he explains the principles of inter-processor communication mechanisms in a very practical way. "Practical" is an adjective that came into my mind frequently while reading through the draft of this book. Yes, the book is about ***practical multi-processor SoC design***. That includes naturally a lot of bits and pieces about Tensilica's Xtensa architecture and its derivatives, but there is a lot more than that.

Shortly, it can be said that by reading the book carefully, the reader will get insight into the history, present and future of processor-based system design. Steve has illustrated all that with a good mixture of words, figures, and tables. This book is one big step ahead in response to the great challenge of transforming the art of SoC design into a more rigorous engineering discipline.

Jari Nurmi
Professor of Digital and Computer Systems
Tampere University of Technology
Finland

PREFACE

In the 21st century, system-on-chip (SOC) design styles are changing. Not because they can but because they must. Moore's law and the side benefits of classical semiconductor scaling (faster transistors running at lower power at each new processing node) parted company when the semiconductor industry hit the 130 nm processing node just after the turn of the century. As a result, on-chip clock rates stopped rising as quickly as they had and power levels stopped falling as quickly as they had. These two side effects of lithographic scaling, usually (and incorrectly) attributed to Moore's law, were actually developed by IBM's Robert Dennard and his team in the early 1970s. On-chip clock rate and power dissipation have tracked Moore's law for nearly 30 years. But no longer.

The net effect of this split between Moore's law and classical scaling is to remove two of SOC design's key assumptions:

- The next processing node promises faster processors.
- Lower energy consumption for increasingly complex systems is just one processing node away.

As a result, SOC designers must become more sophisticated.

This is not an unbearable burden. SOC design first appeared around 1995 when designers started to place processor cores on their ASIC designs. Since then, system architectures developed for SOC designs have, more often than not, closely resembled microprocessor-based, board-level systems that had been designed in the 1980s. However, Moore's law has continued to deliver transistors and interconnect in abundance so there's really no reason that system designers should be burdened with system-design rules of thumb developed under the constraints of packaged processors and layout rules developed for circuit boards.

This book advocates a departure from these decades-old design styles, although not as radical a departure as others may promote. System block diagrams need not change substantially to take full advantage of the resources available to SOC designers in the 21st century. However, many more blocks in these existing block diagrams can now be implemented with processors rather than hand-designed, custom-built logic blocks. The result of this change alone will bring large benefits to system design. Blocks that were previously hard-wired can now become firmware-programmable, which reduces design risk by allowing the SOC's function to change without the need to re-spin the chip.

In addition to advocating a shift to multiple-processor SOC (MPSOC) design, this book also promotes the idea that the almost universal use of globally shared buses should be substantially reduced. Buses are shared resources and came into existence in the 1970s to accommodate the pin limitations of packaged microprocessors. Such limitations don't exist on an SOC and these no-longer-in-effect limitations should therefore cease to hobble system architects. New, more efficient ways of interconnecting on-chip blocks exist and they should be used to improve system performance and reduce power dissipation.

None of these changes should cause experienced system designers any headaches. All of the concepts advocated in this book should be familiar to any system designer. The suggested departure from conventional design is more a shift in perspective that favors one type of existing design style (processor-based block implementation) over another (custom-designed logic). The means to achieving this shift is through the use of configurable and pre-configured processor cores that provide substantially better (orders of magnitude) system performance than widely used, general-purpose processor cores.

To help system designers and architects make this shift, this book is organized as follows:

Chapter 1: Introduction to 21st-Century SOC Design provides an overview of conventional SOC design as it exists today. It discusses the use of processors in SOC architectures and gives a history of processor-based design going back to the introduction of the original microprocessor in 1971. It discusses the tradeoffs and engineering decisions made when architecting an SOC.

Chapter 2: The SOC Design Flow covers the design steps currently used to develop ASICs and SOCs and then proposes a new, system-level design style that should precede the implementation phase. SOC designers currently jump into the implementation phase prematurely (let's get something coded now!) by relying on decades-old assumptions that are no longer valid.

Chapter 3: Xtensa Architectural Basics introduces the concept of a configurable, extensible microprocessor core for SOC design and discusses a specific processor architecture that serves as the foundation for a very broad family of configurable and pre-configured processor cores.

Chapter 4: Basic Processor Configurability lays out the many aspects of processor configurability and discusses how they can be used to boost SOC performance while cutting power dissipation.

Chapter 5: MPSOC System Architectures and Design Tools discusses the many ways, both good and bad, to architect an SOC. It then discusses how to best implement the good ways.

Chapter 6: Introduction to Diamond Cores introduces the six members of the Diamond Standard Series of pre-configured, 32-bit processor cores that have been built using Xtensa configurable-processor technology for specific target applications.

Chapters 7–12 include detailed explanations of each Diamond processor core, including the 108Mini and 212GP controller cores, the 232L and 570T CPUs, and the 330HiFi and 545CK DSPs. Each chapter also illustrates one or more ways to use each processor core in a system.

Chapter 13: Using Fixed Processor Cores in SOC Designs takes the reader deeper into the realm of system design based on multiple processor cores. It also discusses the mixing of pre-configured and configurable cores in system designs.

Chapter 14: Beyond Fixed Cores reintroduces the topic of configurable cores and applies them to the system designs discussed in the previous chapter. It also discusses the mixing of pre-configured and configurable cores in system designs.

Chapter 15: The Future Of Soc Design looks at the design challenges that remain.

Acknowledgements

First and foremost, I thank Pat and Shaina, my wife and daughter, for putting up with me once more while I write yet another book.

Next, I thank Steve Roddy who originally conceived of this book and for allowing me the time to write it.

Thanks to the entire team at Tensilica for making a new way to design SOCs possible in the first place. A large number of people have contributed to Tensilica's ability to revolutionize SOC design. First and foremost is Chris Rowen who is Tensilica's founder, CEO, and president. Rowen has been involved with RISC microprocessors essentially since day one. He was a graduate student under John Hennesy during the early academic development of the MIPS-I RISC microprocessor at Stanford and he went on to become one of the founders of MIPS Technologies, the company that commercialized the MIPS processor architecture.

From this very deep understanding of RISC architectures and of the growing use of such architectures in SOCs, Rowen developed the idea of a post-RISC, configurable microprocessor core that took advantage of rapidly evolving silicon fabrication technology in ways that no fixed-ISA processor cores can. Rowen's ideas evolved into the Xtensa architecture at Tensilica. Many of the ideas and charts that appear in this book originated with him.

Over the next several years, the dozens of members of Tensilica's engineering group have augmented Rowen's initial idea with a huge number of groundbreaking hardware architecture and software contributions. Hardware innovations include the TIE (Tensilica Instruction Extension) language, the FLIX (flexible-length instruction extensions) VLIW (very-long instruction word)-like extensions that give processors the ability to execute multiple independent operations per instruction without VLIW code bloat, and bottleneck-breaking ports and queues. Software innovations include the automatic configuration of entire software-development tool sets from TIE descriptions, fast vectorizing compilers, and fast instruction-set simulators. Without the configurable Xtensa processor, system designers would still be in the Stone Age.

Special thanks to Stuart Fiske for checking and correcting all of my cache-interface diagrams, to Paula Jones for acting as my copy editor (everyone needs a good editor), and to Tomo Tohara for working up the H.264 video decoder block diagram in Chapter 2.

Also, special thanks to my first readers. Besides Paula Jones, these include: Max Baron, Marcus Binning, and Grant Martin. Grant has been a real mentor to me and has taught me a lot about writing books in the last 2 years.

Thanks to photographic and multimedia artist extraordinaire Steve Emery who created a dynamite cover for this book. Believe it or not, I found him on eBay!

And finally, thanks to Chuck Glaser at Elsevier, whose enthusiasm for the book made sure the project advanced when nothing else would.

CHAPTER 1

INTRODUCTION TO 21ST-CENTURY SOC DESIGN

The past is prologue for the future
—common saying,
frequently ignored

Systems-on-chips (SOCs) are, by definition, electronic systems built on single chips. Also by definition, every SOC incorporates at least one microprocessor. Some SOCs use two or three microprocessors to accomplish required tasks and a few SOC designs employ many dozens of processors. To see how SOC design got where it is today and where market and technological forces will take it tomorrow, we start by first looking at how electronic system design has evolved since the microprocessor's introduction.

1.1 THE START OF SOMETHING BIG

The course of electronic systems design changed irreversibly on November 15, 1971, when Intel introduced the first commercial microprocessor, the 4004. Before that date, system design consisted of linking many hardwired blocks, some analog and some digital, with point-to-point connections. After the 4004's public release, electronic system design began to change in two important ways.

First, and most obvious, was the injection of software or firmware into the system-design lexicon. Prior to the advent of the microprocessor, the vast majority of system designers had only analog and digital design skills. If they had learned any computer programming, it was used for developing design-automation aids or simulation programs, not for developing system components. After the Intel 4004's introduction, system developers started to learn software programming skills, first in assembly language and then in high-level languages such as PL/1, Pascal, and C as compilers got better and memory became less expensive.

The other major change to system design caused by the advent of the microprocessor—a change that's often overlooked—is the use of buses to interconnect major system blocks. Figure 1.1 shows a block diagram of a Hewlett-Packard 3440A digital voltmeter that was designed in 1963, eight

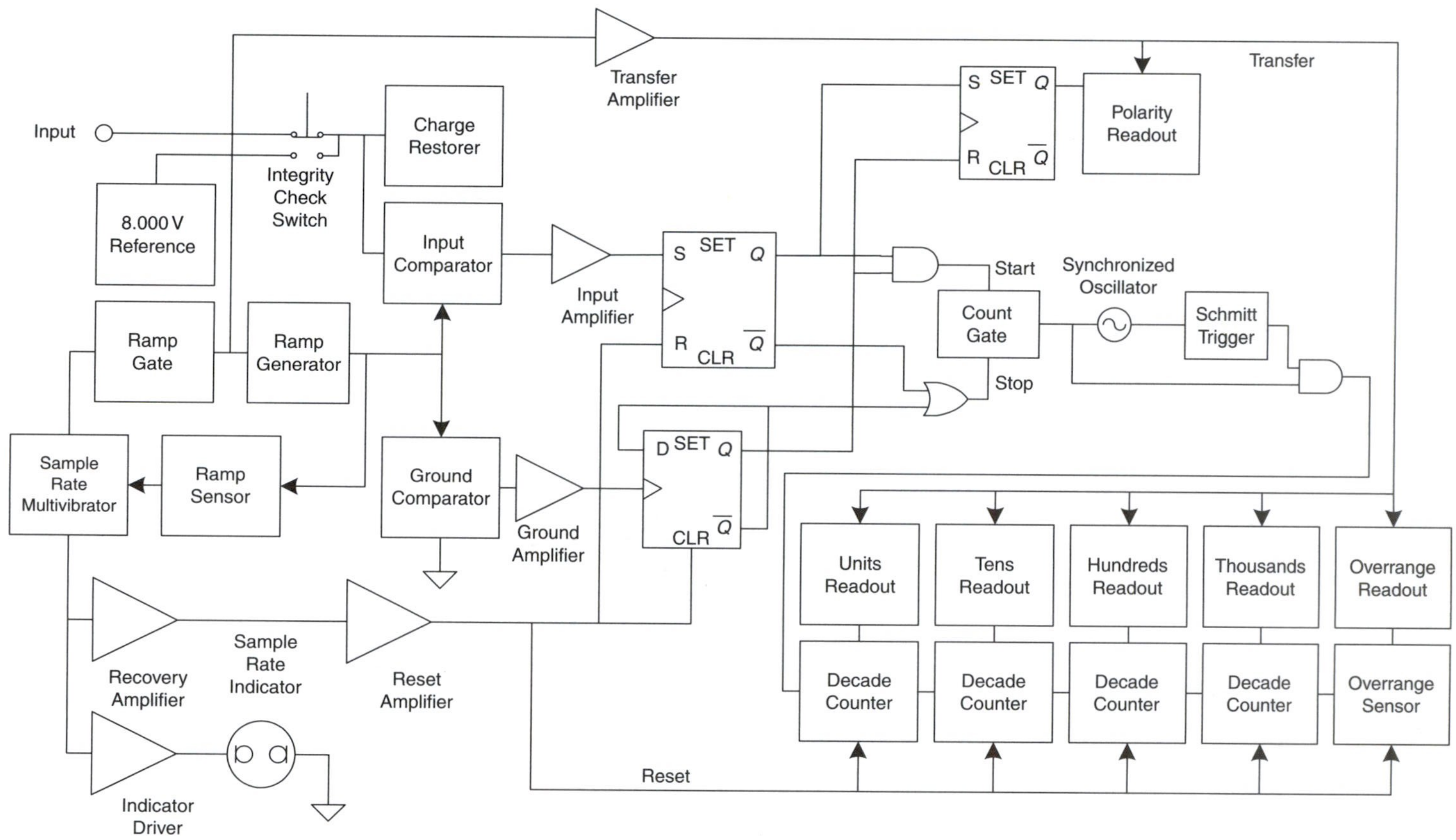

■ **FIGURE 1.1**

A digital voltmeter block diagram adapted from the design of a HP 3440A, circa 1963.

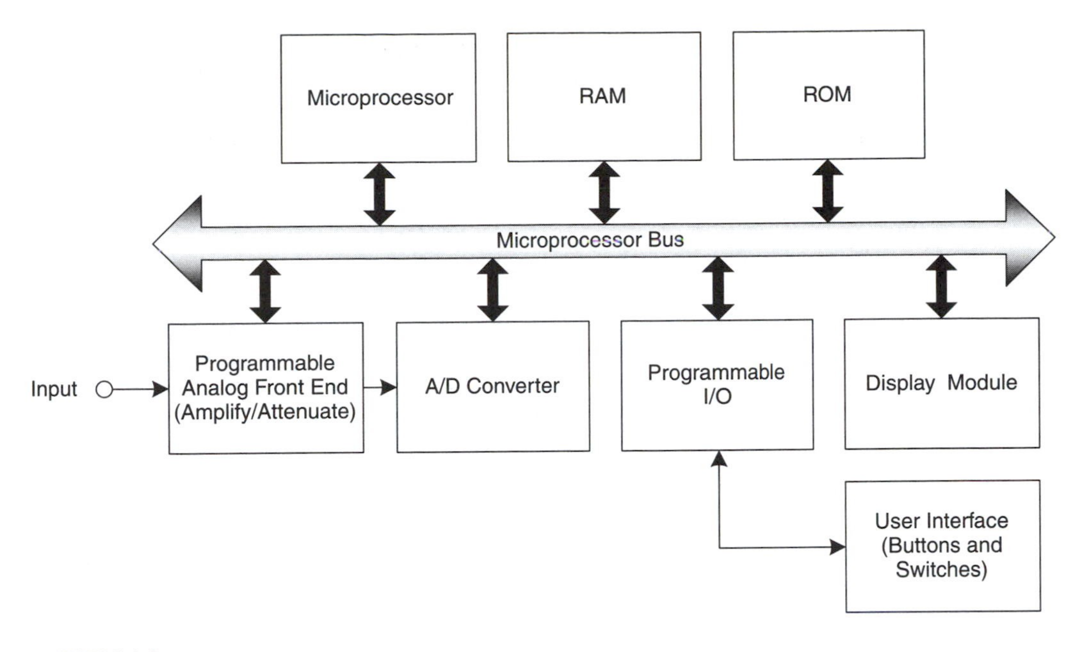

■ **FIGURE 1.2**

A version of the HP 3440A block diagram of Figure 1.1, adapted for microprocessor-based implementation. This system-design style is still quite popular.

years before the microprocessor appeared. This block diagram, typical of the era, shows a mixture of analog and digital elements interconnected with point-to-point connections. Even the all-digital measurement counter, which stores the result of the analog-to-digital conversion, consists of four independent decade counters. Each decade counter communicates to the next counter over one wire and each counter drives its own numeric display. There are no buses in this design because none are needed. Note that there are no microprocessors in this design either. Microprocessors wouldn't appear for another eight years.

Figure 1.2 illustrates how a system designer might implement a digital voltmeter like the HP 3440A today. A microprocessor controls all of the major system components in this modern design implementation. One key change: the processor communicates with the other components over a common bus—the microprocessor's main bus.

From a systems-design perspective, there are significant differences between the digital voltmeter design from the early 1960s and the modern implementation. For the purposes of a systems-level discussion, the most significant difference is perhaps the massive amount of parallelism occurring in the early 1960s design versus the modern design's ability to perform only one operation at a time over the microprocessor bus. For example, if the microprocessor in Figure 1.2 is reading a word from RAM or ROM over the bus, it cannot also be reading a word from the A/D converter at the same time. The processor bus is a shared resource and can only support one operation at a time.

This loss of concurrent operation arose because of the economics of microprocessor packaging and printed-circuit board design. It's less expensive to use buses to move data into and out of packaged microprocessors and it's often much easier to route buses on a circuit board than to accommodate multiple point-to-point connections.

The consequence of creating a shared resource like a microprocessor bus is that the use of the shared resource must be multiplexed in time. As a result of the multiplexed operation, the operating frequency of the shared resource must increase to accommodate the multiple uses of the shared resource—the bus in this case. When the signal frequencies are low to start with, as they are for a voltmeter design that's more than 40 years old, then the final operating frequency of the shared resource places little strain on the system design.

However, as the various tasks performed by a system become more ambitious causing the work to be done every clock cycle to increase, the aggregated requirements for all of the tasks begin to approach the bandwidth limit of the shared resource. As this happens, system-design margins shrink and then vanish. At that point, the system design jumps the threshold from marginal to faulty.

As a reminder: the point-to-point architecture of the original HP 3440A digital voltmeter concurrently operated all of the various systems blocks, which means that there's a lot more design margin in the interconnection scheme than for the microprocessor-based version of the system design. This loss of design margin is an engineering tradeoff and it undoubtedly reduces the implementation cost of the design, as long as no future performance increases are envisioned that would further reduce design margin.

1.2 FEW PINS = MASSIVE MULTIPLEXING

As shown in Figure 1.3, the 4-bit Intel 4004 microprocessor was packaged in a 14-pin dual-inline package (DIP). Consequently, this microprocessor's 4-bit bus not only multiplexed access to the various components in the system, it also had to multiplex the bus-access addresses with the data on the same four wires. It took three clock cycles to pass a 12-bit address out over the bus and two or four more clock cycles to read back an 8- or 16-bit instruction. All instructions came from ROM in a 4004-based system. RAM accesses were even slower because they required one instruction to pass out the target address and then a second instruction to read data from or write data to the selected location. With a maximum operating frequency of 740 kHz and long, multi-cycle bus operations, the Intel 4004

■ **FIGURE 1.3**

The Intel 4004 microprocessor, introduced in 1971, was packaged in a 16-pin package that severely limited the processor's I/O bandwidth and restricted its market accordingly. Photo Courtesy of Stephen A. Emery Jr., www.ChipScapes.com.

microprocessor was far too slow to take on many system control tasks and the electronics design community largely ignored the world's first microprocessor.

The world's second commercial microprocessor, Intel's 8008 introduced in April, 1972, was not much better than the 4004 processor in terms of bus bandwidth. The 8008 microprocessor's 8-bit bus needed two cycles to pass a 14-bit address and one to three cycles to accept an 8- to 24-bit instruction. The 8008 microprocessor had a wider, 8-bit bus that Intel squeezed into an unconventional 18-pin DIP (shown in Figure 1.4). Although the instruction times for the 4004 and 8008 microprocessors were similar (10.5 μsec versus 12.5 to 20 μsec, respectively), the 8008 microprocessor's RAM accesses were faster than those of the Intel 4004 processor because the Intel 8008 processor used a more conventional RAM-access cycle that output an address and then performed the data transaction during the same instruction cycle. The 8008 microprocessor ran at clock rates of 500–800 kHz. Consequently, like its predecessor, it was too slow to fire the imagination of many system designers.

■ **FIGURE 1.4**

Intel got more bus bandwidth from the 8008 microprocessor by squeezing it into an unconventional 18-pin package. Photo Courtesy of Stephen A. Emery Jr., www.ChipScapes.com.

1.3 THIRD TIME'S A CHARM

Intel finally got it right in April, 1974 when the company introduced its third microprocessor, the 8-bit 8080. The 8080 microprocessor had a non-multiplexed bus with separate address and data lines. Its address bus was 16 bits wide, allowing a 64-Kbyte address range. The data bus was 8 bits wide. As shown in Figure 1.5, Intel used a 40-pin DIP to house the 8080 microprocessor. This larger package and the microprocessor's faster 2-MHz clock rate finally brought bus bandwidth up to usable levels. Other microprocessor vendors such as Motorola and Zilog also introduced microprocessors in 40-pin DIPs around this time and system designers finally started to adopt the microprocessor as a key system building block.

1.4 THE MICROPROCESSOR: A UNIVERSAL SYSTEM BUILDING BLOCK

Over the next 30 years, microprocessor-based design has become the nearly universal approach to systems design. Once microprocessors had achieved

■ **FIGURE 1.5**

Intel finally crossed the bus-bandwidth threshold into usability by packaging its third-generation 8080 microprocessor in a 40-pin package. Many competitors swiftly followed suit (shown is Zilog's 8-bit Z80 microprocessor) and the microprocessor quickly became a standard building block for system designers. Photo Courtesy of Stephen A. Emery Jr., www.ChipScapes.com.

the requisite processing and I/O bandwidth needed to handle a large number of system tasks, they began to permeate system design. The reason for this development is simply engineering economics. Standard microprocessors offered as individual integrated circuits (ICs) provide a very economical way to package thousands of logic transistors in standard, testable configurations. The resulting mass-produced microprocessor ICs have become cheap, often costing less than $1 per chip, and they deliver abilities that belie their modest cost.

Microprocessors also dominate modern electronic system design because hardware is far more difficult to change than software or firmware. To change hardware, the design team must redesign and re-verify the logic, change the design of the circuit board (in pre-SOC times), and then re-run any required functional and environmental tests. Software or firmware developers can change their code, recompile, and then burn new ROMs or download the new code into the existing hardware.

In addition, a hardware designer can design a microprocessor-based system and build it before the system's function is fully defined. Pouring the software or firmware into the hardware finalizes the design and this event can occur days, weeks, or months after the hardware has been designed, prototyped, verified, tested, manufactured, and even fielded. As a consequence, microprocessor-based system design buys the design team extra time because hardware and firmware development can occur concurrently, which telescopes the project schedule (at least under ideal conditions).

With the advent of practical 8-bit microprocessors in the mid-1970s, the microprocessor's low cost and high utility snowballed and microprocessor vendors have been under great pressure to constantly increase their products' performance as system designers think of more tasks to execute on processors. There are some obvious methods to increase a processor's performance and processor vendors have used three of them.

The first and easiest performance-enhancing technique used was to increase the processor's clock rate. Intel introduced the 8086 microprocessor in 1978. It ran at 10 MHz, five times the clock rate of the 8080 microprocessor introduced in 1974. Ten years later, Intel introduced the 80386 microprocessor at 25 MHz, faster by another factor of 2.5. In yet another ten years, Intel introduced the Pentium II processor at 266 MHz, better than a ten times clock-rate increase yet again. Figure 1.6 shows the dramatic rise in microprocessor clock rate over time.

Note that Intel was not the only microprocessor vendor racing to higher clock rates. At different times, Motorola and AMD have also produced

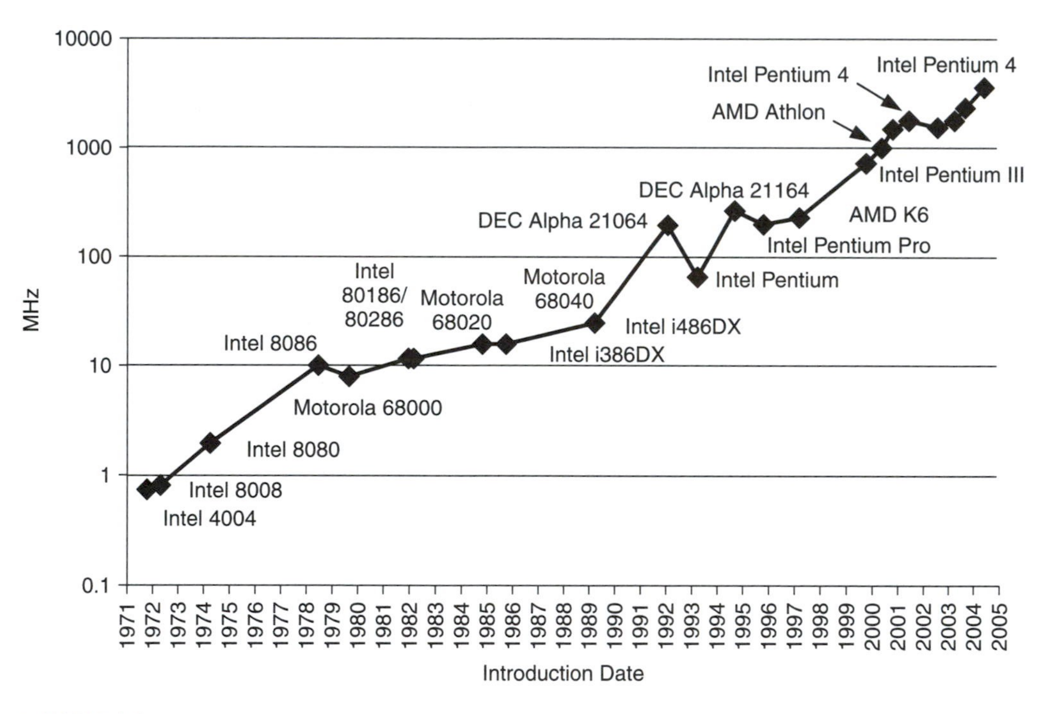

■ **FIGURE 1.6**

Microprocessor clock rates have risen dramatically over time due to the demand of system designers for ever more processor performance.

microprocessors that vied for the "fastest clock rate" title and Digital Equipment Corporation (DEC) hand-tuned its series of Alpha microprocessors to world-beating clock rates. (That is until Compaq bought the company in 1998 and curtailed the Alpha development program. HP then acquired Compaq in late 2001.)

At the same time, microprocessor data-word widths and buses widened so that processors could move more data during each clock period. Widening the processor's bus is the second way to increase processing speed and I/O bandwidth. Intel's 16-bit 8086 microprocessor had a 16-bit data bus and the 32-bit 80386 microprocessor had a 32-bit data bus.

The third way to increase processor performance and bus bandwidth is to add more buses to the processor's architecture. Intel did exactly this with the addition of a separate cache-memory bus to its Pentium II processor. The processor could simultaneously run separate bus cycles to its high-speed cache memory and to other system components attached to the processor's main bus.

As processor buses widen and as processor architectures acquire extra buses, the microprocessor package's pin count necessarily increases. Figure 1.7 shows how microprocessor pin count has increased over the years.

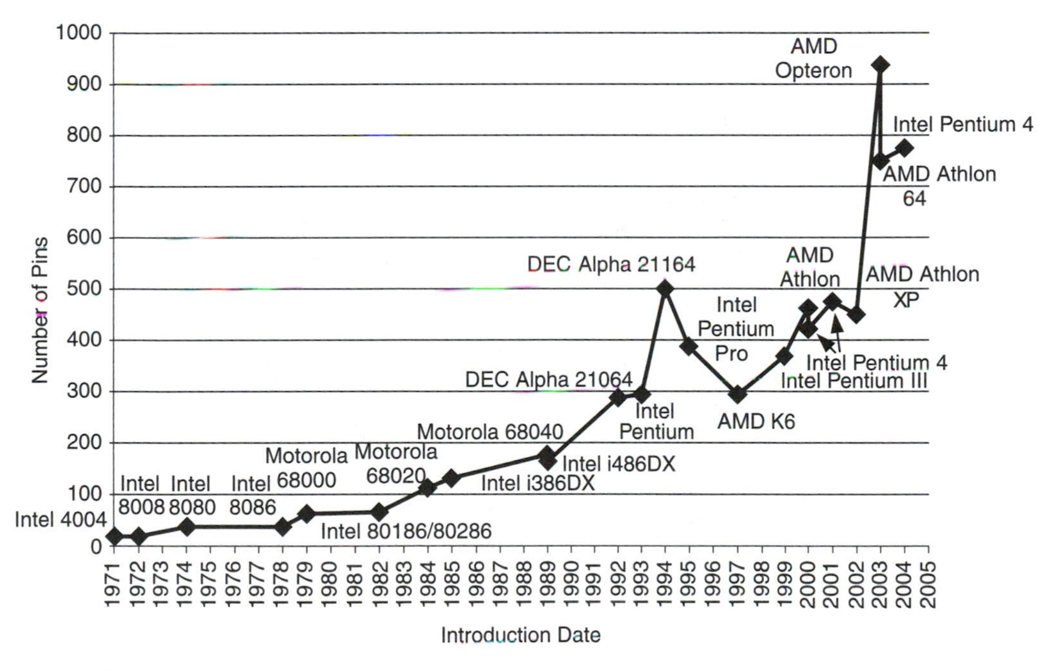

■ **FIGURE 1.7**

Microprocessor pin counts have also risen dramatically over time due to the demand of system designers for ever more processor performance.

Like the rising curve plotted in Figure 1.6, the increasing pin count shown in Figure 1.7 is a direct result of system designers' demand for more processor performance.

1.5 THE CONSEQUENCES OF PERFORMANCE—IN THE MACRO WORLD

Faster clock rates coupled with more and wider buses do indeed increase processor performance, but at a price. Increasing the clock rate extracts a penalty in terms of power dissipation, as shown in Figure 1.8. In fact, power dissipation rises roughly with the square of the clock-rate increase, so microprocessor power dissipation and energy density have been rising exponentially for three decades. Unfortunately, the result is that the fastest

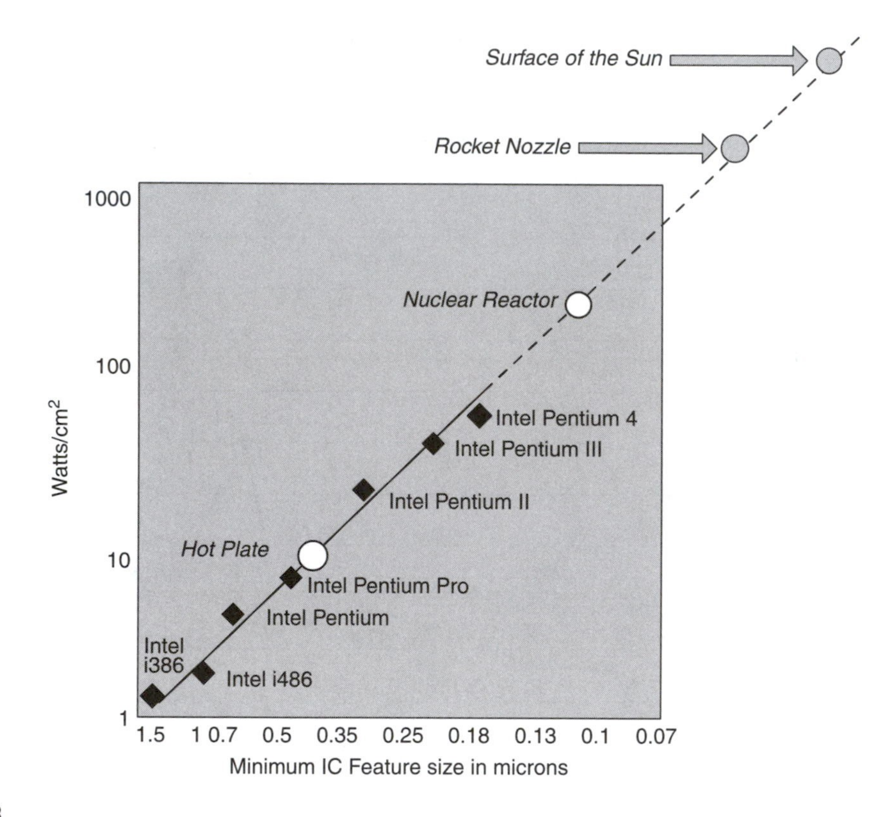

▪ **FIGURE 1.8**

Packaged microprocessor power density has risen exponentially for decades. (*Source*: F. Pollack, keynote speech, "New microarchitecture challenges in the coming generations of CMOS process technologies," MICRO-32, Haifa, Israel, 1999.)

packaged processors today are bumping into the heat-dissipation limits of their packaging and cooling systems. Since their introduction in 1971, cooling design for packaged microprocessors progressed from no special cooling to:

- careful system design to exploit convection cooling,
- active air cooling without heat sinks,
- active air cooling with aluminum and then copper heat sinks,
- larger heat sinks,
- even larger heat sinks,
- dedicated fans directly attached to the processor's heat sink,
- heat pipes,
- heat sinks incorporating active liquid cooling subsystems.

Each step up in heat capacity has increased the cost of cooling, increased the size of required power supplies and product enclosures, increased cooling noise (for fans), and decreased system reliability due to hotter chips and active cooling systems that have their own reliability issues.

SOCs cannot employ the same sort of cooling now used for PC processors. Systems that use SOCs generally lack the PC's cooling budget. In addition, a processor core on an SOC is only a small part of the system. It cannot dominate the cost and support structure of the finished product the way a processor in a PC does. Simple economics dictate a different design approach.

In addition, SOCs are developed using an ASIC design flow, which means that gates are not individually sized to optimize speed in critical paths the same way and to the same extent that transistors in critical paths are tweaked by the designers of packaged microprocessors. Consequently, clock rates for embedded processor cores used in SOC designs have climbed modestly over the past two decades to a few hundred MHz, but SOC processors do not run at multi-GHz clock rates like their PC brethren, and probably never will because the link between smaller transistors and faster clock rates actually broke when the semiconductor industry reached the 130 nm processing node.

Gordon Moore first formulated his famous and long-lasting prediction—that shrinking transistor size and the resulting increase in device density would double roughly every two years—while working at Fairchild Semiconductor's R&D laboratories in 1965. IBM's Robert Dennard, the man who invented and patented the 1-transistor DRAM (dynamic RAM) in 1968, codified the linkages between physical, Moore's law transistor scaling, increasing transistor speed, and decreasing power dissipation. He made this connection in an article titled "Design of Ion-Implanted

MOSFETs With Very Small Physical Dimensions," which was published in the *IEEE Journal of Solid-State Circuits* in 1974, although Dennard and his colleagues published the basic data and conclusion two years earlier at the International Electron Devices Meeting (IEDM). The semiconductor industry has ridden this linkage between Moore's law of transistor scaling and transistor speed and power dissipation for 30 years, until it broke. Now, transistor scaling continues to produce smaller and cheaper transistors but, with each new IC process node, the transistors are only a little faster and they no longer consume considerably less power than the immediately preceding device generation. As a result, SOC designers can no longer depend on the next semiconductor-processing node to solve performance and power-dissipation problems afflicting their system designs. To get better system performance in the 21st century, systems designers must become more sophisticated in the way they architect their systems.

1.6 INCREASING PROCESSOR PERFORMANCE IN THE MICRO WORLD

Lacking the access to the high clock rates previously available to PC processor designers, processor core designers must turn to alternative performance-enhancing strategies. Use of additional buses and wider buses are both good performance-enhancing strategies for SOC-centric processor design. In the macro world of packaged microprocessors, additional processor pins incur a real cost. Packages with higher pin counts are more expensive, they're harder to test, and they require more costly sockets. However, in the micro world of SOC design, additional pins for wider buses essentially cost nothing. They do incur some additional routing complexity, which may or may not increase design difficulty. However, once routed, additional pins on a microprocessor core do not add much to the cost of chip manufacture, except for a fractionally larger silicon footprint.

In much the same way, additional microprocessor buses also incur very little cost penalty but provide a significant performance benefit. Processor cores for SOCs often have many more buses than their packaged counterparts, as shown in Figure 1.9.

Figure 1.9 shows a hypothetical processor core with eight buses. One of those buses, the main one, is the traditional multimaster bus that all microprocessors have had since 1971. Four of the buses communicate directly with local data and instruction memories. Two more buses communicate with instruction and data caches, respectively. The remaining bus is a fast local bus used for high-speed communications with closely coupled devices such as FIFOs and high-bandwidth peripherals.

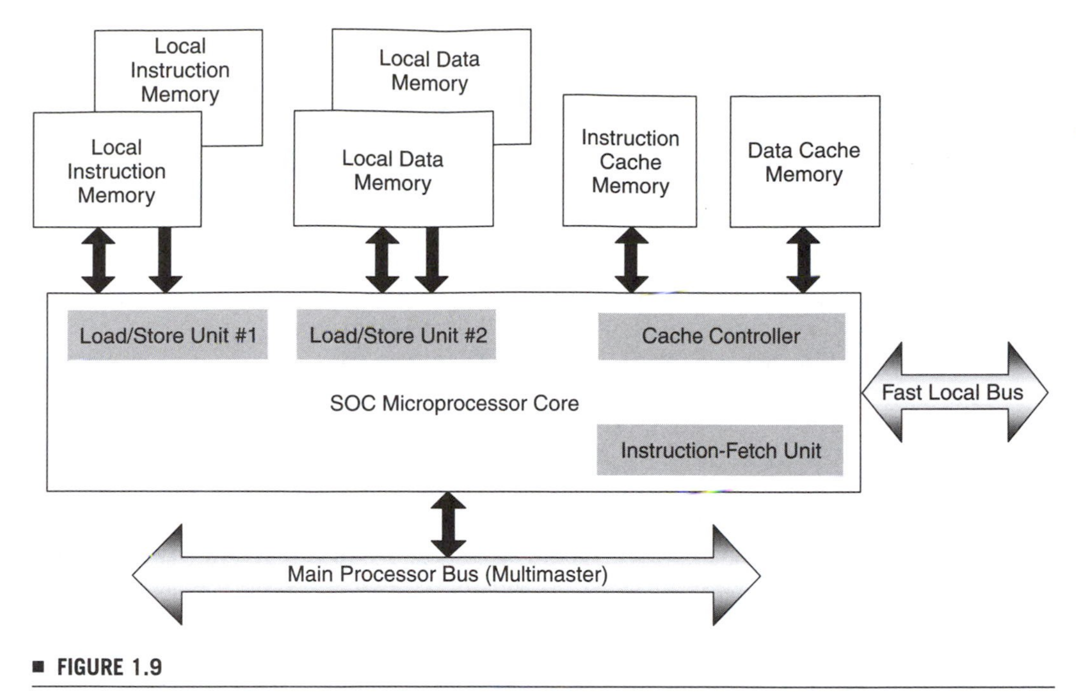

■ **FIGURE 1.9**

SOC processor cores can incorporate several buses to increase performance without incurring the pin limitations and costs of packaged processors.

The processor shown in Figure 1.9 has two load/store units plus an instruction-fetch unit, which can operate three of the processor's buses simultaneously. In addition, the processor's cache-control unit can independently use the cache buses and the memory buses. As a consequence, several of the buses on the processor shown in Figure 1.9 can be operating simultaneously. This sort of I/O concurrency is precisely what's needed to keep the processor core's operating frequency low, and thus keep operating power low as well. The number of pins required to implement eight buses is cost-prohibitive for a packaged processor IC, but is not at all costly for a processor core.

1.7 I/O BANDWIDTH AND PROCESSOR CORE CLOCK RATE

This dichotomy is a significant point of difference between processor chips and processor cores. A processor core's ability to support multiple simultaneous I/O transactions on several buses profoundly expands the possibilities for high-performance system architectures and topologies that would be uneconomical or impossible using packaged processor ICs for board-level system designs. Consequently, SOC designers should not feel the same pressures to pursue processors with high clock rates that PC designers use to achieve performance goals.

However, entrenched system-design habits and rules of thumb developed from the industry's 35 years of collective, microprocessor-based, board-level system-design experience frequently prompt even experienced system designers to compare processor cores solely on clock frequency, as they might for packaged processor chips. One of the goals of this book is to convince you that SOC processor cores are not as interchangeable as once thought and that high processor clock rates are no longer as important for SOC design as they were for high-performance, board-level system designs. In fact, as discussed above, the quest for high clock rates carries severely negative consequences with respect to power dissipation, system complexity, reliability, and cost. Achieving system performance goals at lower clock rates virtually always results in superior system designs.

Note that this does not mean that high processor-core clock rates are never important. Sometimes, just there is no alternative. However, if more task concurrency at lower clock rates can achieve system-performance goals, then that design approach will almost always prove superior. Ideally, the best SOC processor core will therefore be one that both:

- delivers high performance at low clock rates,
- can achieve high clock rates if needed.

1.8 MULTITASKING AND PROCESSOR CORE CLOCK RATE

Multitasking is another system-design choice that tends to increase processor clock rates. Processor multitasking predates the introduction of microprocessors by at least a decade. Early computers of the 1940s, 1950s, and 1960s were very expensive. Consequently, computer time was also very expensive. One way to distribute the high hardware costs was to give each computer user a share of the computer's time—timesharing. Timeshared operating systems started to appear on computers by 1961. Multitasking is timesharing, recast. Multitasking operating systems queue multiple tasks (rather than users) and give each task a time-multiplexed share of the computer. Multitasking makes one processor appear to be doing the work of several. When computers were big and expensive, multitasking made perfect sense.

Initially, microprocessors were also expensive. The first production units of the earliest processor chips cost several hundred dollars throughout the 1970s, so there was significant financial incentive for the expensive processor to execute as many concurrent tasks as possible to amortize the processor's cost across tasks rather than using many expensive processors to implement the multiple tasks. An entire industry has grown up around the development of real-time operating systems for the specific purpose of making microprocessors execute multiple concurrent tasks.

Microprocessor multitasking encourages clock-rate escalation. A faster clock rate allows a processor to execute more concurrent tasks and more complex tasks. As long as processors are expensive, the system-design scales tip toward multitasking because larger power supplies and cooling components (incurred when running a processor at a higher clock rate) are probably not as expensive as a second processor. However, when processors are cheap, the scales tip against multitasking.

In 1968, a dollar bought one packaged transistor and you needed thousands of transistors to build a computer. In the 21st century, a dollar buys several million transistors on an SOC. It takes roughly 100,000 transistors to build a 32-bit RISC processor, which now cost less than a penny. Moore's law has made transistors—and therefore processors—cheap, but conventional system-design techniques that conserve processors in exchange for increased clock rates are based on habits and rules of thumb developed when processors cost many dollars instead of less than a penny.

1.9 SYSTEM-DESIGN EVOLUTION

Semiconductor advances achieved through the relentless application of Moore's law have significantly influenced the evolution of system design since microprocessors became ubiquitous in the 1980s. Figure 1.10 shows how minimum microprocessor feature size has tracked Moore's law since the introduction of the Intel 4004, which used 10-micron (10,000-nm) lithography. The figure also incorporates ITRS 2005 (International Technology Roadmap for Semiconductor) projections to the year 2020, when the minimum feature size is expected to be an incredibly tiny 14 nm. Each reduction in feature size produces a corresponding increase in the number of transistors that will fit on a chip. Presently, Intel's dual-core Itanium-2 microprocessor holds the record for the largest number of transistors on a microprocessor chip at 1.72 billion. Most of the Itanium-2's on-chip transistors are devoted to memory. In the 21st century, SOCs routinely contain tens of millions to several hundred million transistors.

A series of system-level snapshots in 5-year intervals illustrates how system design has changed, and how it also has clung to the past. Figure 1.11 shows a typical electronic system, circa 1985. At this point in the evolution of system design, microprocessors have been available for nearly 15 years and microprocessor-based system design is now the rule rather than the exception. Packaged microprocessor ICs are combined with standard RAM, ROM, and peripheral ICs and this collection of off-the-shelf LSI chips is arrayed on a multilayer printed-circuit board. Glue logic provides the circuits needed to make all of the standard LSI chips work together as a system.

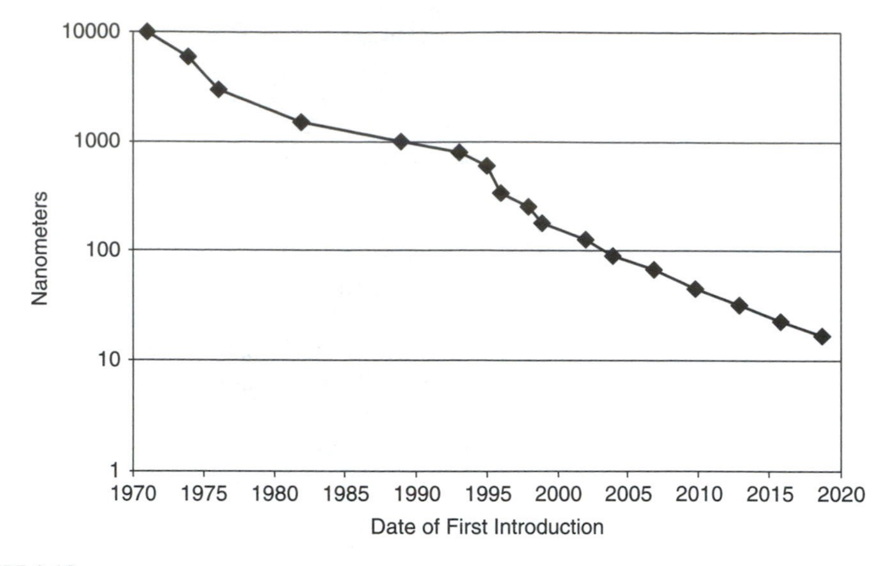

■ **FIGURE 1.10**

The relentless decrease in feature size that has slavishly followed Moore's law for decades fuels rapid complexity increases in SOC designs.

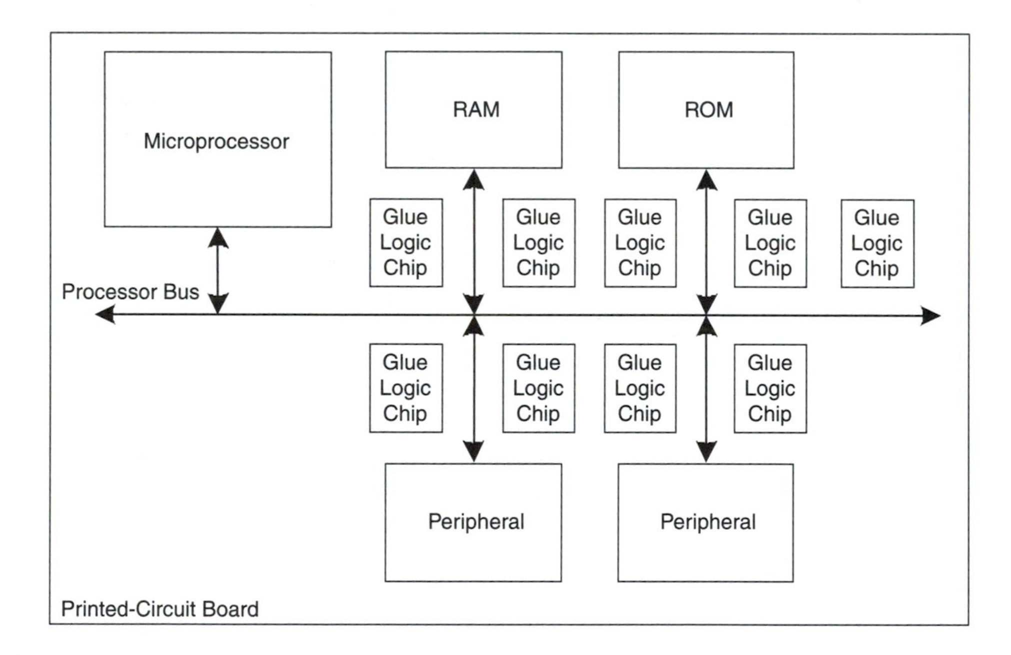

■ **FIGURE 1.11**

By 1985, microprocessor-based system design using standard LSI parts and printed-circuit boards was common.

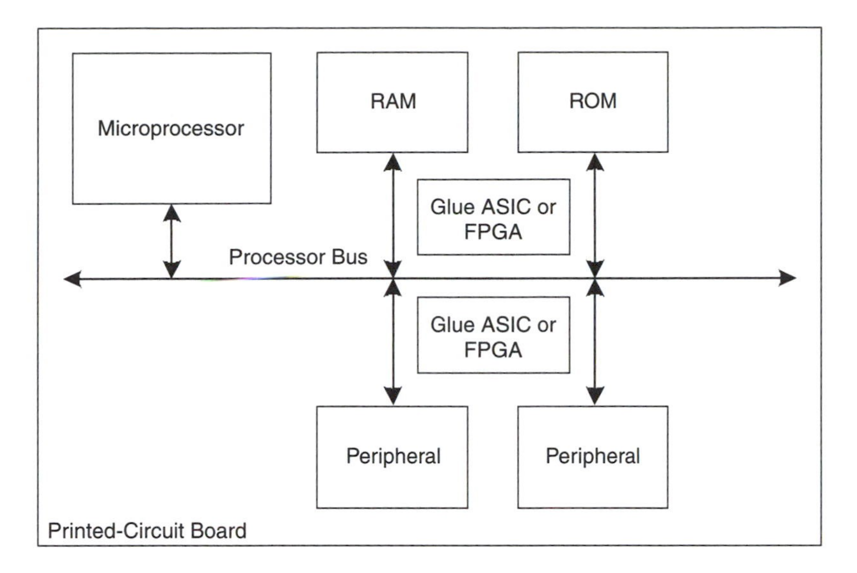

■ **FIGURE 1.12**

By 1990, system design was still mostly done at the board level but system designers started to use ASICs and FPGAs to consolidate glue logic into one or two chips.

In this design era, the glue-logic chips are probably bipolar 74LS series TTL chips but more advanced designers are using small, field-programmable chips (usually fast, bipolar PROMs or PALs made by Monolithic Memories) for glue. Note the single-processor bus that links all of the chips together. Even though microprocessors have been available for nearly a decade and a half, the block diagram shown in Figure 1.11 could easily be representative of systems designed with Intel's original 4004 microprocessor.

Five years later, in 1990, system designers were still largely working at the board level with standard packaged microprocessor ICs. However, much of the glue logic has migrated into one or more ASICs (for high-volume systems) or FPGAs (for lower volume systems), as shown in Figure 1.12. These ASICs were usually too small (had insufficient gate count) for incorporating microprocessor cores along with the glue logic.

ASIC capacities had advanced enough to include a processor core by 1995, initiating the SOC design era. Despite the additional potential flexibility and routability afforded by on-chip system design, system block diagrams continued to look much like their board-level predecessors, as illustrated in Figure 1.13. In general, system designers used the new silicon SOC technology in much the same way they had used printed-circuit boards.

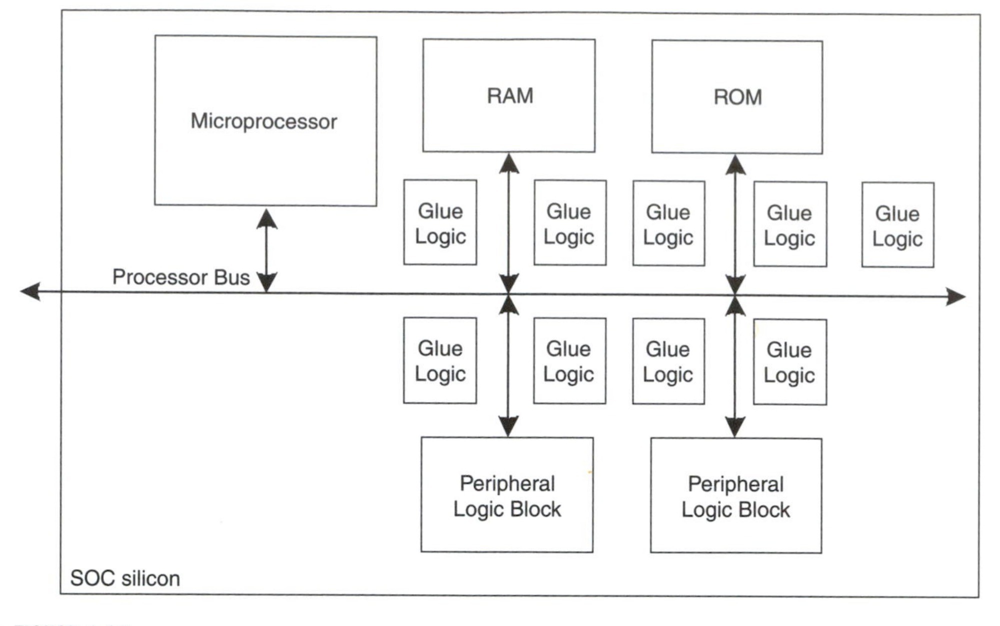

■ **FIGURE 1.13**

Although some system designs had fully migrated onto one chip by 1995, system block diagrams continued to closely resemble earlier board-level designs.

By the year 2000, increasingly advanced IC lithography permitted the incorporation of processor cores with increasingly high clock rates, which caused a mismatch between the processor's bus speed and slower peripheral devices. To uncouple the fast processor-memory subsystem from the slower peripheral sections of the SOC, system designers started to use on-chip bus hierarchies with fast and slow buses separated by a bus bridge, as shown in Figure 1.14. This system topology allows the high-speed bus to shrink in size, which reduces its capacitance and allows the processor's memory bus to keep up with the processor's high clock rates. Logically, the system block diagram of a system designed in the year 2000 still closely resembles one designed in 1985.

Present-day SOC design has started to break with the 1-processor system model that has dominated since 1971. Figure 1.15 shows a 2-processor SOC design with a control-plane processor and a data-plane processor. Each processor has its own bus and shares a peripheral device set by communicating over bus bridges to a separate peripheral bus. This arrangement is an extension of the bus hierarchy discussed in connection with Figure 1.14.

The terms "control plane" and "data plane" came into use during the Internet boom of the late 1990s and early part of the 21st century. At first, these terms referred largely to the design of multiple-board networking systems. High-speed network data passed through high-performance

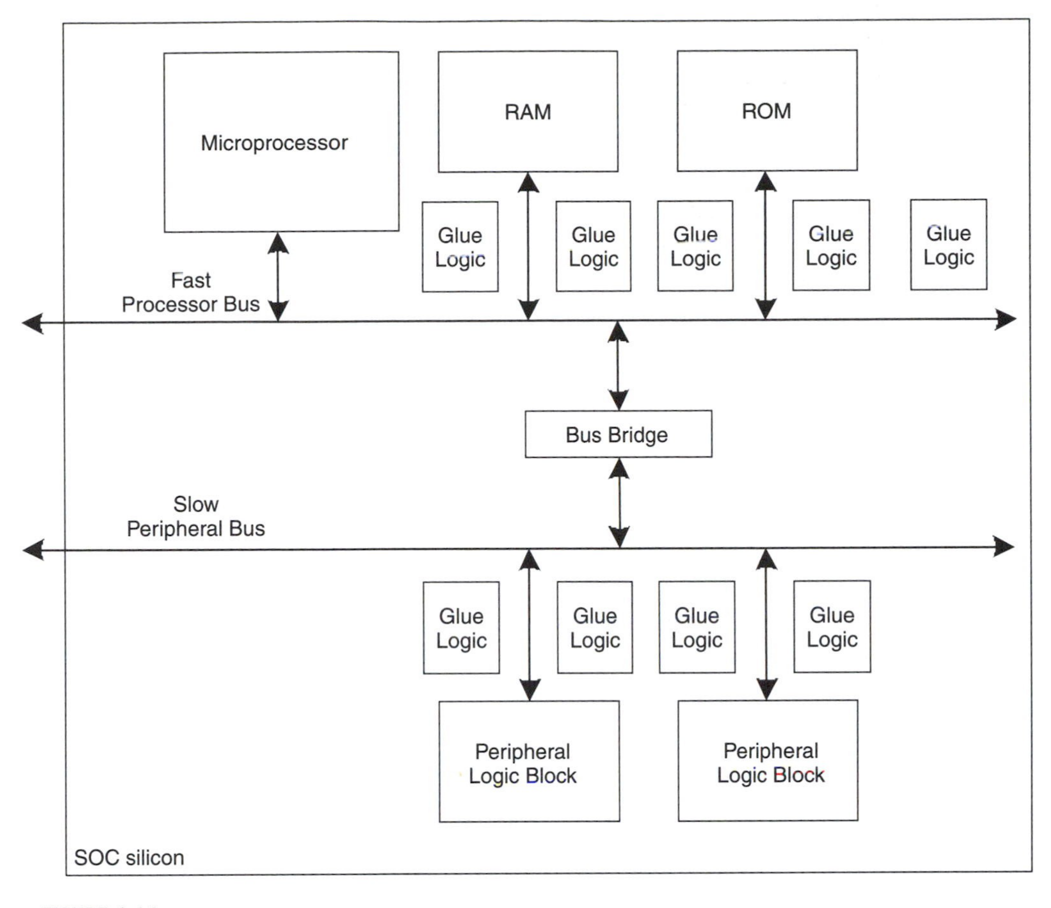

FIGURE 1.14

By the year 2000, system designers were splitting the on-chip bus into a small, fast processor-memory bus and a slower peripheral bus. A hierarchical bus topology allowed chip designers to greatly reduce the capacitance of the high-speed bus by making it physically smaller. Even so, the SOC's logical block diagram continued to strongly resemble single-processor, board-level systems designed 15 years earlier.

processors and hardware accelerators on a high-speed circuit board—called the data plane. Overall system control did not require such high performance so the control task was given to a general-purpose processor on a separate circuit board—called the control plane. These terms have now become universal because they suitably describe many processing systems such as video-encoding and video-decoding designs that must handle high-speed data and perform complex control.

A 2-processor design approach has also become very common in the design of voice-only mobile-telephone handsets. A general-purpose processor (almost universally an ARM RISC processor due to legacy-software and type approval considerations) handles the handset's operating system, user interface, and protocol stack. A DSP (digital signal processor) handles the

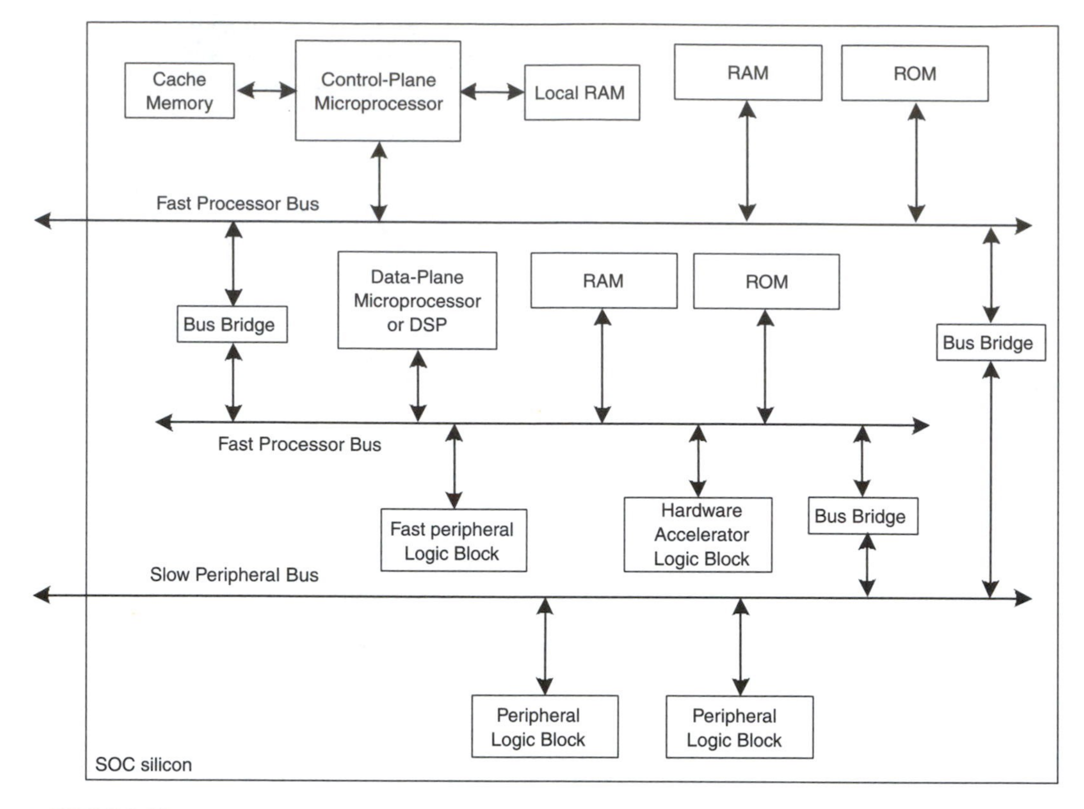

■ **FIGURE 1.15**

Present-day SOC design has started to employ multiple processors instead of escalating clock rates to achieve processing goals.

mobile phone's baseband and voice processing (essentially DSP functions such as Fast Fourier Transformations (FFTs) and inverse FFTs, symbol coding and decoding, filtering, etc.). The two processors likely run at different clock rates to minimize power consumption. Processing bandwidth is finely tuned to be just enough for voice processing—which minimizes product cost (mobile-phone handset designs are sensitive to product-cost differentials measured in fractions of a penny because they sell in the hundreds of millions of units per year) and also minimizes power dissipation—which maximizes battery life, talk time, and standby time.

1.10 HETEROGENEOUS- AND HOMOGENEOUS-PROCESSOR SYSTEM-DESIGN APPROACHES

Figure 1.15 shows the use of two different microprocessor cores, one general-purpose processor and one DSP. Such a system is called a heterogeneous-multiprocessor system. A heterogeneous-multiprocessor design approach has the advantage of matching processor cores with application-appropriate features to specific on-chip tasks.

Selecting just the right processor core or tailoring the processor core to a specific task has many benefits. First, the processor need have no more abilities than required by its assigned task set. This characteristic of heterogeneous-multiprocessor system design minimizes processor gate counts by trimming unneeded features from each processor.

One of the key disadvantages of heterogeneous-multiprocessor design is the need to use a different software-development tool set (compiler, assembler, debugger, instruction-set simulator, real-time operating system, etc.) for each of the different processor cores used in the system design. Either the firmware team must become proficient in using all of the tool sets for the various processors or—more likely—the team must be split into groups, each assigned to different processor cores.

However, this situation is not always the case for heterogeneous-processor system designs, as we'll see later in this book. Each instance of a configurable processor core can take on exactly the attributes needed for a specific set of tasks and all of the variously configured processor cores, which are based on a common ISA (instruction-set architecture), can still use the same software-development tool suite so that software team members can use familiar tools to program all of the processors in an SOC.

Some SOC designs, called homogeneous multiprocessor systems, use multiple copies of the same processor core. This design approach can simplify software development because all of the on-chip processors can be programmed with one common set of development tools. However, processor cores are not all created equally able. General-purpose processors are not generally good at DSP applications because they lack critical execution units and memory-access modes. A high-speed multiplier/accumulator (MAC) is essential to efficient execution of many DSP algorithms but MACs require a relatively large number of gates so few general-purpose processors have them.

Similarly, the performance of many DSP algorithms can benefit from a processor's ability to fetch two data words from memory simultaneously, a feature often called XY memory addressing. Few general-purpose processors have XY memory addressing because the feature requires the equivalent of two load/store units and most general-purpose processors have only one such unit.

Although basic, voice-only mobile-telephone handsets generally have only two processors, incorporation of multimedia features (music, still image, and video) has placed additional processing demands on handset system designs and the finely tuned, cost-minimized 2-processor system designs for voice-only phones simply lack processing bandwidth for these additional functions. Consequently, the most recent handset designs with new multimedia features are adding either hardware-acceleration blocks or "application processors" to handle the processing required by the additional features. The design of multiple-processor SOC systems is now sufficiently common to prompt a new term that describes the system-design style that employs more than one processor; multiple-processor SOCs are called MPSOCs. Figure 1.16 is a die layout of one such device, the MW301 media processor for camcorders and other video and still-image devices. The MW301 incorporates five processor cores. Four identical cores share the work of MPEG4 video coding and decoding. A fifth core handles the audio processing.

The MediaWorks' MW301 media processor is a good example of an SOC that transcends the special-purpose nature of early SOC designs. With five firmware-programmable processors, the MW301 is flexible enough to serve many applications in the audio/video realm. Consequently, an SOC like the MW301 media processor is often called a "platform" because the SOC's hardware design can be used for a variety of applications and its function can be altered, sometimes substantially, simply by changing the code it runs. There's no need to redesign the SOC hardware for new applications.

Platform SOCs are far more economical to design because they sell in larger volumes due to their flexible nature. Design and NRE (non-recurring engineering) costs can therefore be amortized across a larger chip volume, which reduces the design and NRE burden placed on the total cost of each chip sold.

The current record holder for the maximum number of microprocessor cores placed on one chip is Cisco's SPP (silicon packet processor) SOC, designed for the company's CRS-1 92-Tbit/sec router. The massively parallel SPP network processor chip incorporates 192 packet-processing elements (PPEs) organized in 16 clusters. Each cluster contains 12 PPEs and each PPE incorporates one of Tensilica's 32-bit Xtensa RISC processor cores, a very small instruction cache, a small data memory, and a DMA (direct memory access) controller for moving packet data into and out of the PPE's data memory. Figure 1.17 shows a diagram of the SPP network processor's design.

Cisco's SPP chip measures 18 mm on a side and is manufactured in a 130 nm process technology by IBM. Although there are 192 PPEs on the SPP die, four of the processors are spares used for silicon yield enhancement so the SPPs operate as 188-processor devices when fielded. There are approximately 18 million gates and 8 Mbits of SRAM on the SPP chip. Each of the PPEs on the chip consumes about 0.5 mm^2.

Cisco's earlier router chips employed hard-wired logic blocks to implement the router algorithms. Because of the number of features Cisco wanted to build into the CRS-1 router, and because networking standards change continuously, it was clear that the CRS-1 architecture required a firmware-programmable architecture. This requirement was brought into sharp focus as the effort to migrate from Internet protocol IPv4 to IPv6 escalated. The firmware-programmable SPP architecture allowed the CRS-1 design team to accommodate this major change in the protocol specification.

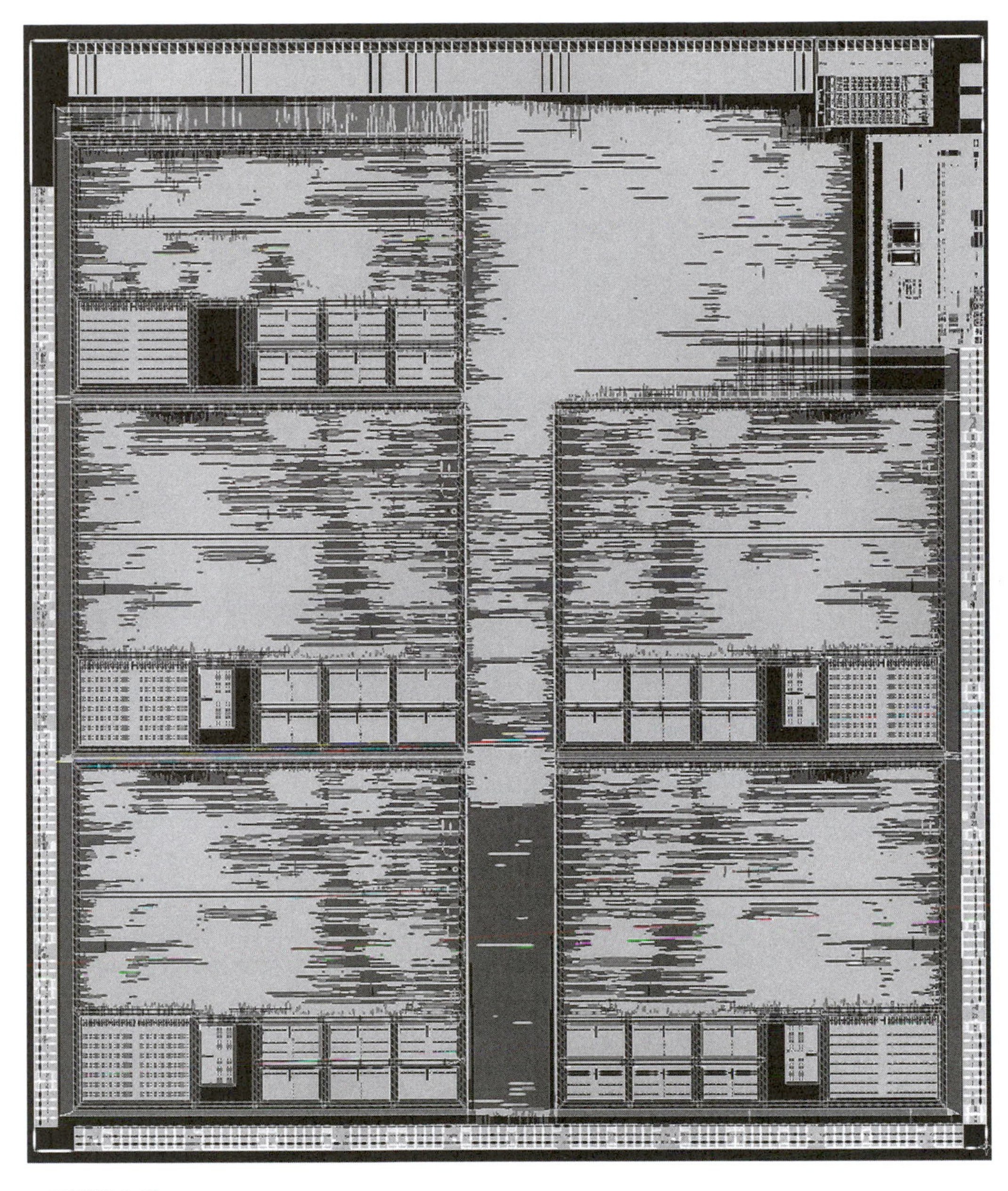

■ **FIGURE 1.16**

The MediaWorks MW301 media processor chip is a DVD resolution MPEG video and audio encoder/decoder system on a chip for solid state camcorder and portable video products. The chip contains a set of five loosely coupled heterogeneous processors. This is the highest performance fully C-code programmable media processor ever built. Photo courtesy of MediaWorks.

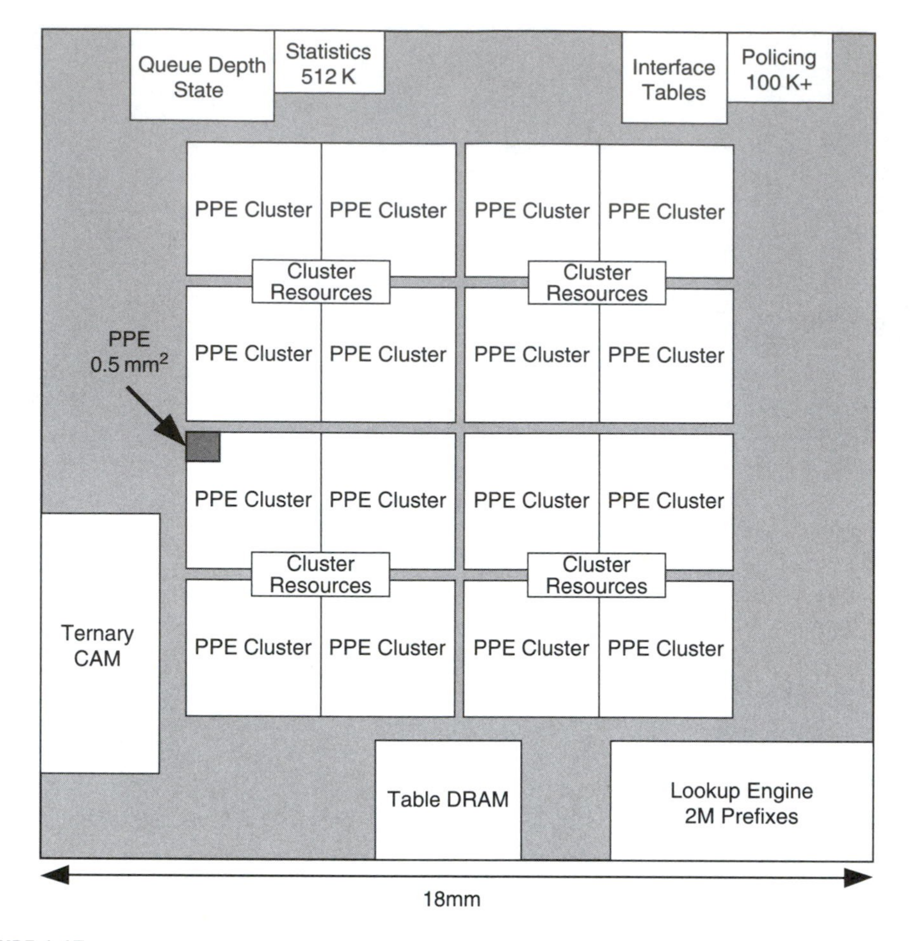

■ **FIGURE 1.17**

Cisco's SPP network processor SOC incorporates 192 packet-processing elements, each built with an Xtensa 32-bit RISC processor core with instruction enhancements specifically for packet processing.

1.11 THE RISE OF MPSOC DESIGN

MediaWorks' MW301 and Cisco's SPP are noteworthy examples of 21st-century MPSOC design. Both system designs employ processor cores extensively, preferring them to large blocks of manually designed, custom RTL (register transfer level). This preference reflects a change in SOC-design philosophy from the design style in vogue just a few years before. In the late 1990s, most SOC designs employed just one processor core. Any additional processing was performed by RTL blocks. The reason for using this older design style was, primarily, a scarcity of gates on chip.

ASIC technology at the time could scarcely accommodate one processor core and its memory. Squeezing several processors onto a chip was simply out of the question.

Consequently, any additional processing to be performed had to be assigned to custom-designed logic. When there were few such blocks of custom logic, chip verification wasn't too difficult. However, with climbing gate counts driven by Moore's law, the number of gates available to the system design has rocketed into the tens of millions. Consequently, chip-level verification has become a burden—onerous in fact. Verification now consumes approximately 70% of an SOC project's development time and budget. This design style, with its reliance on large numbers of manually designed custom-logic blocks, is not sustainable in the 21st century.

Whenever system-design styles are forced to change by technological advances, a few early adopters jump on the bandwagon. However, many design engineers remain skeptical of the need to change even when confronted by both the need and the historical facts surrounding previous changes. In the 1980s, for example, systems and chips were designed using schematics. Synopsys and its competitors started to sell logic-synthesis tools in the latter part of that decade, but it was an uphill battle for years. Most chip designers would not abandon their existing design styles, which were based on schematic capture. By the late 1990s, 10 years later, the schematic-capture design style for digital IC design had essentially disappeared and all SOCs are now designed using hardware description languages—either Verilog or VHDL.

In 1990, very few ASICs incorporated even one processor core. There simply weren't enough gates available on economically sized chips to make on-chip processors practical. ASICs designed during the early 1990s primarily served as glue-logic chips that connected packaged processor ICs to memories and peripheral devices, as discussed previously. Ten years later, at the turn of the century, nearly every ASIC incorporated at least one on-chip processor core because the semiconductor technology permitted it and because it was more efficient and less expensive for the processor to reside on the chip with other system components.

By 1995, RISC microprocessor cores were just starting to emerge as ASIC building blocks because their compact nature (relatively low gate count) gave them a very attractive capability-per-gate quotient and their programmability provided useful design flexibility. As semiconductor technology and design styles evolved over the next 10 years, processor cores became sufficiently pervasive in ASIC design to give rise to a new name, the system on chip or SOC. Of course, once systems designers started using one on-chip processor core, it was only a matter of time before they started using two, three, or more processor cores to achieve processing goals. The high-water mark to date for Tensilica's customers is 192 processor cores.

Despite numerous dire predictions over the past several decades, Moore's law has not stopped nor will it for many years. Semiconductor process technology will continue to advance. As it does, the increasing practicality of SOC designs like the MediaWorks MW301 and Cisco's SPP will cause all system designers to adopt multiple-processor system design. This change is inevitable because the semiconductor technologies, tool flows, and existing technical skills of working design engineers all support it. In fact, it is so predictable that the rapid rise in the number of on-chip processors is now built into the ITRS semiconductor roadmap.

1.12 VEERING AWAY FROM PROCESSOR MULTITASKING IN SOC DESIGN

The contemporary design trend toward increasing the number of on-chip processor cores is a significant counter trend to decades-old multitasking. Where multitasking loads multiple tasks onto one processor—which increases software complexity, forces processor clock rates up, and therefore increases processor power dissipation—use of multiple on-chip processors takes system design in a different direction by reducing the number of tasks each processor must execute. This design style simplifies the software by reducing the possibility of intertask interference, cutting software overhead (it takes extra processor cycles to just schedule and track multiple software tasks), and thus moderating the rise of processor clock rates.

Multitasking was developed when microprocessors were expensive; when every processor cycle was precious; and when the use of multiple processors was completely out of the question for reasons of engineering economics, circuit-board real estate, and power dissipation. However, Moore's law has now reduced the cost of silicon for an on-chip processor to mere pennies (or less) and these costs will further decline as the relentless application of Moore's law continues to advance the semiconductor industry's ability to fit more transistors on a chip. Microprocessor cores on SOCs are no longer expensive—and they get cheaper every year. All system-design techniques based on the old assumptions about processor costs must be rethought just as system-design techniques had to be rethought when logic-synthesis tools started to appear. Decades old, pre-SOC system-design techniques that conserve processors (which are now cheap) in the face of increasing power dissipation, software complexity, and development cost are clearly obsolete in the 21st century.

1.13 PROCESSORS: THE ORIGINAL, REUSABLE DESIGN BLOCK

Microprocessors became successful because they were the first truly universal, reusable block of logic to become available. With firmware reprogramming, microprocessors could be made to perform a very wide range of tasks with no changes to the hardware. This characteristic allowed system designers to use fixed-ISA, packaged processor ICs in an ever-expanding number of systems. As the popularity of these universal system building blocks grew, an entire software-development tool industry grew up around packaged microprocessors. Large numbers of compiler and RTOS (real-time operating system) vendors popped into existence in the 1980s, the decade when microprocessors became firmly established in the system designers' lexicon.

When systems design began to migrate from the board level to the chip level, it was a natural and logical step to continue using fixed-ISA processor cores in SOCs. Packaged processors had to employ fixed ISAs to achieve economies of scale in the fabrication process. System designers became versed in the selection and use of fixed-ISA processors and the related tool sets for their system designs. Thus, when looking for a processor to use in an SOC design, system designers first turned to fixed-ISA processor cores. RISC microprocessor cores based on processors that had been designed for personal computers and workstations from ARM and MIPS Technologies were early favorites due to their low gate count.

However, when you're designing custom silicon, there's no technical need to limit a design to fixed-ISA microprocessor cores as there is for board-level systems based on discrete, pre-packaged microprocessors. If there's legacy software to reuse, there's certainly a reason to retain a particular microprocessor ISA from one system design to the next. However, if there is no legacy code or if the legacy code is written in C, system designers have a freer hand in the selection of a processor with a different ISA if such a processor improves the system's performance, power dissipation, or manufacturing cost.

The first practical configurable microprocessor cores started to appear in the late 1990s. A configurable processor core allows the system designer to custom tailor a microprocessor to more closely fit the intended application (or set of applications) on the SOC. A "closer fit" means that the processor's register set is sized appropriately for the intended task and that the processor's instructions also closely fit the intended task. For example, a processor tailored for digital audio applications may need a set of 24-bit registers for the audio data and a set of specialized instructions that operate on 24-bit audio data using a minimum number of clock cycles.

Processor tailoring offers several benefits. Tailored instructions perform assigned tasks in fewer clock cycles. For real-time applications such as audio processing, the reduction in clock cycles directly lowers operating clock rates, which in turn cuts power dissipation. Lower power dissipation extends battery life for portable systems and reduces the system costs associated with cooling in all systems. Lower clock rates also allow the SOC to be fabricated in slower and therefore less expensive IC-fabrication technologies.

Even though the technological barriers to freer ISA selection were torn down by the migration of systems to chip-level design, system-design habits are hard things to break. Many system designers who are well versed in comparing and evaluating fixed-ISA processors from various vendors elect to stay with the familiar, which is perceived as a conservative design approach. When faced with designing next-generation systems, these designers immediately start looking for processors with higher clock rates that are just fast enough to meet the new system's performance requirements. Then they start to worry about finding batteries or power supplies with extra capacity to handle the higher power dissipation that accompanies operating these processors at higher frequencies. They also start to worry about finding ways to remove the extra waste heat from the system package. In short, the design approach cited above is not nearly as conservative as it is perceived; it is merely old fashioned. In Valley Speak (that is, San Fernando Valley), "It is so last century."

It is not necessary for SOC designers to incur these system-design problems if they use the full breadth of the technologies available instead of limiting themselves to older design techniques developed when on-chip transistors were not so plentiful. Moore's law has provided new ways to deal with the challenges of rising system complexity, market uncertainties, and escalating performance goals. This book focuses on one such system technology: configurable and preconfigured, application-specific processor cores.

1.14 A CLOSER LOOK AT 21st-CENTURY PROCESSOR CORES FOR SOC DESIGN

Microprocessor cores used for SOC design are the direct descendents of Intel's original 4004 microprocessor. They are all software-driven, stored-program machines with bus interconnections. Just as packaged microprocessor ICs vary widely in their attributes, so do microprocessors packaged as IP cores. Microprocessor cores vary in architecture, word width, performance characteristics, number and width of buses, cache interfaces, local memory interfaces, and so on. Early on, when transistors were somewhat scarce, many SOC designers used 8-bit microprocessor cores to save silicon real estate. In the 21st century however, some high-performance 32-bit RISC microprocessor cores consume less than 0.5 mm^2 of silicon, so there's no longer much reason to stay with lower performance processors. Indeed, the vast majority of SOC designers now use 32-bit processor cores.

In addition, microprocessor cores available as IP have become specialized just like their packaged IC brethren. Thus you'll find 32-bit, general-purpose processor cores and DSP cores. Some vendors offer other sorts of very specialized microprocessor IP such as security and encryption processors, media processors, and network processors. This architectural diversity is accompanied by substantial variation in software-development tools, which greatly complicates the lives of developers on SOC firmware-development teams.

The reason for this complication is largely historic. As microprocessors evolved, a parallel evolution in software-development tools also occurred. A split between the processor developers and tool developers opened and grew. Processor developers preferred to focus on hardware architectural advances and tool developers focused on compiler advancements. Processor developers would labor to produce the next great processor architecture and, after the processor was introduced, software-tool developers would find ways to exploit the new architectural hardware to produce more efficient compilers.

In one way, this split was very good for the industry. It put a larger number of people to work on the parallel problems of processor architecture and compiler efficiency. As a result, it's likely that microprocessors and software tools evolved more quickly than if the developments had remained more closely linked.

However, this split has also produced a particular style of system design that is now limiting the industry's ability to design advanced systems. SOC designers compare and select processor cores the way they previously compared and selected packaged microprocessor ICs. They look at classic, time-proven figures of merit such as clock rate, main-bus bandwidth, cache-memory performance attributes, and the number of available third-party software-development tools to compare and select processors. Once a processor has been selected, the SOC development team then chooses the best compiler for that processor, based on other figures of merit. Often, the most familiar compiler is the winner because learning a new software-development tool suite consumes precious time more economically spent on actual development work.

If the SOC requires more than one processor, it's often because there's specialized processing to be done. In the vast majority of such cases, there is some sort of signal or image processing to be performed and a general-purpose processor isn't an efficient choice for such work. Usually, this situation leads to another processor selection, this time for a DSP and an associated software-development tool suite.

The big problem with this selection method is that it assumes that the laws of the microprocessor universe have remained unchanged for decades. This assumption is most definitely flawed in the 21st century. Processor cores for SOC designs can be far more plastic than packaged microprocessor ICs for board-level system designs. Shaping these processor cores for specific applications produces much better processor efficiency and much lower system clock rates. In short, using the full abilities of microprocessor core IP to more closely fit the application problems produces better system designs.

Certainly, many microprocessor cores are not at all plastic. They lack configurability. However, a relatively new class of microprocessor core, the configurable microprocessor, exploits the plastic nature of the SOC's underlying silicon to permit the creation of processors that perform much better in SOC designs than older microprocessor cores, which are based on processor architectures that were originally developed to fit into DIPs and thus inherit some of the limitations placed on packaged microprocessor ICs.

Tensilica's Xtensa processor family is an example of such a configurable core. The Xtensa architecture was specifically designed for SOC use. It was designed to fully exploit the plastic nature of nanometer silicon. The original Xtensa processors were available as fully configurable processor cores. These processor cores have now been joined by a family of preconfigured microprocessor cores called the Diamond Standard Series, which is based on and compatible with Xtensa processors. Members of the Diamond Standard Series of processor cores have already been tailored for specific SOC applications and are not further configurable.

Together, the Xtensa and Diamond processor cores constitute a family of software-compatible microprocessors covering an extremely wide performance range from simple control processors, to DSPs, to 3-way superscalar processors. The configurability of Xtensa processors allows the performance range to grow even further, to the limits supported by the underlying silicon. Yet all of these processors use the same software-development tools so that programmers familiar with one processor in the family can easily switch to another.

As a consequence of this microprocessor core family's wide performance range, it's entirely possible to develop entire SOCs using only Xtensa and Diamond microprocessor cores, barring issues of legacy software. In fact many SOCs have already been designed in just this manner. The benefit of this system-design approach is that it boosts both hardware- and firmware-development productivity in a way that cannot be approached when using old system-design methods that employ different processor architectures for different tasks on the SOC—and therefore different processor-interface schemes and different software-development tools.

This book emphasizes a processor-centric MPSOC design style shaped by the realities of the 21st-century and nanometer silicon. It advocates the assignment of tasks to firmware-controlled processors whenever possible to maximize SOC flexibility, cut power dissipation, reduce the size and number of hand-built logic blocks, shrink the associated hardware-verification effort, and minimize the overall design risk. The design examples in this book employ members of the Xtensa and Diamond Standard processor families because of the extremely broad performance range and extended I/O abilities that these processor cores offer as a group. The advanced SOC design styles discussed in this book can be used with other microprocessor cores to create MPSOCs, but it will be more difficult to achieve the same results.

BIBLIOGRAPHY

Dennard, R.H., Gaensslen, F.H., Yu, H-N., Rideout, V.L., Bassous, E. and LeBlanc, A.R. "Design of ion-implanted MOSFETs with very small physical dimensions," *IEEE Journal of Solid-State Circuits*, 9(5), October 1974, pp. 256–268.

Leibson, S. "The end of Moore's law," *Embedded Systems Design*, December 2005, pp. 40–47.

Moore, G. "Cramming more components onto integrated circuits," *Electronics*, 38(8), April 19, 1965, pp. 114–117.

Moore, G.E., "Progress in Digital Integrated Electronics," *Technical Digest 1975*, International Electron Devices Meeting, IEEE, 1975, pp. 11–13.

CHAPTER 2

The SOC Design Flow

Our primary tools for comprehending and controlling complex objects are structure and abstraction
—Niklaus Wirth

SOC design techniques are layered upon ASIC design methods. Therefore, any proposed system-design style intended for SOCs must be compatible with ASIC design flows. A truly practical SOC design style must, in fact, seamlessly dovetail with the ASIC design flow. Such a design style should not require the SOC design team to learn and use radically new design tools. Some newly proposed system-design approaches such as so-called "C-to-gates," behavioral synthesis, reconfigurable-logic, and design methods based on the use of reconfigurable processors attempt to overcome shortcomings in existing SOC design techniques that have developed over the last 30 years. However, these design styles have not achieved wide acceptance because they either have significant shortcomings of their own or they require SOC designers to learn new and radically different ways to design systems.

The processor-centric MPSOC (multi-processor SOC) design style advocated in this book does indeed dovetail nicely with the existing ASIC design flow. It does not require SOC designers to learn radically new ways to design systems. It does ask SOC designers to unshackle their thinking from the many limitations previously imposed by 20th-century microprocessors and processor cores discussed in Chapter 1.

2.1 SYSTEM-DESIGN GOALS

All system designers have three major tasks:

- achieve system-design goals
- minimize project costs
- reduce design risks

At the same time, rising system complexity conspires to make these tasks increasingly difficult. For example, a system designed in 1995 might have simple I/O requirements such as a few parallel and asynchronous serial ports. Today's electronic systems bristle with USB, IEEE 1394, and Ethernet ports that employ many complex protocol layers. Consequently, system-processing requirements have grown and are growing dramatically. Increasingly, system designers turn to SOCs to achieve the system complexity and cost goals. However, due to its very nature, SOC design is inherently more costly and riskier than board-level design.

Because SOCs are physically smaller, run at lower power, and have greatly expanded abilities compared to board-level systems based on off-the-shelf ICs, system designers can tackle far more challenging designs that involve processing tasks such as complex data encoding and decoding (for audio and video compression), encryption, and signal processing. The greatly increased systems complexity incurred by these additional processing loads increases design risk—complicated systems are harder to design and debug. SOCs are inherently riskier and more costly to develop than board-level designs because respinning a circuit board takes a day or a week. Respinning a chip takes months.

2.2 THE ASIC DESIGN FLOW

SOC design is quite unlike board-level system design. There's much less physical prototyping in the SOC world. Instead, SOC design verification relies almost exclusively on simulations. In fact, the SOC EDA design toolkit and ASIC design flow differ substantially from board-level design tools and flow. Figure 2.1, adapted from the *Reuse Methodology Manual for System-on-a-Chip Design*, presents a simplified view of the ASIC design flow called the "waterfall model" that was already showing its age when the manual was published in 1998.

The first box of the simplified ASIC design flow in Figure 2.1 shows the creation of the system specification. For simpler ASICs, this is possibly a straightforward task. However, this simple box belies the complexity of developing system specifications for complex SOCs. We will ignore this issue, for now.

Once the system specification is complete, the simplified ASIC design flow in Figure 2.1 shows conversion of the system specification into RTL. Again, it may have been possible to compose an entire ASIC in RTL in earlier times, but SOCs aren't so simple. At the very least, complex SOCs require the integration of numerous IP blocks including processors, memories, and other pre-coded hardware. Such IP integration may be implied by the "RTL code development" block in Figure 2.1, but it's not obvious.

The third box in the simplified ASIC design flow is functional verification. In the simplified flow of Figure 2.1, this box represents the first time the system is simulated. Realistically, if an SOC design team waits until

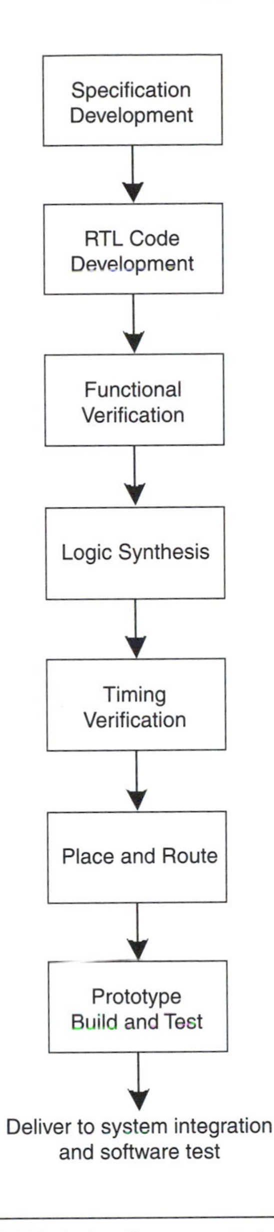

■ **FIGURE 2.1**

This simplified representation of the ASIC design flow enumerates some of the steps required to transform a system specification into a chip layout.

after the RTL coding for the entire SOC is finished, large design problems are very likely to erupt. RTL code is a hand-translated version of a system specification and converting a system specification into an hardware-description language (HDL) description is a time-consuming and error-prone manual process. If system verification only occurs after RTL coding, as proposed in Figure 2.1, errors from the overall system design will only be discovered after considerable time and effort has been expended on the coding phase. A realistic design flow for complex SOCs simply cannot postpone initial system verification until after RTL coding. To be effective and helpful, system verification must start much earlier.

After functional verification, Figure 2.1 shows logic synthesis occurring. Timing verification follows. As with functional system verification, the simple ASIC design flow of Figure 2.1 shows timing verification occurring far too late in the design flow for complex SOC design. Trying to achieve timing closure for an entire system becomes extremely difficult as system complexity and clock rates climb. Postponing all timing simulation until after logic synthesis is a prescription for disaster when designing complex SOCs.

Figure 2.1 shows placement and routing following timing verification. Once the chip is placed and routed, it can be taped out and fabricated. Figure 2.1 then shows system integration and software testing almost as an afterthought. This step is where most of the big design disasters materialize. If system integration and software testing strictly follow chip fabrication, the probability of getting a system that fails to meet system requirements are very high indeed. Many SOC design teams have faced this dismal outcome.

The *Reuse Methodology Manual* notes that the waterfall model worked very well for designs with a complexity of as many as 100K gates. However, the next sentence in the *Manual* starkly contradicts this claim: "It produced chips that worked right the first time, although often the systems that were populated with them did not." Systems that don't work right even if the key component chips meet spec still qualify as design disasters. Ultimately, the *Manual* concludes that the simple waterfall-development model simply does not work for more complex SOC designs.

Figure 2.2 shows an expanded ASIC design flow. This figure provides more detail than Figure 2.1 and shows the people involved at each design step, the tasks they perform, and the tools used to perform each of the tasks. This figure provides a better view into the steps needed to convert a system definition into a placed and routed netlist that is ready for chip fabrication. Double-ended arrows in the first column of the figure show that communication passes continually up and down the design chain so the expanded design flow has feedback paths, which the simple flow shown in Figure 2.1 lacked.

However, Figure 2.2 still falls short of showing an SOC design flow (as opposed to an ASIC design flow) because the entire SOC design occupies just one box—a task to be performed by the system architect. System design is inherently more complex than the simple single-box representation in Figure 2.2. At the very least, the ASIC design flow shown in Figure 2.2 is dedicated to producing a chip, not a system. In particular, Figure 2.2 shows no software developers or software development. Similarly, no system-integration or system-testing tasks appear in this diagram. ASIC

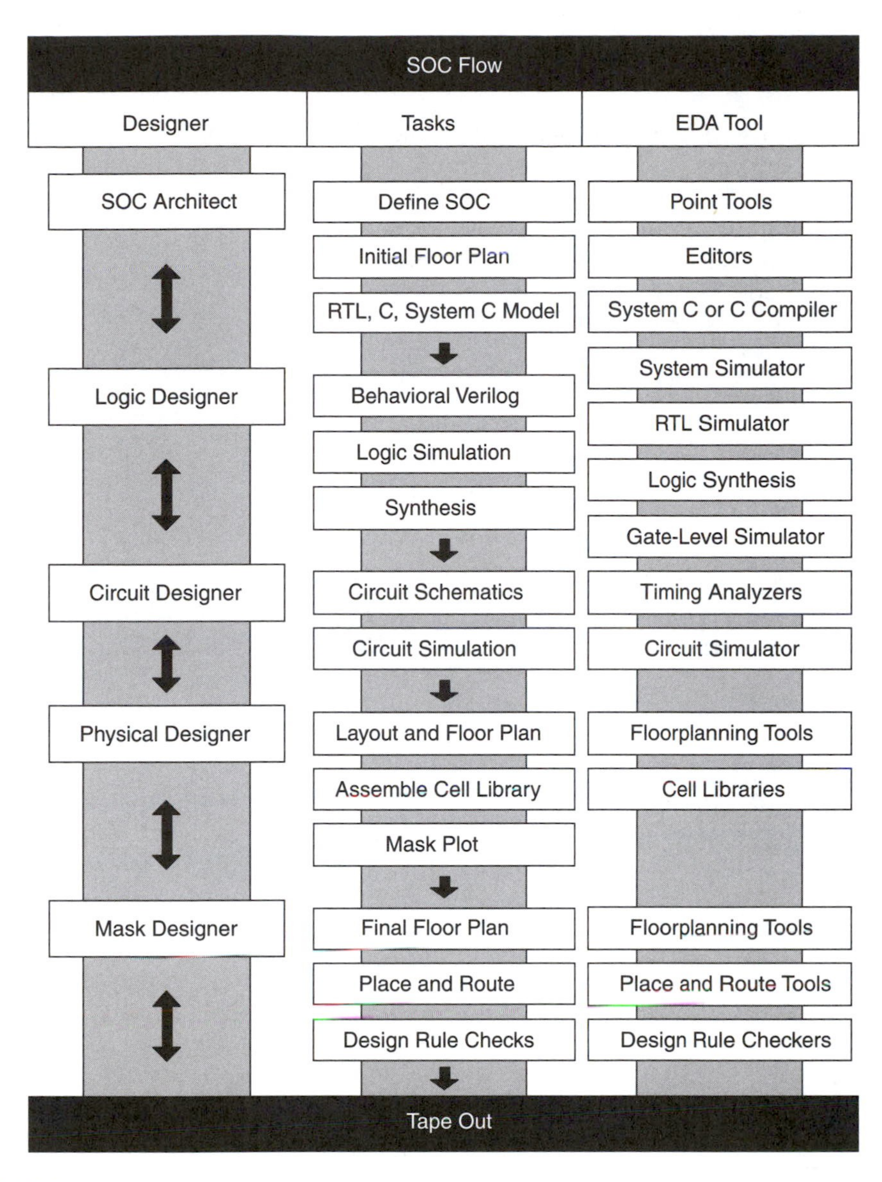

■ **FIGURE 2.2**

An expanded ASIC design flow details the people involved at each design step, the tasks they perform, and the tools used to perform each of the tasks.

design flows do not adequately describe system-design tasks that must preface the work of translating RTL descriptions into silicon. Ultimately, ASIC design flows focus on producing chips, not systems. If the system isn't right, the silicon won't be right either.

2.3 THE ad-hoc SOC DESIGN FLOW

The absence of a well-accepted, industry-standard SOC design flow has produced numerous ad-hoc approaches. The first step in one very popular approach is to design a simple microprocessor system like the one shown in Figure 2.3. This system closely resembles every simple microprocessor-based system designed since the Intel 4004 microprocessor became available in 1971.

The design of the microprocessor-based system shown in Figure 2.3 is quite generic. It merely connects the processor to memory and peripherals. If this simple system design happens to also meet the needs of the target system, it's purely serendipitous. However, if this simple design is sufficient, then the system design is finished. Most likely, if this simple design meets the needs of the overall system, then a standard, off-the-shelf microcontroller would likely serve the system design just as well, for a lot less money.

In fact, the ad-hoc system-design style that produces generic systems such as the one shown in Figure 2.3 routinely produces ASSPs (application-specific standard products) such as the OMAP (Open Media Applications Platform) chips from Texas Instruments. The block diagram of the OMAP chip in Figure 2.4 shows a 2-processor system with a shared DMA controller, a shared-memory controller, and a collection of peripheral blocks. Study the block diagram alone and you will not discern the OMAP chip's intended function.

The OMAP chip's system design doesn't at all resemble the actual application problem's block diagram. Firmware must perform the transformation

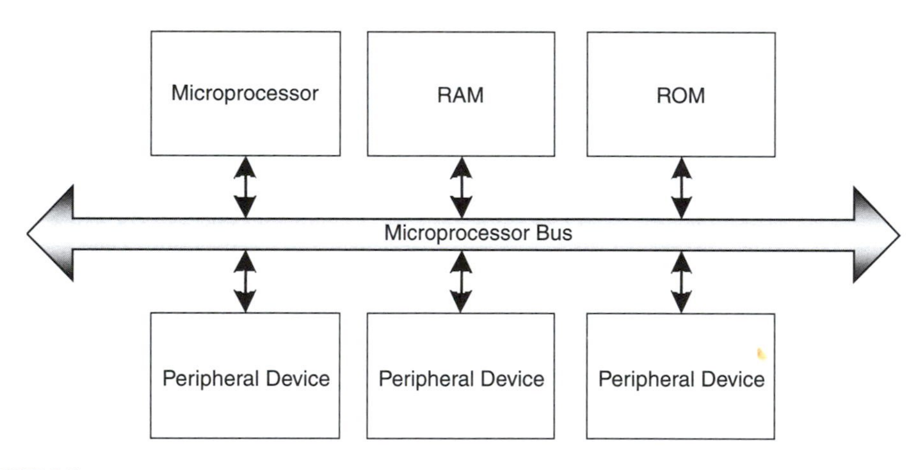

■ **FIGURE 2.3**

This generic, microprocessor-based system is representative of many system designs. It is so successful because of the microprocessor's inherent flexibility, not because this system-design matches the actual needs of a system.

from generic to specific. As a system's performance goals expand, it becomes increasingly difficult for firmware to transform generic hardware into a high-performance system.

The OMAP chip demonstrates two additional "rules" of the ad-hoc approach to system design:

- If one processor lacks sufficient processing power, add another.
- If another processor won't do, add an application accelerator hardware block.

The OMAP chip uses both of these approaches. The general-purpose control processor is not adequately efficient for signal processing, so the OMAP chip adds a DSP (digital signal processor) core to the mix. The chip also has a DMA control block that performs data movement to offload this task from either of the two on-chip processor cores because neither processor can move data as efficiently as a DMA controller.

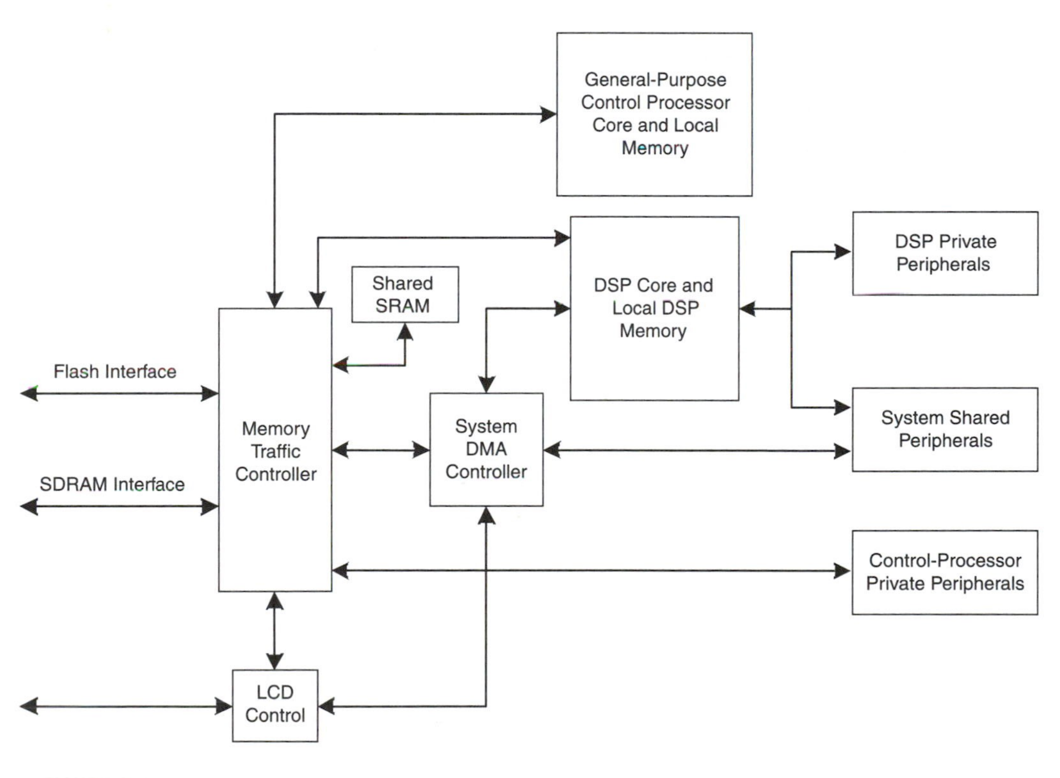

■ **FIGURE 2.4**

This OMAP chip from Texas Instruments epitomizes the ad-hoc approach to system design, which produces a generic platform chip. Firmware gives the chip a specific function.

The two processors and the DMA controller all communicate with most of the other system blocks on the chip, which makes the design generic and therefore less than optimal for any specific application because the likelihood of resource conflict is very high, especially on buses shared by the two processors and the DMA controller.

There is nothing wrong with using multiple processors and hardware accelerators to improve an SOC's performance, but an ad-hoc approach to combining these system components is not likely to provide an optimal design because of unnecessary shared-resource conflicts. In fact, the ad-hoc design approach often produces a design that fails to meet system-performance goals because of these shortcomings.

For examples of system architectures directly derived from the problem being solved, see Figures 2.5–2.8. These figures show, respectively, a HiperLAN/2 decoder, a UMTS decoder, a Mondiale digital radio decoder, and an H.264 digital video decoder. Each of these four block diagrams shows a variety of different tasks to be performed and the expected bit rates required of the communications paths linking these tasks. This detailed level of understanding helps the system designer create successful, efficient SOCs.

None of the decoder systems shown in these four block diagrams looks remotely like a microprocessor-based system. All four of the systems shown in Figures 2.5–2.8 consist of a series of blocks communicating with each other over point-to-point connections defined by the overall application and the various tasks to be performed, not by the underlying implementation technology.

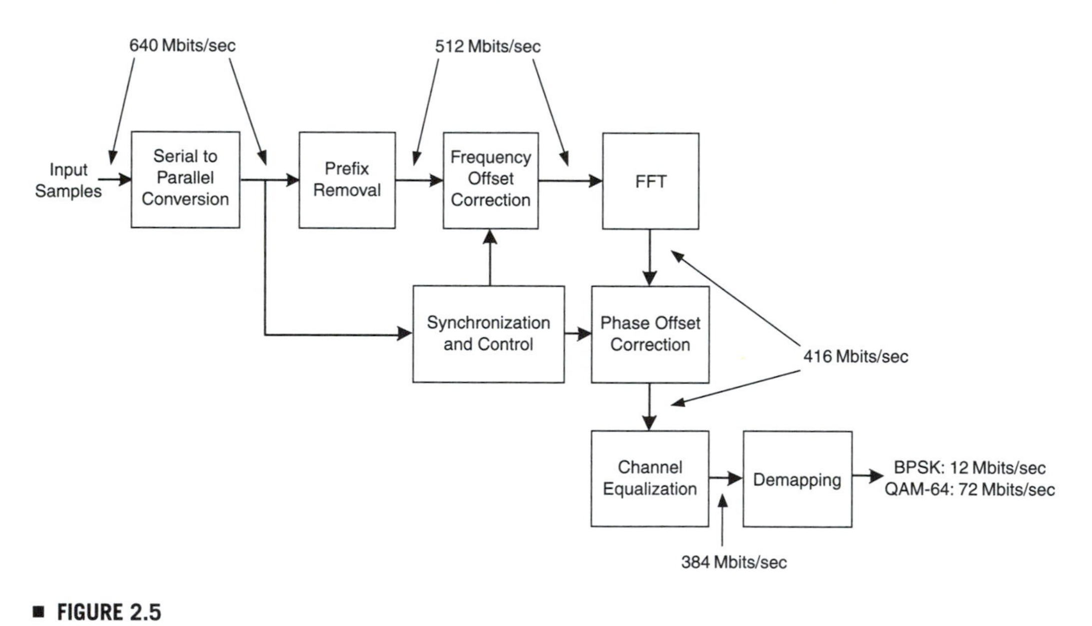

■ **FIGURE 2.5**

HiperLAN/2 wireless LAN decoder.

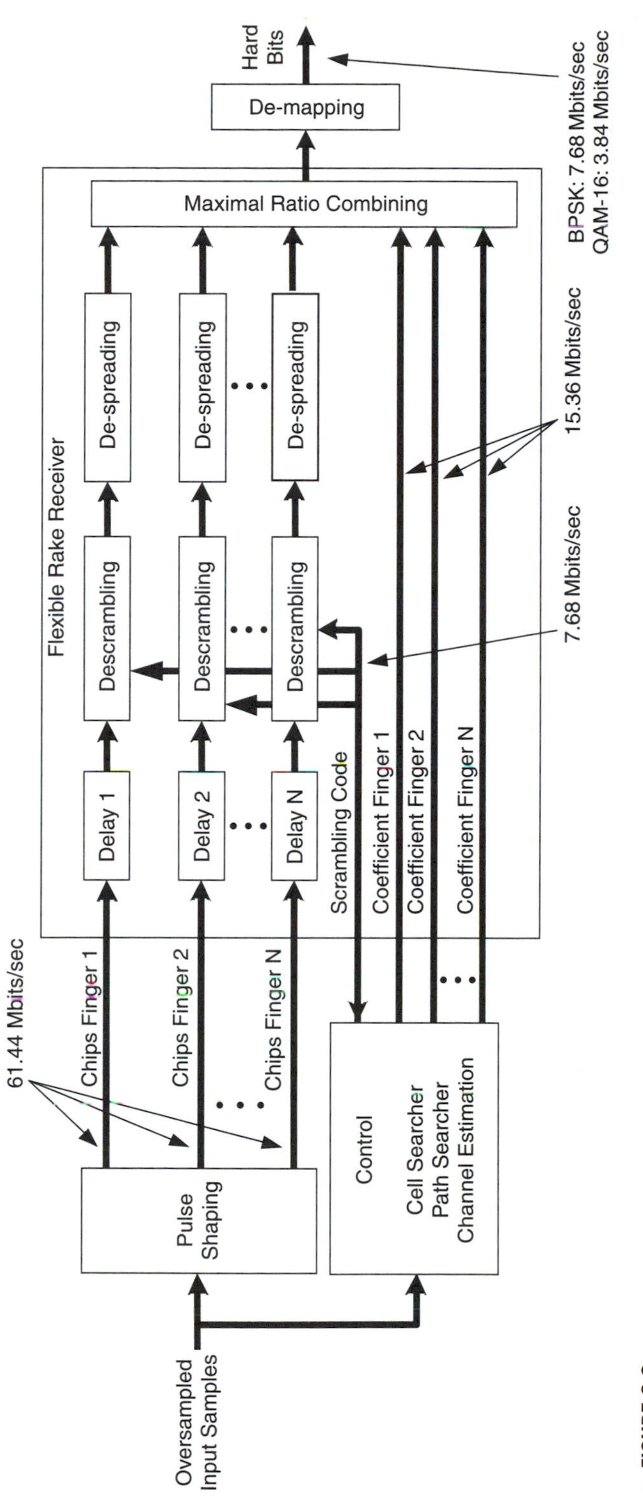

■ **FIGURE 2.6**

UMTS digital cellular telephone decoder.

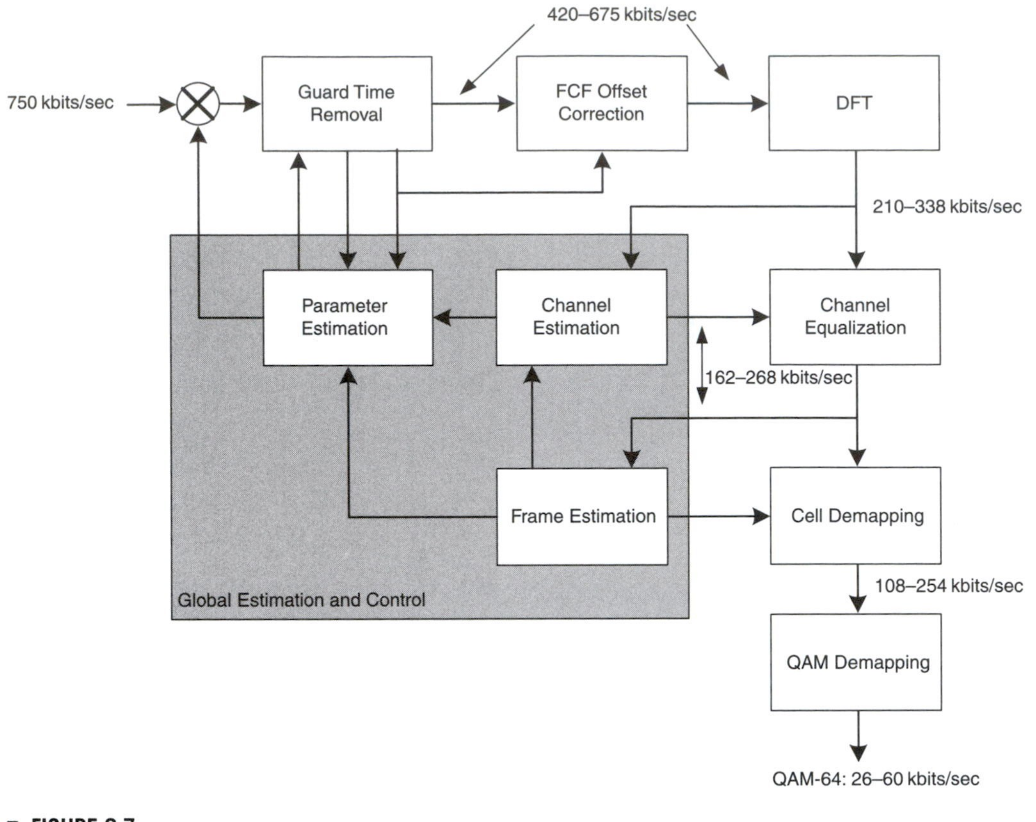

■ **FIGURE 2.7**

Mondiale digital radio decoder.

Systems that perform digital decoding and encoding, encryption and decryption, and other data-transformation algorithms are optimally organized or configured when the system block diagram matches the problem at hand. It is our familiarity with 30 years of microprocessor-based system design that starts us thinking along the lines of force-fitting task-specific architectural structures to traditional, bus-based microprocessor systems.

For example, look at Figure 2.9, which shows the mapping of a set-top box system design into a traditional bus-based microprocessor system as it appeared in Chapter 4 of the book *Surviving the SOC Revolution: A Guide to Platform-Based Design*. After 30 years of microprocessor-based system-design experience, SOC designers simply take such re-mappings for granted. They're usually not even consciously aware that remapping has occurred. However, such mappings are not the only way to design

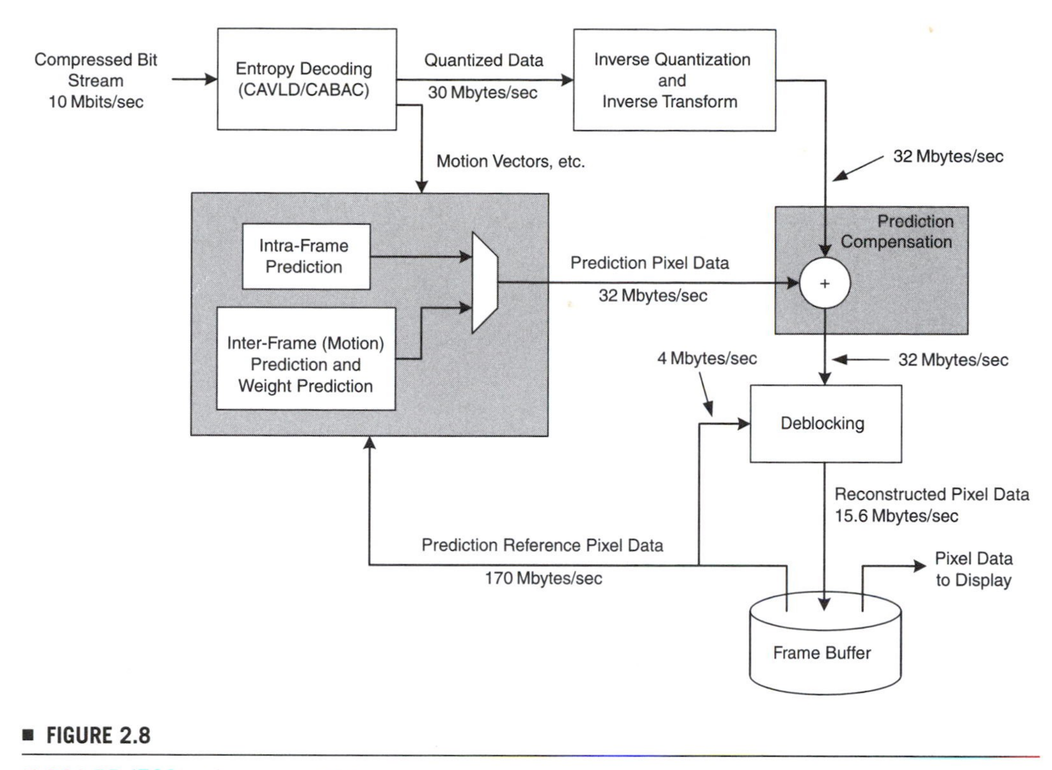

■ **FIGURE 2.8**

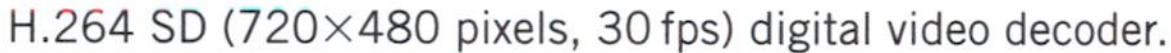

H.264 SD (720×480 pixels, 30 fps) digital video decoder.

SOCs. In many cases, these mappings no longer represent the best way to develop 21st-century system architectures. In the system shown in Figure 2.9, for example, all communications in the system must traverse the main system bus to reach any of the system's computational blocks. This requirement immediately forces the bus to become a critical shared resource that can easily be overwhelmed in high-performance operating modes.

After many decades of microprocessor design experience, we have become so acclimated to treating processors as scarce system resources that we automatically try to minimize the number of processors without sufficient engineering analysis of processor and bus loading. If it's discovered well into the design project that the processor has too much work to perform in the allotted time, a higher clock rate is often the most expedient remedy sought although the cure may require use of a more expensive IC process technology and it will likely incur higher—perhaps even unacceptable—power-dissipation levels.

Too often, we blindly accept the inefficiencies that result from using outdated system-design styles to force-fit task-oriented systems into

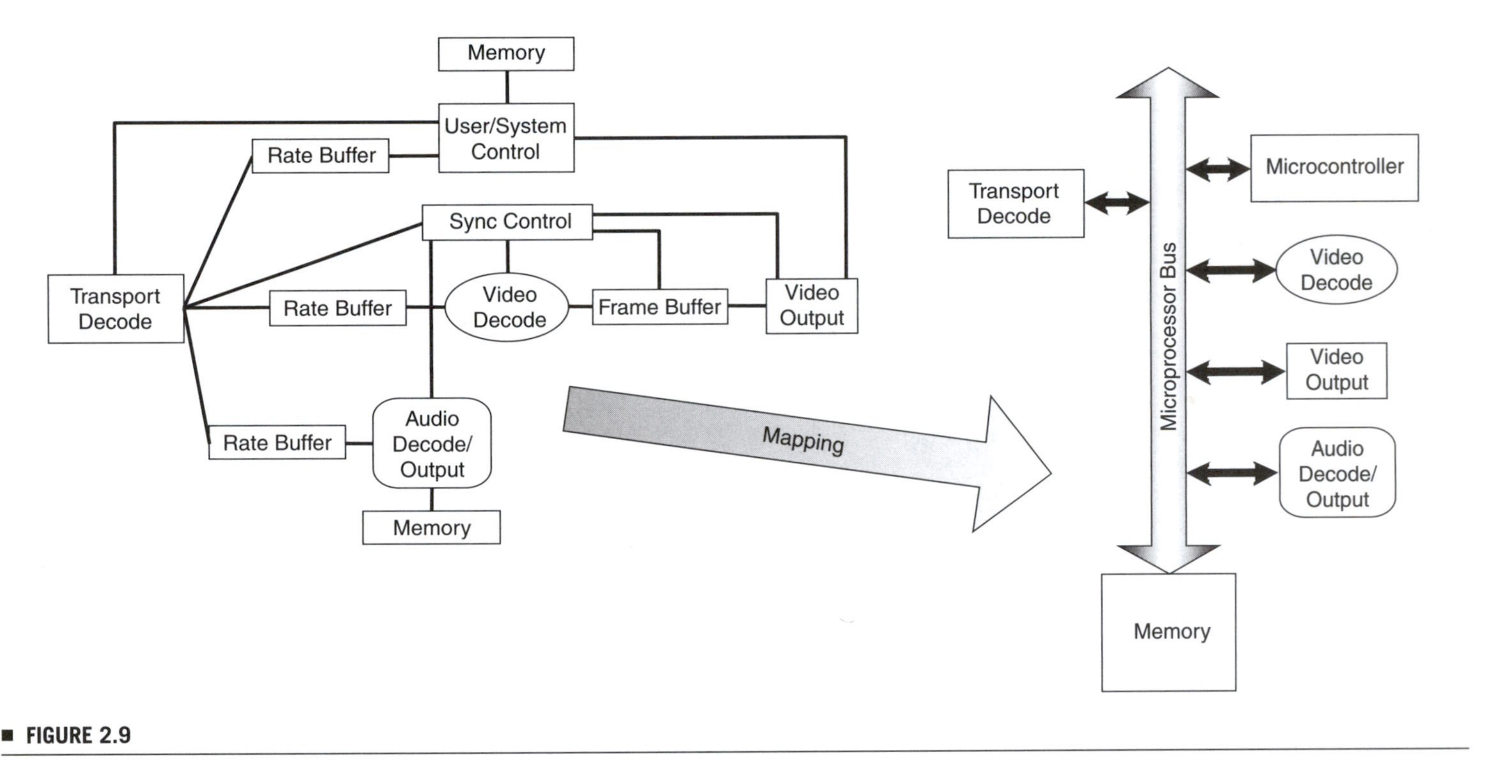

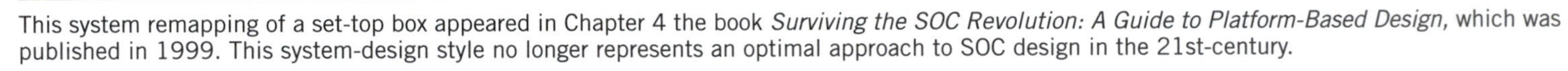

■ **FIGURE 2.9**

This system remapping of a set-top box appeared in Chapter 4 the book *Surviving the SOC Revolution: A Guide to Platform-Based Design*, which was published in 1999. This system-design style no longer represents an optimal approach to SOC design in the 21st-century.

hierarchical, bus-oriented hardware architectures that bear no structural resemblance to the problems being tackled. The inefficiencies of generic microprocessor-based architectures have become camouflaged through their extreme familiarity; they simply blend into today's system-design landscape.

The ceaseless advance of Moore's Law, the advent of nanometer silicon, and the appearance of mega-gate SOCs make on-chip processors far more available so that the "processor-as-scarce-resource" mentality is now thoroughly outdated. Today's silicon supports and encompasses a broader and richer universe of practical system designs that meet performance goals while running at lower clock rates and with lower power requirements. Companies designing SOCs must change their system-design styles to accommodate this shift and to stay competitive.

2.4 A SYSTEMATIC MPSOC DESIGN FLOW

Just as an ad-hoc architectural approach to skyscrapers and other large civil structures produces jumbled, inefficient buildings, the ad-hoc approach to system architectures produces inefficient electronic systems for complex applications. Figure 2.10 presents a detailed, systematic flow for efficient MPSOC design. This diagram dovetails with the ASIC design flow shown in Figure 2.2. Instead of corralling the system-design process into one small process box, the MPSOC design flow encapsulates the ASIC implementation flow in one box and details the tasks required to create an overall design that meets system-design goals. Because SOC prototyping is next to impossible, this systematic approach to MPSOC design focuses on extensive system and subsystem simulation, performed early and often. Simulation should start during architectural design, long before any RTL is written.

The systematic design approach illustrated in Figure 2.10 starts with the definition of the two essential system attributes: the computational requirements and the I/O interface requirements. These two requirement sets are then combined and used to write a set of abstract task models that define a system. The most likely candidate language for these models is SystemC, now codified as IEEE Standard 1666. Nearly all major EDA tool vendors now offer SystemC simulators.

System-level simulation at such a high abstraction level allows the SOC design team to quickly explore candidate system architectures because these high-level simulations run quickly. High-level simulation results of the abstracted system allow the design team to see, analyze, and understand the critical data flows in each candidate architecture, refine the overall system architecture to balance computational and data-transfer loads, and relieve the inevitable and unforeseen bottlenecks that every complex system seems to have. At this early phase of the MPSOC

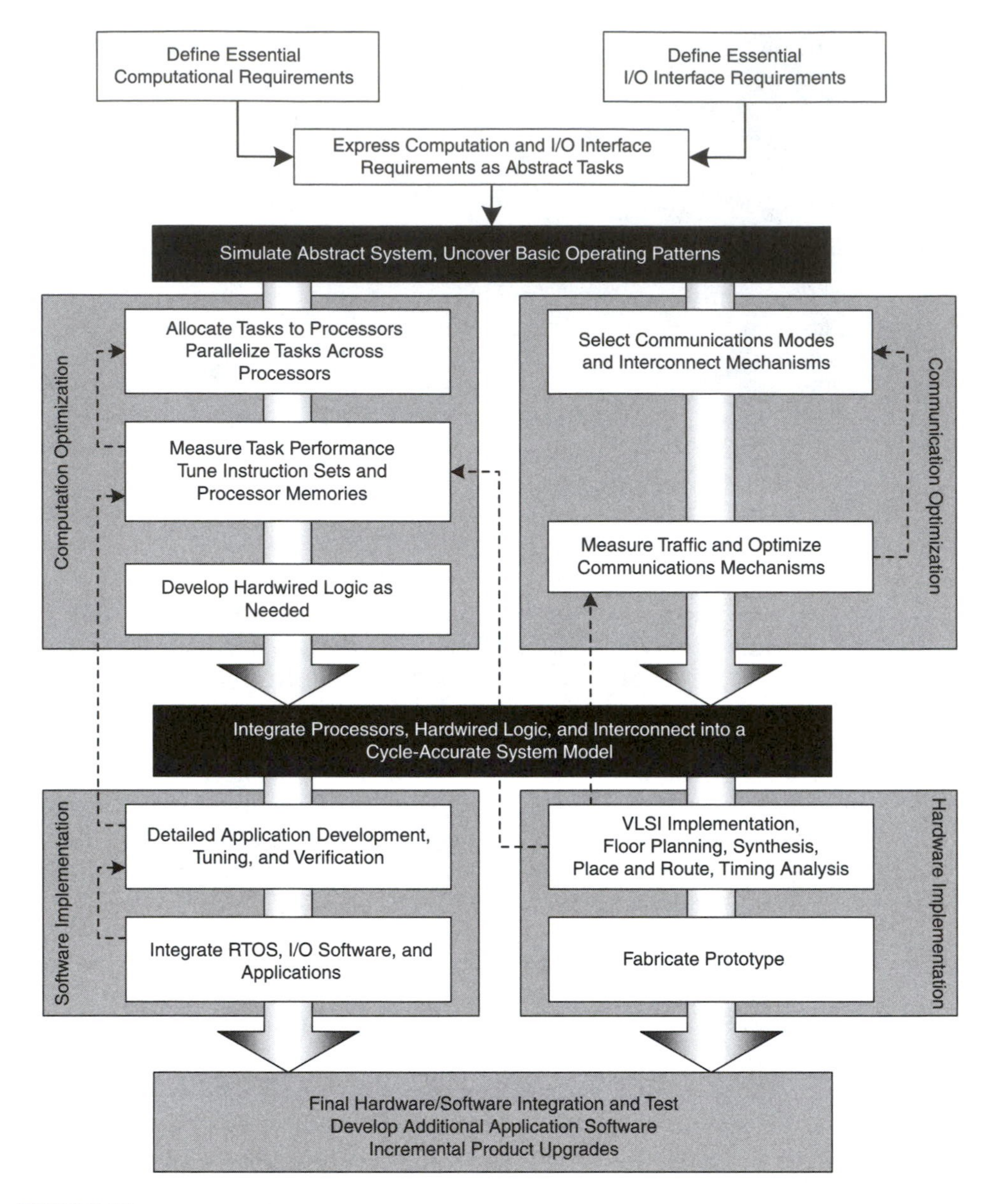

■ **FIGURE 2.10**

A systematic approach to MPSOC design will produce chips that more closely match the target application. In addition, this approach to system design will significantly reduce the number of problems encountered during design verification and hardware/software integration.

design, the design team should have selected the best of the candidate architectural designs and should have developed a good "gut feel" for how the system behaves under a variety of typical and worst-case operating conditions.

At this point, the design process splits into two subprocesses:

1. Optimization of computational tasks by mapping tasks to processors and other hardware blocks.
2. Optimization of inter-task communications by selecting appropriate interconnection structures (buses, shared memories, FIFO queues, or simple wires).

Because the abstract computation models are already written in SystemC, the systems represented by these abstract models can be readily simulated. Those tasks with low to moderate computational requirements can probably run on fixed-ISA (instruction-set architecture) and pre-configured processor cores without the need for excessive clock rates. Computational tasks with high-performance requirements such as media processing (audio and video) may not comfortably fit on one fixed-ISA processor without requiring that processor to run at wholly unrealistic or unreachable clock rates. SOC processor cores cannot achieve the multi-GHz clock rates of packaged PC processors, nor do most SOC designers wish to deal with the 50–100 W power dissipation that accompanies such clock rates.

2.5 COMPUTATIONAL ALTERNATIVES

There are several alternatives for executing high-performance tasks without resorting to stratospheric clock rates. These alternatives include tuning a configurable processor, splitting the task across more than one processor, and using or implementing a hardware block to perform the task. Each of these alternatives is quite viable and each should be carefully evaluated to make the most appropriate choice given overall system requirements.

Tuning a configurable processor allows the design team to retain the programmability and flexibility of a firmware-based computing element while adding parallel, programmable execution units to the processor to boost performance without raising clock rate. The industry is starting to use the term ASIP (application-specific instruction-set processor) to refer to such an augmented processor. In essence, developing an ASIP superficially resembles RTL hardware development in its addition of hardware execution units using a hardware-description language. But it differs markedly from RTL design by retaining the processor's firmware programmability, which permits state machines to be implemented in firmware. The next chapter in this book discusses this design approach in much more detail.

Splitting the high-performance task across multiple processors is another viable approach. Most complex, high-performance tasks have easily identifiable subtasks that can be distributed to multiple processors and executed in parallel. For example, H.264 video decoding consists of subtasks that include bitstream parsing and decoding and pixel reconstruction using macroblocks. One PC processor running at more than 1 GHz can implement the entire H.264 decoding algorithm. However, most consumer applications that would use an H.264 decoder lack the cooling fans or power sources needed to support such a processor in a system.

The single, complex video-decoding task can be split across two tailored ASIPs: a bitstream processor and a pixel processor. This approach results in a 2-processor, firmware-programmable hardware block that performs the desired function, with the desired performance, but without the excessive clock rate or power dissipation.

The third alternative is to either reuse an existing hardware block or develop a new one to implement the desired function. If such a hardware block already exists, and if it's unlikely that the function will change in the future, then this alternative may well be the best one.

2.6 COMMUNICATION ALTERNATIVES

As discussed earlier, the classic method used to interconnect SOC blocks is to use one or more microprocessor-style buses. This design style is a holdover from decades of microprocessor-based system-design experience. Buses are still useful for conveying inter-block traffic, but they are by no means the SOC designer's only alternative. A partial list of alternative interconnection schemes includes:

- Buses
- Shared memories
- FIFO queues
- Simple wires (point-to-point connections)
- On-chip networks

Buses and bus hierarchies are familiar to all system designers. Figure 2.11 shows a system that uses a bus hierarchy to move data around a system that contains two processors and a hardware accelerator block.

Shared or dual-port memories are common in systems that transfer large blocks of data between successive computational elements. The shared memory can hold data blocks such as network packets or video frames. Figure 2.12 shows a shared-memory system. Shared-memory subsystems are good for implementing systems where multiple computational blocks need random access to data within the data block.

Figure 2.13 illustrates a system that employs FIFO memories for inter-processor communications. Unlike shared memories, FIFOs are

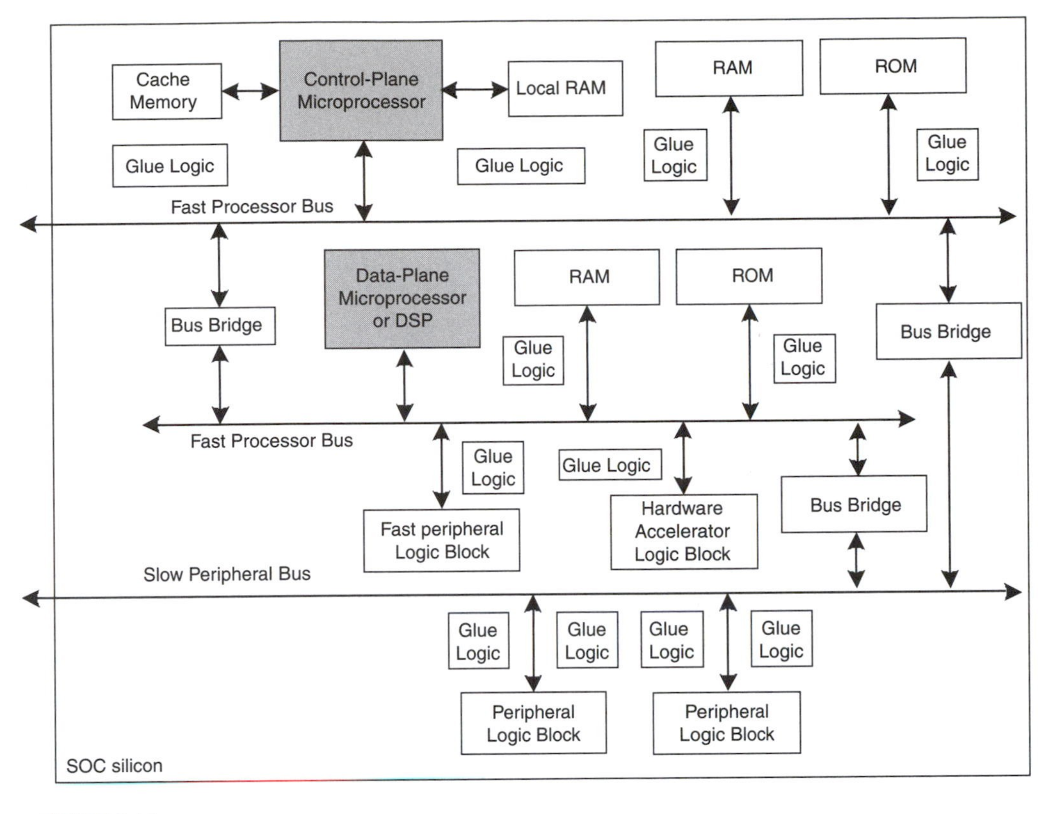

■ **FIGURE 2.11**

System design using a bus hierarchy.

unidirectional, so the system needs two FIFOs if large amounts of data must be passed back and forth between the computational elements. If the amount of data to be transferred is not equal in both directions, the two FIFOs need not even be the same size. However, FIFOs are fast because there's no contention for the port to shared-memory. One computational element controls the FIFO's head and the other controls its tail. The FIFO's head and tail are separate resources.

If the amount of information to be transferred between computational blocks is small, FIFOs and shared memories may be too complex. Simple wires may suffice. Often, these wires originate from a writeable register attached to one computational element and terminate in a readable register attached to the other computational element. Figure 2.14 illustrates such a system. When data-transfer requirements are relatively simple, wires represent a low-cost way of connecting computational elements.

Finally, considerable research is currently going into developing on-chip networks. Although this work is largely beyond this book's scope, you'll find some on-chip networking concepts discussed in the final chapter.

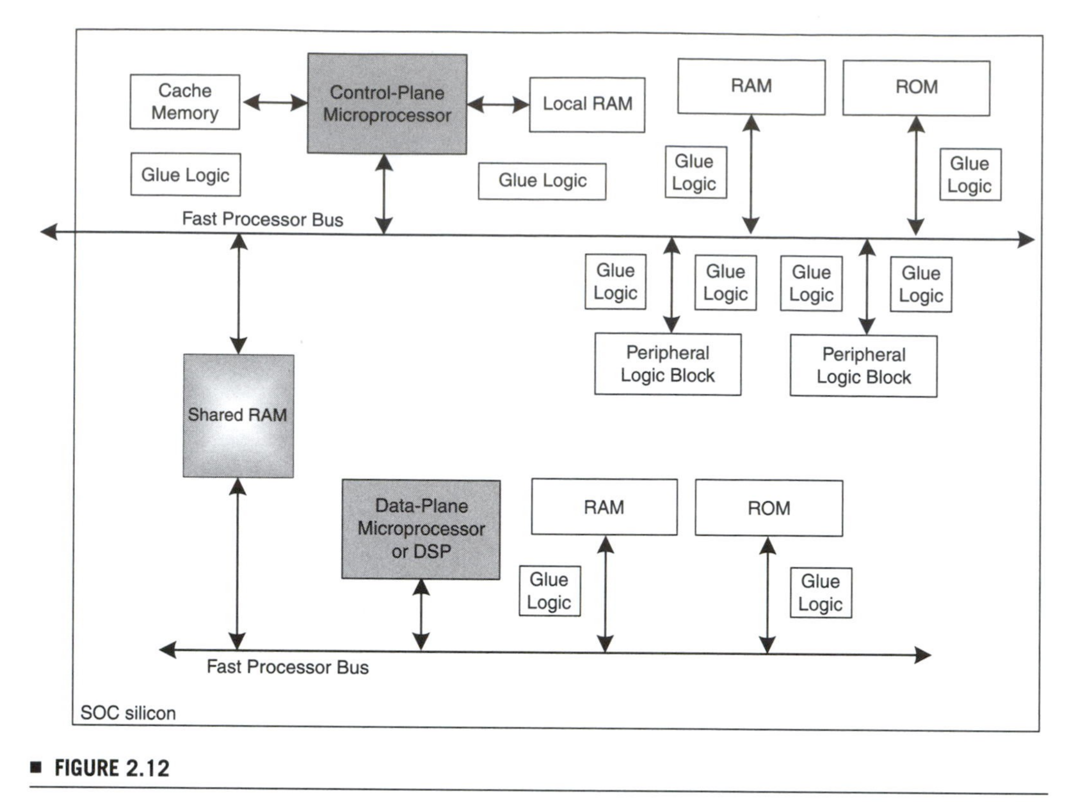

■ **FIGURE 2.12**

System design using shared memory.

One of the advantages of the MPSOC design style advocated in this book is that the methodology allows the SOC design team to experiment with the various interconnection schemes listed above and to measure, through simulations of the incrementally refined system model, how the system architecture responds to the use of the different interconnection methods. In general, the design team should be looking for the lowest-cost interconnection method that meets system-performance goals. However, it's always a good idea to provide a healthy amount of design margin for the unanticipated specification changes or next-generation goals that appear unexpectedly, and often in the middle of the project cycle—often at the least convenient moment.

2.7 CYCLE-ACCURATE SYSTEM SIMULATION

Once the abstracted system model produces the desired-performance results in system simulation, it's time to expand the abstract computational

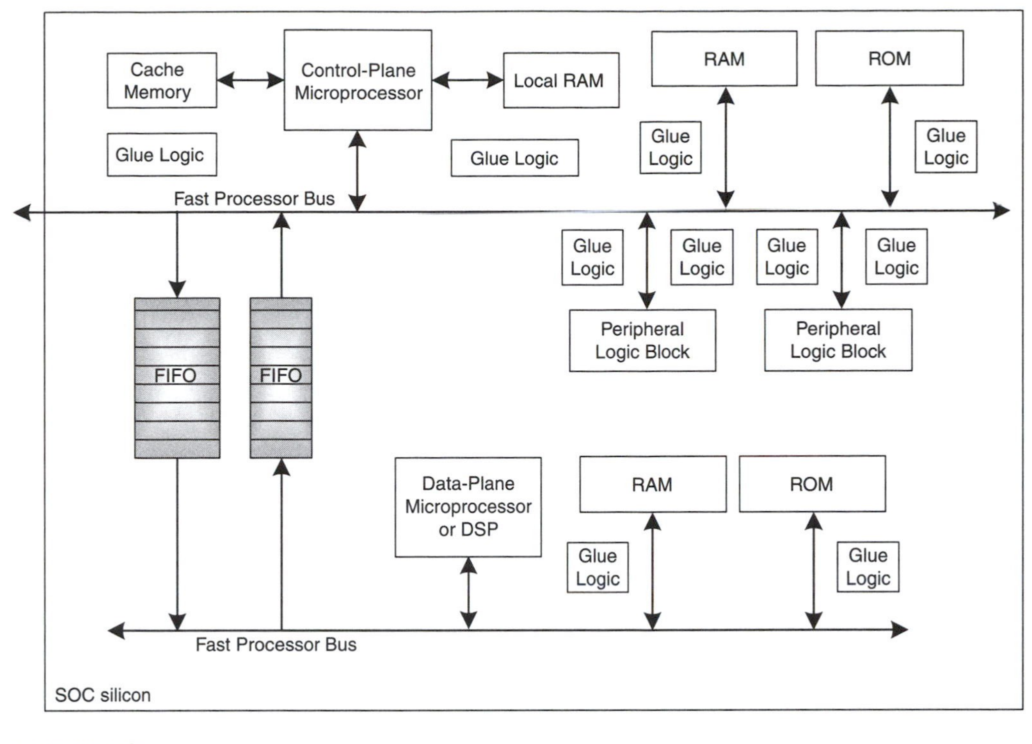

■ **FIGURE 2.13**

System design using FIFOs.

and communication models into cycle-accurate ones. For processors, such models run the actual system firmware on a cycle-accurate instruction-set simulator (ISS). For RTL blocks, a cycle-accurate transaction-level model (TLM) is used. This simulation level provides a more detailed, less abstracted look at overall system performance and serves as a critical checkpoint in the system's design. It validates the prior abstractions.

2.8 DETAILED IMPLEMENTATION

After the cycle-accurate simulations provide the desired system-performance results, the detailed implementation can begin. Again, work splits in to two. The software team now has a set of compilers, assemblers, debuggers, and instruction-set simulators for the processors selected in prior design steps. These tools allow software team to implement the computational and communication tasks in firmware.

Note that this design phase should not be the first time that the software-development team has been involved in the design. Software developers

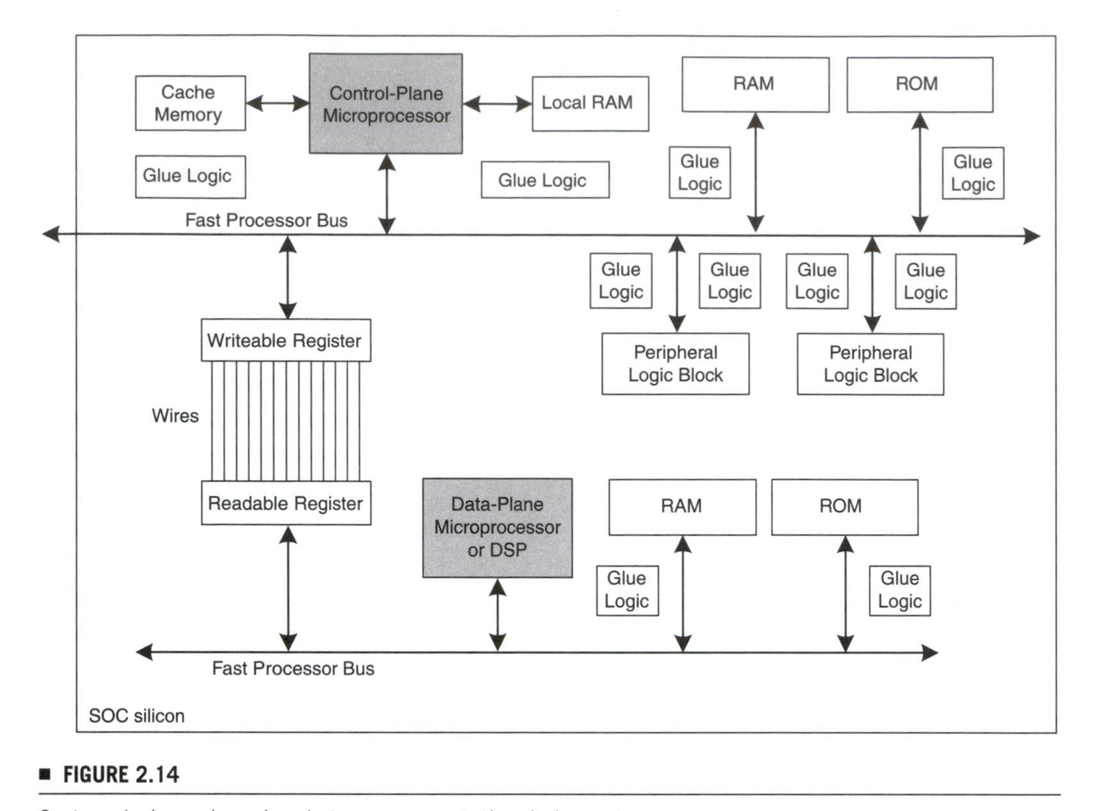

■ **FIGURE 2.14**

System design using wires between computational elements.

should have been involved in the process all along the way—during model development, abstract system simulation, selection of communication mechanisms, and during processor selection and ASIP development. The software team's participation during these design steps will be a major factor that determines the success of the design project.

At the same time, the hardware team now has a set of hardware blocks to integrate. These blocks include processors, memories, pre-built hardware IP blocks, and specifications for RTL blocks requiring detailed implementation. The hardware team also has a set of selected communications mechanisms to put into place. These mechanisms are chosen using quantitative simulation data for alternative on-chip communications schemes. Still to be performed is the assembly of these various blocks into an SOC and RTL coding for blocks yet to be implemented, followed by floor planning, synthesis, placement and routing, and timing simulation. In short, the hardware-implementation phase of the MPSOC design style is the ASIC design flow followed by chip fabrication.

When the hardware team and the software team have finished their detailed implementations, the SOC-development team is ready for final hardware/software integration and testing. In simplistic 20th-century design flows, final hardware/software integration is a time of great trepidation because no prior project work has given any quantitative indication that the integration will succeed. In the MPSOC design flow advocated in this chapter, substantial simulation modeling at multiple abstraction levels gives confidence of successful integration.

2.9 SUMMARY: HANDLING SOC COMPLEXITY

Simply put, SOC design costs and design risk are rising because global competitive pressures are causing system complexity to grow exponentially. Moore's law places relentless pressure on system design from another direction by making more transistors available to system designers every year. Moore's law has been in place since 1960 and it will continue to be enforced for many more years, according to the ITRS. Where "system" designers of the early 1960s were designing 5- and 7-transistor radios and board-level designers in the 1980s put a few LSI (large-scale integration) chips and a few dozen medium-scale integration (MSI) components on a board to make a system, today's SOC designers funnel tens or hundreds of millions of transistors into each chip. Increased design cost and risk inevitably accompany increased system performance and complexity.

Rising design costs and risks are not the only problem. A recent IC/ASIC functional verification study performed by Collett International Research suggests that two-thirds of all chip designs that reach volume production require two or more silicon respins. More than 80% of these designs are respun because of design errors, incomplete or incorrect specifications, or specification changes made after design begins. Functional and logic flaws are the biggest problems. Respinning a design increases design costs through the direct cost of the respin and through missed market windows that result in sales lost to competitors or simply the vagaries of time.

The MPSOC design style advocated in this chapter is a 21st-century process that reflects the realities of today's nanometer silicon. It produces firmware-programmable SOCs that can serve as platforms. These hardware platforms can satisfy the needs of multiple products across multiple product generations through reprogrammability. The MPSOC design style does not require design teams to learn new techniques or to understand radically new component types. Further, this design style has been successfully used by dozens of companies to produce hundreds of SOC designs. Thus the advocated design style is based in practical experience. It is not academic or hypothetical.

The systematic SOC design style discussed in this chapter produces working, flexible silicon with low risk and minimal cost. That result is substantially different from the dismal state of the SOC projects discussed in the IC/ASIC functional verification report from Collett International Research. What remains is for your SOC-development team to consciously select a design style for its next project. Which will it be?

BIBLIOGRAPHY

Chang, H., Cooke, L., Hunt, M., Martin, G., McNelly, A. and Todd, L., *Surviving the SOC Revolution: A Guide to Platform-Based Design*, Kluwer Academic Publishers, 1999.

Donlin, A., "Transaction Level Modeling: Flows and Use Models," *International Conference on Hardware/Software Codesign and System Synthesis*, 2004, pp. 75–80.

Gries, M. and Keutzer, K. (eds), *Building ASIPS: The MESCAL Methodology*, Springer, 2005.

Keating, M. and Bricaud, P., *Reuse Methodology Manual for System-on-a-Chip Designs*, Kluwer Academic Publishers, 1999.

Leibson, S. and Kim, J., "Configurable processors: a new era in chip design," *IEEE Computer*, July 2005, pp. 51–59.

Martin, G. and Chang, H., *Winning the SOC Revolution: Experiences in Real Design*, Kluwer Academic Publishers, 2003.

Panda, P. R., Dutt, N. and Nicolau, A., *Memory Issues in Embedded Systems-On-Chip*, Kluwer Academic Publishers, 1999.

Pieter van der Wolf, et al, "Design and programming of embedded multiprocessors: an interface-centric approach," International Conference on *Hardware/Software Codesign and System Synthesis*, 2004, pp. 206–217.

Rowen, C. and Leibson, S., *Engineering the Complex SOC*, Prentice-Hall, 2004.

Smit, G., "An Energy Efficient-Reconfigurable Circuit-Switched Network-on-Chip," *International SOC Conference*, Tampere, Finland, November 2005.

CHAPTER 3

XTENSA ARCHITECTURAL BASICS

Anyone can build a fast CPU.
The trick is to build a fast system.
—Seymour Cray

Academic researchers, system-on-chip (SOC) designers, and ASIC and EDA vendors are in a fair amount of agreement as to what must be done to reduce SOC design risks. SOC designs must become flexible enough to accommodate design changes brought on by design errors, spec changes, standards changes, and competitive market forces. Designing additional flexibility into an SOC allows one chip design to serve several products and multiple product generations.

One way to add flexibility to an SOC is to add firmware programmability through microprocessor cores. However, firmware programmability alone does not provide the requisite flexibility because general-purpose microprocessor cores with fixed ISAs (instruction-set architectures) often cannot meet system performance requirements.

As discussed in Chapter 1, the popular way to extract higher performance from a fixed-ISA microprocessor is to boost its clock rate. This approach worked in the PC market, as x86 processor clock rates rocketed from 4.77 MHz to more than 3 GHz over two decades. However, high clock rates incur a power-dissipation penalty that increases exponentially with clock rate, which was also discussed in Chapter 1. This chapter and subsequent chapters provide an introduction to a processor ISA that's more appropriate for SOC design: Tensilica's Xtensa microprocessor ISA.

In its basic form, Tensilica's Xtensa processor core is a small, fast 32-bit processor core with industry-leading performance. Unlike most other processor cores available to SOC designers, the Xtensa core can be tuned to a very wide range of applications through the use of automated tools and a processor generator that produce correct-by-construction RTL for the processor and an automatically tailored set of software-development tools for the tuned processor. The automated generation of processor and tool set takes an hour or so. Thus the Xtensa processor ISA and its associated processor generator constitute an SOC-design tool that has been tailored from the start to create application-specific instruction-set processors (ASIPs).

This chapter serves as an introduction to the base Xtensa ISA. The next chapter, Chapter 4, covers several topics associated with basic ISA configurability. Chapter 5 discusses some specific features of the Xtensa ISA that support SOC and multiple-processor SOC (MPSOC) design. Then, Chapters 6–11 discuss a series of pre-configured processor cores that have been tuned for specific application spaces. These processor cores, called Diamond Standard cores, are all based on the configurable Xtensa ISA. Therefore, the discussions of the Diamond Standard cores serve two purposes in this book. They demonstrate the use of configurability for tuning the Xtensa ISA to specific applications and they are optimized SOC processor cores (ASIPs) in their own right, useful for many SOC tasks in their out-of-the-box form. The Diamond Standard cores need no additional tuning to deliver very high performance in several application spaces common to modern SOC design.

3.1 INTRODUCTION TO CONFIGURABLE PROCESSOR ARCHITECTURES

The customary design method for achieving high performance at low clock rates—both at the board and chip level—is to design custom accelerator blocks. On SOCs, these are called RTL blocks and are most often manually developed using Verilog or VHDL. However, custom-designed RTL blocks are not usually firmware programmable because it takes a lot of extra design work to make them programmable. Consequently, most custom-designed logic blocks created for specific acceleration purposes are relatively inflexible. This inflexibility elevates design risk because an SOC that incorporates such blocks must be redesigned to accommodate any design changes brought on by bugs, specification changes, or other market forces.

It is possible to boost a microprocessor's performance for specific applications by adding execution resources directly to the processor's ISA. This is the underlying premise behind the development of configurable processor cores or ASIPs. ISA extension is not new. Processor vendors have long added ISA extensions to enhance application performance. The best-known ISA extensions are probably the enhancements made to Intel's 8086 processor by the 8087 floating-point coprocessor and the MMX and SSE multimedia extensions added to Intel's Pentium processor family. These examples demonstrate that it's quite possible to significantly improve a processor's execution performance for specific tasks through ISA extension.

Note that ISA extension should be used to make a good ISA better, not to shore up an ISA that's poorly suited to SOC design in the first place. This chapter introduces just such an SOC-friendly ISA, that of Tensilica's Xtensa microprocessor core. The Xtensa architecture was designed from the start to serve as an on-chip microprocessor core. Consequently, it was designed to be a small, fast 32-bit processor core.

In those characteristics, the Xtensa ISA follows the RISC heritage traced back to IBM's 801 project in the 1970s and the RISC microprocessor work of John Hennessy at Stanford University and David Patterson at the University of California at Berkeley in the 1980s. The result of this research produced small, fast processor architectures with several characteristic features:

- Load/store architecture
  - no memory references except for load and store instructions
- 3-operand instruction orientation
  - two operand sources, one result destination
- Large general-purpose register file
  - supports the load/store architecture
- Single-cycle instructions
  - for simplicity and speed
- Pipelined operation
  - produces the single-cycle instruction throughput.

The Xtensa ISA shares all of these characteristics with the earlier RISC processor architectures. However, the architects of the Xtensa ISA realized that memory footprint would be critically important for on-chip processors using on-chip SOC memory (on-chip SOC memory is much more expensive than memory contained in standard memory chips), so the Xtensa architecture deviates from the traditional RISC fixed-size instruction to reduce the memory footprint of firmware. The basic Xtensa ISA contains a mix of 16- and 24-bit instructions. These 16- and 24-bit instructions all perform 32-bit operations, so they are just as powerful as the 32-bit instructions of the older RISC architectures—they're merely smaller, which reduces program size and therefore reduces on-chip memory costs.

The original RISC processors employed fixed-size instructions of 32 bits to simplify the processor's fetch/decode/execute circuitry but the mechanism that converts incoming instruction words into 16- and 24-bit instructions in the Xtensa processor is not complex. The amount of memory saved through the use of 16- and 24-bit instructions more than compensates for the gates used to implement the processor's mixed-size instruction-fetch and instruction-decode unit.

Because an Xtensa processor's instruction-fetch cycle retrieves more than one instruction per cycle (including fractions of an instruction word) and because a single Xtensa instruction can cross an aligned fetch (word) boundary, the Xtensa processor stores fetched words from the instruction stream in a FIFO holding buffer. For the base Xtensa ISA, the holding buffer is 32 bits wide and two entries deep. It can be deeper and wider for certain Xtensa processor configurations.

In addition to supporting the 16- and 24-bit instructions in the Xtensa processor's base ISA, this mixed-size instruction-fetch and instruction-decode unit also supports extended 32- and 64-bit instructions that can be added to the configurable Xtensa processor. Some of the pre-configured Diamond Standard series processor cores have been extended in this manner. These processor cores are discussed in later chapters of this book.

3.2 XTENSA REGISTERS

The Xtensa processor's base ISA incorporates the following register files and registers:

- A 32-bit, general-purpose register file that employs register windows
- A 32-bit program counter
- Various special registers.

The Xtensa processor's general-purpose 32-bit register file is shown in Figure 3.1. This file, called the AR register file, has either 32 or 64 entries (a configurable attribute). Xtensa instructions access this physical register file through a sliding 16-register window. Register windowing allows the Xtensa processor to have a relatively large number of physical registers while restricting the number of bits needed to encode a source or destination operand address to four bits each. Thus the 3-operand Xtensa processor instructions need only 12 bits to specify the registers holding the instruction's three operands.

3.3 REGISTER WINDOWING

Register windowing is a key ISA feature that allowed the Xtensa architects to achieve the small instruction sizes needed to minimize the application code's footprint while allowing for a large general-purpose register file that boosts compiler efficiency and improves the performance of compiled code.

Xtensa function calls and call returns slide the register window up and down the physical register file. The window can slide by 4, 8, or 12 register entries on each call, as determined by the call instruction's opcode. Because window movement is restricted to a maximum of 12 register entries per call, there are always some register entries in the physical

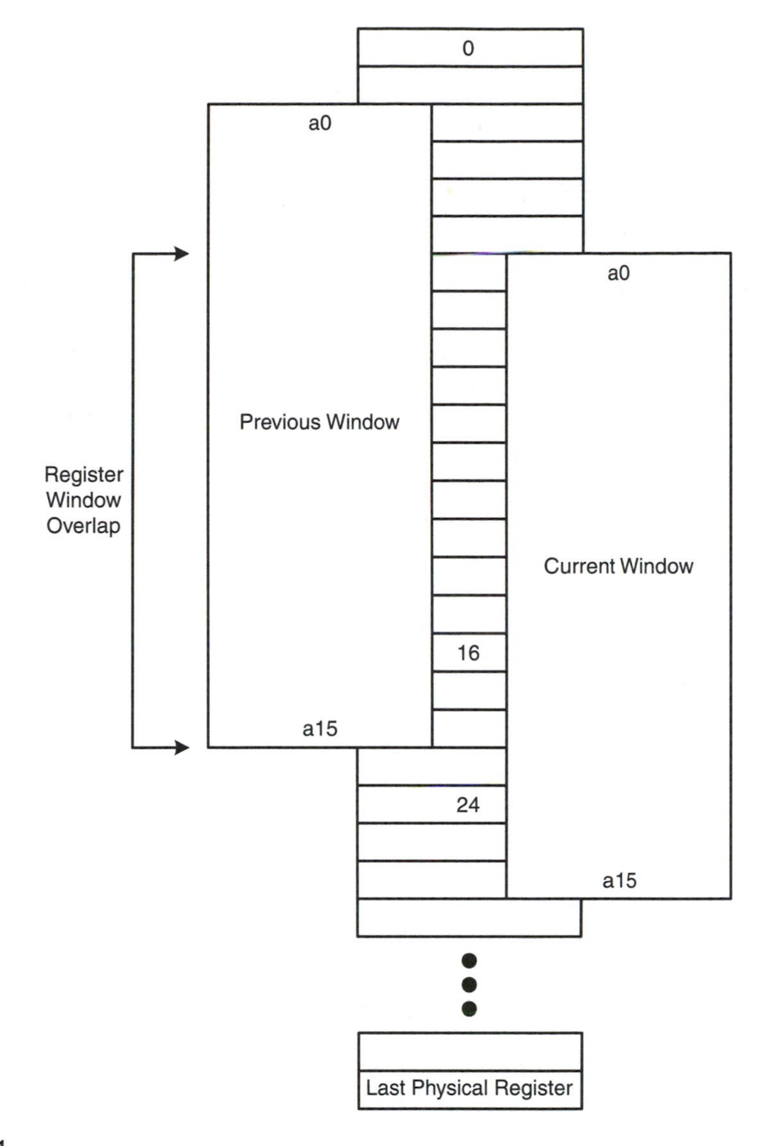

■ **FIGURE 3.1**

The Xtensa ISA incorporates a 32- or 64-entry register file with register windows. Each entry in the file is a 32-bit, general-purpose register.

general-purpose register file that are shared between the current register window and the previous window. This overlap provides a set of register entries that can be used to automatically pass argument values and return function values between a calling function and the called function. XCC—the Xtensa C and C++ compiler—automatically uses the features of this register-windowing scheme to minimize the number of instructions used in function calls.

After a function call moves the register window, the application code has a fresh set of registers that it can use for operand storage. This scheme eliminates the need to save and restore register entries in the general-purpose register file at function-call boundaries, as long as the application code stays within the physical boundaries of the register file. This characteristic also minimizes the number of instructions needed to implement function calls and improves code performance.

Eventually, function calls become sufficiently nested so that the moving register window must reuse some of the previously allocated register-file entries. When such an event takes place, it generates a window-overflow exception and triggers an exception handler that spills register-file entries to a pre-allocated stack area in memory. A function return that causes the window to shift back to a windowed area that has been spilled to the stack generates a window-underflow exception and triggers a window-underflow exception handler that restores register-file entries from the overflow stack to the register file.

3.4 THE XTENSA PROGRAM COUNTER

The Xtensa ISA employs a 32-bit program counter, which encompasses a 4-Gbyte address space. During a function call, only the lower 30 bits of the return address are saved, which restricts nested subroutines using function calls to a 1-Gbyte address space. (The other two address bits are used to store the amount of register-window translation: 0, 4, 8, or 12 entries.)

A function-call return restores the saved 30-bit return address and leaves the upper two bits of the program counter untouched. Thus related sets of function calls can operate in any of the four 1-Gbyte address spaces that comprise the overall 4-Gbyte space. Jump instructions, which use target addresses stored as 32-bit values in register-file entries, can shift program execution across the 1-Gbyte boundaries. This scheme reduces the number of memory accesses required to implement function calls, which improves the overall performance of the Xtensa ISA.

The resulting 1-Gbyte address restriction that the Xtensa ISA imposes on linked subroutines puts no real practical limits on the design of current-generation SOCs, which almost universally employ per-processor code footprints much smaller than 1 Gbyte. Per-processor SOC code footprints are unlikely to approach 1 Gbyte for many, many years. That's simply too much code to run on any one embedded processor. From a system-design perspective, the inherent parallelism latent in such a tremendous amount of program code nearly always demands that the code be partitioned and distributed across multiple on-chip processors to reduce code complexity and thus diminish the occurrence of programming bugs, improve SOC performance, and reduce the on-chip processor clock rate.

3.5 MEMORY ADDRESS SPACE

Data-space requirements for SOCs seem to grow without limit, especially for media-oriented chips. Consequently, the Xtensa ISA provides full, unrestricted 32-bit data-space addressing for its load and store instructions. The Xtensa ISA employs a Harvard architecture that physically separates instruction and data spaces, although they share the same 4-Gbyte address space. An Xtensa processor's local memories are divided into instruction and data memories and the processor employs separate instruction and data caches.

The existence of local memories and caches, the address spaces allocated to the local memories, and the width of the bus interfaces to these local memories are all configuration options for Xtensa processors. Load, store, and fetch operations to addresses that are not allocated to local memories (as defined in the processor's configuration) are directed to the Xtensa processor's main bus interface, called the PIF (processor interface). The existence and width of a processor's PIF bus is another Xtensa configuration option. (*Note*: All of these configuration options are pre-configured for Diamond Standard processor cores.)

3.6 BIT AND BYTE ORDERING

As a configuration option, Xtensa processors support either big- or little-endian byte ordering, shown in Figures 3.2 and 3.3. Diamond Standard series processor cores are pre-configured as little-endian machines. Little-endian byte ordering stores a number's low-order byte in the right-most byte location in memory and the high-order byte at the left-most byte

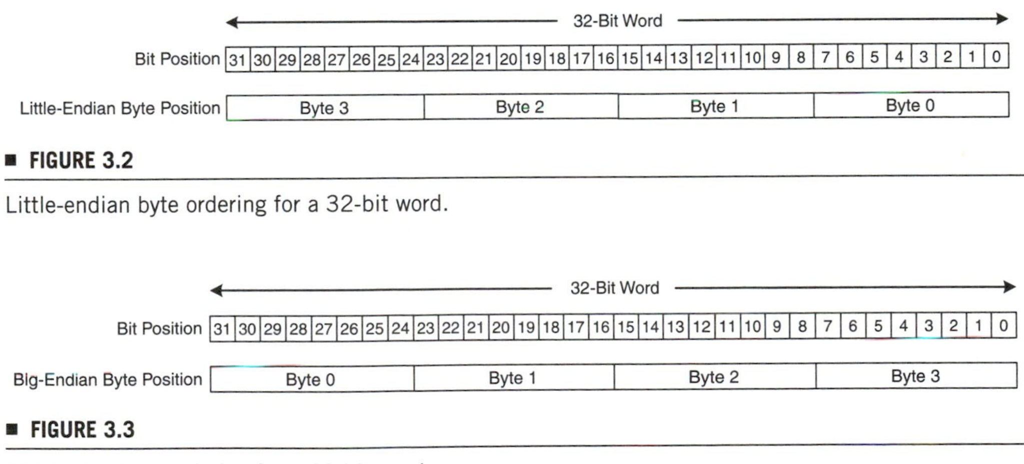

FIGURE 3.2

Little-endian byte ordering for a 32-bit word.

FIGURE 3.3

Big-endian byte ordering for a 32-bit word.

location. (The least significant or "little" end comes first.) For example, a 4-byte long integer is stored using the little-endian format in a 32-bit memory word as shown in Figure 3.2.

Little-endian byte addressing is:

Base Address+0	Byte 0
Base Address+1	Byte 1
Base Address+2	Byte 2
Base Address+3	Byte 3

The big-endian byte-ordering format stores the 4-byte integer's high-order byte in the memory location's right-most byte location and the low-order byte at the left-most byte location. (The most significant or "big" end comes first.) The long integer would then appear as shown in Figure 3.3.

Big-endian byte addressing is:

Base Address+0	Byte 3
Base Address+1	Byte 2
Base Address+2	Byte 1
Base Address+3	Byte 0

There are a seemingly endless number of arguments regarding the relative merits of big- and little-endian byte ordering. Many of these arguments stem from religious discussions about the merits of the IBM PC versus the (pre-Intel) Apple Macintosh computers because the underlying Intel and Motorola processors in those computer systems used different byte orderings. In reality, both byte-ordering formats have advantages.

Little-endian assembly-language instructions pick up a multi-byte number in the same manner for all number formats: first pick up the low-order byte at offset 0 and then work up to the most-significant byte. Due to the one-to-one relationship between address offset and byte number (offset 0 is byte 0), multiple precision math routines are correspondingly easy to write for little-endian processors.

Numbers stored in big-endian form, with the high-order byte first, can easily be tested as positive or negative by looking at the byte value stored at offset zero, no matter the number of bytes contained in the number format. Software doesn't need to know how long the number is nor does it need to skip over any bytes to find the byte containing the sign information. Numbers stored in big-endian format are also stored in the order in which they are printed out, so binary-to-decimal conversion routines are particularly efficient.

The existence of "endian" issues means is that whenever multi-byte values are written to a file, the software must know whether the underlying processor is a big- or little-endian machine and whether the target file format is big or little endian. For example, if a big-endian machine operates on a standard-format, little-endian file, the software must first reverse the byte order of all the multi-byte values in the file or the file format will not adhere to the file-format standard.

SOC designers developing chips that must deal with standard file formats are most concerned about byte ordering because many of these formats have an established byte ordering assigned to multi-byte values embedded in the files. Any software that deals with these file formats must correctly deal with the format's byte ordering. Some common file formats and their endian orders are:

- Adobe Photoshop—big endian
- GIF—little endian
- JPEG—big endian
- PostScript—not applicable (byte-oriented format)
- QTM (QuickTime movies)—little endian (Macintosh native format)
- TGA (Targa)—little endian
- TIFF—both, endian identifier encoded into file.

Audio and video media streams come in both endian formats as well. In essence, nearly any SOC dealing with multimedia files will need to work with both byte-ordering formats, so the choice of a processor's endian orientation is usually not critical. Figure 3.4 illustrates how bytes are swapped in a 32-bit word to convert from one endian format to the other.

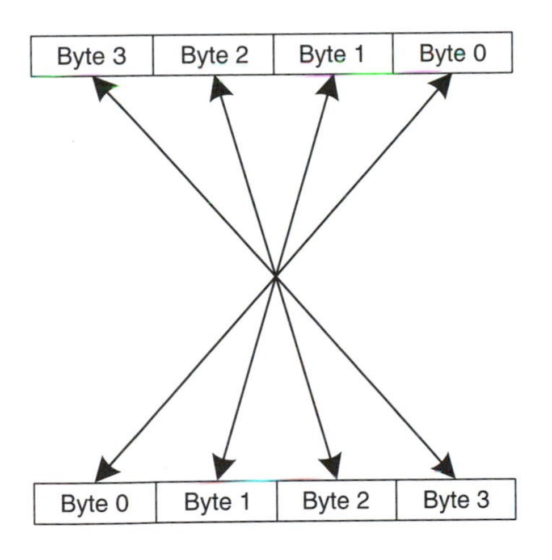

FIGURE 3.4

Conversion from one endian format to the other for a 32-bit word.

The following line of C code converts from one endian format to the other:

```
unsigned ss = (s<<24) | ((s<<8)&0xff0000) | ((s>>8)&0xff00) | (s>>24);
```

The XCC compiler for the Xtensa processor translates the line of C above into nine base Xtensa assembly-language instructions as follows (the input word is passed in register a14 and the endian-converted result ends up in register a10):

slli a9, a14, 24	Form intermediate result bits 24–31 in register a9
slli a8, a14, 8	Shift 32-bit word left by 8 bits, save in register a8
srli a10, a14, 8	Shift 32-bit word right by 8 bits, save in register a10
and a10, a10, a11	Form intermediate result bits 9–15 in register a10
and a8, a8, a13	Form intermediate result bits 16–23 in register a8
or a8, a8, a9	Form intermediate result bits 16–31, save in register a8
extui a9, a14, 24, 8	Extract result bits 0–7, save in register a9
or a10, a10, a9	Form result bits 0–15, save in register a10
or a10, a10, a8	Form final result in register a10

This conversion code could execute nine times faster if the base Xtensa ISA had an instruction that performed a 32-bit endian conversion. (One instruction would replace nine, so the code footprint would also be smaller.) However, one of the basic tenets of ISA development for configurable processors is to minimize the processor's gate count by keeping the base ISA as lean as possible and not overly stuffing it with instructions that might only be useful in some applications. For example, an application might require an instruction to perform 64- or 128-bit endian conversions and, in such cases, an instruction that performed 32-bit endian conversions would not be helpful.

An ISA-design philosophy that advocates the addition of task-specific instructions differs significantly from the current lean-and-mean approach to ISA design for fixed-ISA processors. Designers of such processors are always under pressure from existing customers to add specialized instructions that will improve the processor's performance for specific applications. While such specialized instructions may improve the processor's performance for a few application programs, they increase the processor's size for everyone using that processor. ISA design for configurable processors avoids adding specialized instructions to the base ISA by allowing each design team to add the task-specific instructions it needs to meet project-specific design goals.

Specialized instructions easily can be added to an Xtensa configurable processor by the SOC designer when the target application is known. Chapter 4 discusses the creation of such ISA extensions to Xtensa processor cores and illustrates the use of the TIE (Tensilica Instruction Extension) language to add an endian-conversion instruction called BYTESWAP to the base instruction set. Pre-configured Diamond Standard processor core ISAs have been extended with format-conversion instructions appropriate to the applications targeted by those cores.

3.7 BASE XTENSA INSTRUCTIONS

The base Xtensa ISA consists of more than 70 instructions in ten groups, listed in Table 3.1. Tables 3.2–3.9 give expanded explanations for the base instruction groups. These base instructions are sufficient to cover the needs of any program written in assembly language, C, or C++. For more performance, TIE instruction extensions can introduce nearly any instruction imaginable, as discussed in Chapter 4. Detailed explanations of the base Xtensa instructions appear in the Xtensa and Diamond Standard processor core data books.

TABLE 3.1 ■ Base Xtensa ISA instruction summary

Instruction group	Instructions
Load	L8UI, L16SI, L16UI, L32I, L32R
Store	S8I, S16I, S32I
Memory Ordering	MEMW, EXTW
Jump, Call	CALL0, CALLX0, RET J, JX
Conditional Branch	BALL, BNALL, BANY, BNONE BBC, BBCI, BBS, BBSI BEQ, BEQI, BEQZ BNE, BNEI, BNEZ BGE, BGEI, BGEU, BGEUI, BGEZ BLT, BLTI, BLTU, BLTUI, BLTZ
Move	MOVI, MOVEQZ, MOVGEZ, MOVLTZ, MOVNEZ
Arithmetic	ADDI, ADDMI, ADD, ADDX2, ADDX4, ADDX8, SUB, SUBX2, SUBX4, SUBX8, NEG, ABS
Bitwise Logical	AND, OR, XOR
Shift	EXTUI, SRLI, SRAI, SLLI SRC, SLL, SRL, SRA SSL, SSR, SSAI, SSA8B, SSA8L
Processor Control	RSR.SAR, WSR.SAR

TABLE 3.2 ■ Base Xtensa load and store instructions

Instruction	Definition
L8UI	8-bit unsigned load (8-bit offset)
L16SI	16-bit signed load (8-bit shifted offset)
L16UI	16-bit unsigned load (8-bit shifted offset)
L32I	32-bit load (8-bit shifted offset)
L32R	32-bit load PC-relative (16-bit negative word offset)
S8I	8-bit store (8-bit offset)
S16I	16-bit store (8-bit shifted offset)
S32I	32-bit store (8-bit shifted offset)

TABLE 3.3 ■ Base Xtensa call, return, and jump instructions

Instruction	Definition
CALL0	Call subroutine, PC-relative
CALLX0	Call subroutine, address in register
RET	Subroutine return. Jump to return address. Used to return from a routine called by CALL0/CALLX0
J	Unconditional jump, PC-relative
JX	Unconditional jump, address in register

TABLE 3.4 ■ Base Xtensa conditional branch instructions

Instruction	Definition
BEQZ	Branch if Equal to Zero
BNEZ	Branch if Not Equal to Zero
BGEZ	Branch if Greater than or Equal to Zero
BLTZ	Branch if Less than Zero
BEQI	Branch if Equal Immediate
BNEI	Branch if Not Equal Immediate
BGEI	Branch if Greater than or Equal Immediate
BLTI	Branch if Less than Immediate
BGEUI	Branch if Greater than or Equal Unsigned Immediate
BLTUI	Branch if Less than Unsigned Immediate
BBCI	Branch if Bit Clear Immediate
BBSI	Branch if Bit Set Immediate
BEQ	Branch if Equal
BNE	Branch if Not Equal
BGE	Branch if Greater than or Equal
BLT	Branch if Less than
BGEU	Branch if Greater than or Equal Unsigned
BLTU	Branch if Less than Unsigned
BANY	Branch if Any of Masked Bits Set
BNONE	Branch if None of Masked Bits Set (All Clear)
BALL	Branch if All of Masked Bits Set
BNALL	Branch if Not All of Masked Bits Set
BBC	Branch if Bit Clear
BBS	Branch if Bit Set

TABLE 3.5 ■ Base Xtensa move instructions

Instruction	Definition
MOVI	Load Register with 12-bit Signed Immediate Constant
MOVEQZ	Conditional Move if Zero
MOVNEZ	Conditional Move if Non-Zero
MOVLTZ	Conditional Move if Less than Zero
MOVGEZ	Conditional Move if Greater than or Equal to Zero

TABLE 3.6 ■ Base Xtensa arithmetic instructions

Instruction	Definition
ADD	Add two registers AR[r]←AR[s] + AR[t]
ADDX2	Add register to register shifted by 1 AR[r]←(AR[s]$_{30..0}$ ll 0) + AR[t]
ADDX4	Add register to register shifted by 2 AR[r]←(AR[s]$_{29..0}$ ll 02) + AR[t]
ADDX8	Add register to register shifted by 3 AR[r]←(AR[s]$_{28..0}$ ll 03) + AR[t]
SUB	Subtract two registers AR[r]←AR[s] − AR[t]
SUBX2	Subtract register from register shifted by 1 AR[r]←(AR[s]$_{30..0}$ ll 0) − AR[t]
SUBX4	Subtract register from register shifted by 2 AR[r]←(AR[s]$_{29..0}$ ll 02) − AR[t]
SUBX8	Subtract register from register shifted by 3 AR[r]←(AR[s]$_{28..0}$ ll 03) − AR[t]
NEG	Negate AR[r]←0 − AR[t]
ABS	Absolute value AR[r]←if AR[s]$_{31}$ then 0 − AR[s] else AR[s]
ADDI	Add signed constant to register AR[t]←AR[s] + (imm8$_7^{24}$ ll imm8)
ADDMI	Add signed constant shifted by 8 to register AR[t]←AR[s] + (imm8$_7^{16}$ ll imm8ll0^8)

Note: AR[ ] designates one of the entries visible through the register window to the general-purpose register file.

TABLE 3.7 ■ Base Xtensa bitwise logical instructions

Instruction	Definition
AND	Bitwise logical AND AR[r]←AR[s] and AR[t]
OR	Bitwise logical OR AR[r]←AR[s] or AR[t]
XOR	Bitwise logical exclusive OR AR[r]←AR[s] xor AR[t]

Notes:
1. AR[ ] designates one of the entries visible through the register window to the general-purpose register file.
2. The Extract Unsigned Immediate instruction (EXTUI) performs many of the most commonly used bitwise-AND operations.

TABLE 3.8 ■ Base Xtensa shift instructions

Instruction	Definition
EXTUI	Extract Unsigned field Immediate Shifts right by 0..31 and ANDs with a mask of 1..16 ones
SLLI	Shift Left Logical Immediate by 1..31 bit positions
SRLI	Shift Right Logical Immediate by 0..15 bit positions
SRAI	Shift Right Arithmetic Immediate by 0..31 bit positions
SRC	Shift Right Combined (a funnel shift with shift amount from SAR) The two source registers are concatenated, shifted, and the least significant 32 bits of the result is returned.
SRA	Shift Right Arithmetic (shift amount from SAR)
SLL	Shift Left Logical (Funnel shift AR[s] and 0 by shift amount from SAR)
SRL	Shift Right Logical (Funnel shift 0 and AR[s] by shift amount from SAR)
SSA8B	Set Shift Amount Register (SAR) for big-endian byte align
SSA8L	Set Shift Amount Register (SAR) for little-endian byte align
SSR	Set Shift Amount Register (SAR) for Shift Right Logical This instruction differs from WSR.SAR in that only the four least significant bits of the register are used.
SSL	Set Shift Amount Register (SAR) for Shift Left Logical
SSAI	Set Shift Amount Register (SAR) Immediate

Notes:
1. SAR is the "Shift Amount Register," one of the Xtensa ISA's special registers.
2. There is no SRLI instruction for shifts of 16 bits or greater. Use the EXTUI instruction instead.

TABLE 3.9 ■ Base Xtensa processor control instructions

Instruction	Definition
RSR.SAR	Read SAR (Shift Amount Register) Special Register
WSR.SAR	Write SAR (Shift Amount Register) Special Register

3.8 BENCHMARKING THE XTENSA CORE ISA

A processor's ISA efficiency is best measured by specific target application code. Often, that code doesn't exist when the SOC designers select a processor for the target tasks. Consequently, benchmark programs often stand in for the target application code.

The original definition of a benchmark was literally a mark on a workbench that provided some measurement standard. Eventually, early benchmarks were replaced with standard measuring tools such as yardsticks. Processor benchmarks provide yardsticks for measuring processor performance.

In one sense, the ideal processor benchmark is the actual application code that the processor will run. No other piece of code can possibly be as representative of the actual task to be performed as the actual code that executes that task. No other piece of code can possibly replicate the instruction-use distribution, register and memory use, or data-movement patterns of the actual application code. In many ways, however, the actual application code is less than ideal as a benchmark.

First and foremost, the actual application code may not exist when candidate processors are benchmarked because benchmarking and processor selection usually occur early in the project cycle. A benchmark that doesn't exist is worthless because it cannot aid in processor selection.

Next, the actual application code serves as an overly specific benchmark. It will indeed give a very accurate prediction of processor performance for a specific task, and for no other task. In other words, the downside of a highly specific benchmark is that the benchmark will give a less-than-ideal indication of processor performance for other tasks. Because on-chip processor cores are often used for a variety of tasks, the ideal benchmark may well be a suite of application programs and not just one program.

Yet another problem with application-code benchmarks is their lack of instrumentation. The actual application code has almost always been written to execute the task, not to measure a processor core's performance. Appropriate measurements may require modification of the application code. This modification consumes time and resources, which may not be readily available. Even with all of these issues, the target application code (if it exists) provides invaluable information on processor core performance and should be used to help make a processor core selection whenever possible.

EDN editor Markus Levy founded the non-profit, embedded-benchmarking organization called EEMBC (EDN Embedded Benchmark Consortium) in 1997. (EEMBC—pronounced "embassy"—later dropped the "EDN" from its name but not the corresponding "E" from its abbreviation.) EEMBC's stated goal was to produce accurate and reliable metrics based on real-world embedded applications for evaluating embedded processor performance. EEMBC drew remarkably broad industry support from microprocessor and DSP (digital signal processor) vendors including

Advanced Micro Devices, Analog Devices, ARC, ARM, Hitachi, IBM, IDT, Lucent Technologies, Matsushita, MIPS, Mitsubishi Electric, Motorola, National Semiconductor, NEC, Philips, QED, Siemens, STMicroelectronics, Sun Microelectronics, Texas Instruments, and Toshiba.

EEMBC spent nearly three years working on a suite of benchmarks for testing embedded microprocessors and introduced its first benchmark suite at the Embedded Processor Forum in 1999. EEMBC released its first certified scores in 2000 and, during the same year, announced that it would start to certify benchmarks run on simulators so that processor cores could be benchmarked. EEMBC has more than 50 corporate members.

EEMBC's benchmark suites are based on real-world application code. As such, they provide some of the industry's best measuring tools for comparing the performance of various processor cores. Descriptions of EEMBC's consumer, networking, and telecom benchmark suites appear in Tables 3.10–3.12.

EEMBC's consumer benchmarking suite, shown in Table 3.10, consists of four image-processing algorithms commonly used in digital cameras

TABLE 3.10 ■ EEMBC consumer benchmark programs

EEMBC Consumer benchmark name	Benchmark description	Example applications
High Pass Grey-Scale Filter	2-D array manipulation and matrix arithmetic	CCD and CMOS sensor signal processing
JPEG	JPEG image compression and decompression	Still-image processing
RGB to CMYK conversion	Color-space conversion at 8 bits/pixel	Color printing
RGB to YIQ conversion	Color-space conversion at 8 bits/pixel	NTSC video encoding

TABLE 3.11 ■ EEMBC networking benchmark programs

EEMBC networking 2.0 benchmark name	Benchmark description	Example applications
IP Packet Check	IP header validation, checksum calculation, logical comparisons	Network router, switch
IP Network Address Translator (NAT)	Network-to-network address translation	Network router, switch
OSPF version 2	Open shortest path first/Djikstra shortest path first algorithm	Network routing
QoS	Quality of service network bandwidth management	Network traffic flow control

and printers. The networking benchmark suite, shown in Table 3.11, consists of four networking algorithms used in networking equipment such as routers. The telecom suite, shown in Table 3.12, consists of algorithms used to develop wired and wireless telephony equipment.

Figures 3.5–3.7, respectively show the performance of Tensilica's base ISA Xtensa compared with processor ISAs offered by MIPS and ARM as

TABLE 3.12 ■ EEMBC telecom benchmark programs

EEMBC telecom benchmark name	Benchmark description	Example applications
Autocorrelation	Fixed-point autocorrelation of a finite-length input sequence	Speech compression and recognition, channel and sequence estimation
Bit allocation	Bit-allocation algorithm for DSL modems using DMT	DSL modem
Convolutional encoder	Generic convolutional coding algorithm	Forward error correction
Fast Fourier Transform (FFT)	Decimation in time, 256-point FFT using Butterfly technique	Mobile phone
Viterbi decoder	IS-136 channel decoding using Viterbi algorithm	Mobile phone

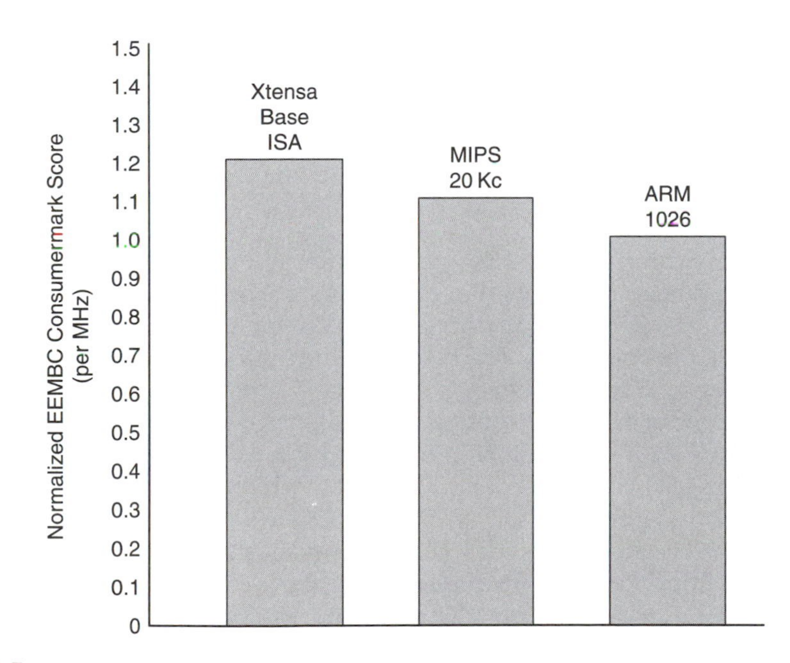

■ **FIGURE 3.5**

EEMBC "out-of-the-box" consumer benchmark scores.

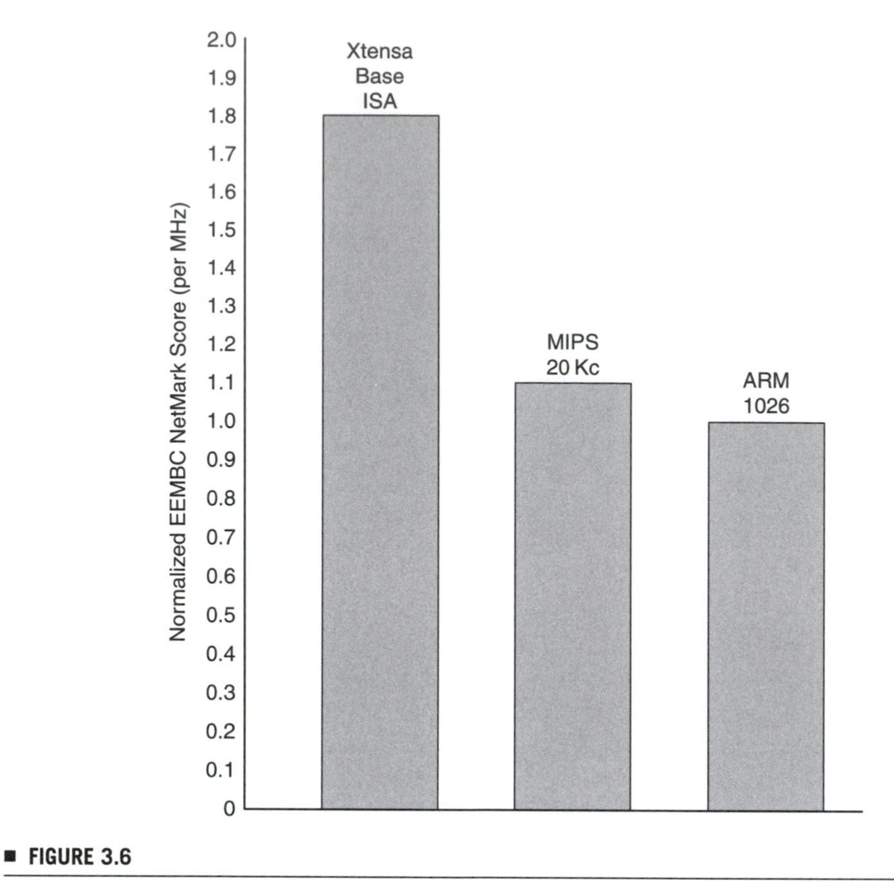

■ **FIGURE 3.6**

EEMBC "out-of-the-box" networking benchmark scores.

published on the EEMBC Web site for the consumer, networking, and telecom benchmark results. The results in Figures 3.5–3.7 are all normalized to the ARM 1026 processor core, which is the slowest of the three processors compared. The results in the figures are also normalized with respect to clock frequency so that the figures give a true impression of work performed per clock. If a processor core performs more work per clock, then it can perform more work overall and it can perform time-constrained tasks at lower clock rates, which in turn reduces the SOC's power dissipation.

The Xtensa processor's reduced instruction size, large register file, and tuned compiler all contribute to the superior EEMBC benchmark results achieved by the Xtensa ISA, when compared to other commonly used microprocessor cores. These attributes are shared with the Diamond Standard processor cores, which are all built on the Xtensa processor's base ISA.

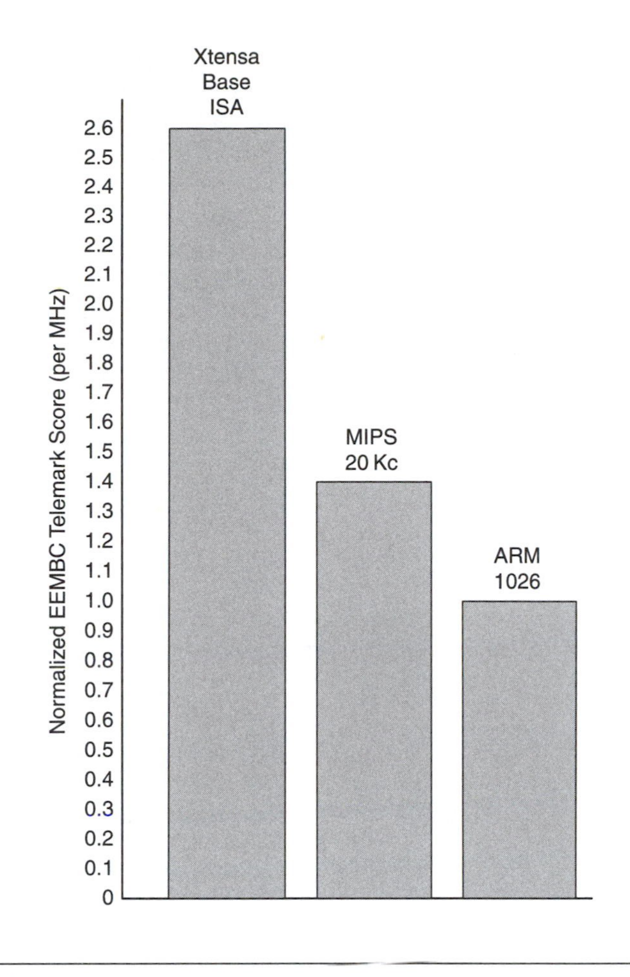

■ **FIGURE 3.7**

EEMBC "out-of-the-box" telecom benchmark scores.

EEMBC rules allow for two levels of benchmarking. The lower level produces the "out-of-the-box" scores shown in Figures 3.5–3.7. Out-of-the-box EEMBC benchmark tests can use any compiler (in practice, the compiler selected has affected the performance results by as much as 40%) and any selection of compiler switches, but the benchmark source code cannot be modified. EEMBC's "out-of-the-box" results therefore give a fair representation of the abilities of the processor/compiler combination without adding programmer creativity as a wild card.

The higher benchmarking level is called "full-fury." Processor vendors seeking to improve their full-fury EEMBC scores (posted as "optimized" scores on the EEMBC Web site) can use hand-tuned code, assembly-language subroutines, special libraries, special CPU instructions, coprocessors, and other hardware accelerators. As a result, full-fury scores tend to be much better than out-of-the-box scores, just as application-optimized production code generally runs much faster than code that has merely been run through a compiler.

As will be discussed in Chapter 4, the Xtensa configurable processor core delivers considerably better performance than fixed-ISA processor cores on EEMBC's full-fury benchmarks because of its ability to incorporate specialized instructions that collapse large portions of benchmark code into individual instructions, as was demonstrated in the discussion of endian-conversion routines above. EEMBC's full-fury benchmarking process provides SOC designers with a more realistic assessment of a configurable processor core's capability because it precisely mirrors the way that the configurable processor core will be applied to a target application.

BIBLIOGRAPHY

Gries, M. and Keutzer, K. (eds), *Building ASIPS: The MESCAL Methodology*, Springer, 2005.

Xtensa LX Microprocessor Data Book, Tensilica, Inc., August 2004.

EEMBC benchmarking consorium web site, www.eembc.org

CHAPTER 4

BASIC PROCESSOR CONFIGURABILITY

They always say time changes things, but you actually have to change them yourself.
—Andy Warhol

Processor configuration is a new concept to most SOC designers. For decades, the industry has worked with fixed-ISA (instruction-set architecture) processors and the idea that processors can be tailored to specific tasks is rather novel. Tensilica has developed a processor-configuration design flow that allows SOC developers to enhance the 32-bit Xtensa processor's ISA (instructions, register files, and registers) so that the processor executes a specific algorithm or set of algorithms faster. This approach results in a processor that executes the target task in fewer cycles. Reducing the required number of execution cycles eases the need for high processor clock rates and thus drops SOC operating power. The process of configuring an Xtensa processor closely resembles the now-common practice of optimizing performance through manual translation of critical routines from C to assembly language, so the specifics of the design flow may be different but the concept is somewhat familiar to any experienced system-development team.

4.1 PROCESSOR GENERATION

To start the process of developing an enhanced processor core, the developer profiles target code by compiling it and identifying critical loops with a profiler. Then, instead of recoding these critical loops in assembly language, the developer describes new data types, new registers and register files, and new machine instructions that better fit the needs of the target application's algorithms. Figure 4.1 shows the tool flow for developing an enhanced Xtensa processor core.

The processor-development process starts with the compilation of the available application code using a compiler built for a base Xtensa processor ISA. Code profiling then reveals the expected performance from the base processor architecture and identifies hot spots in the code that should be scrutinized for acceleration. So far, this process is no different

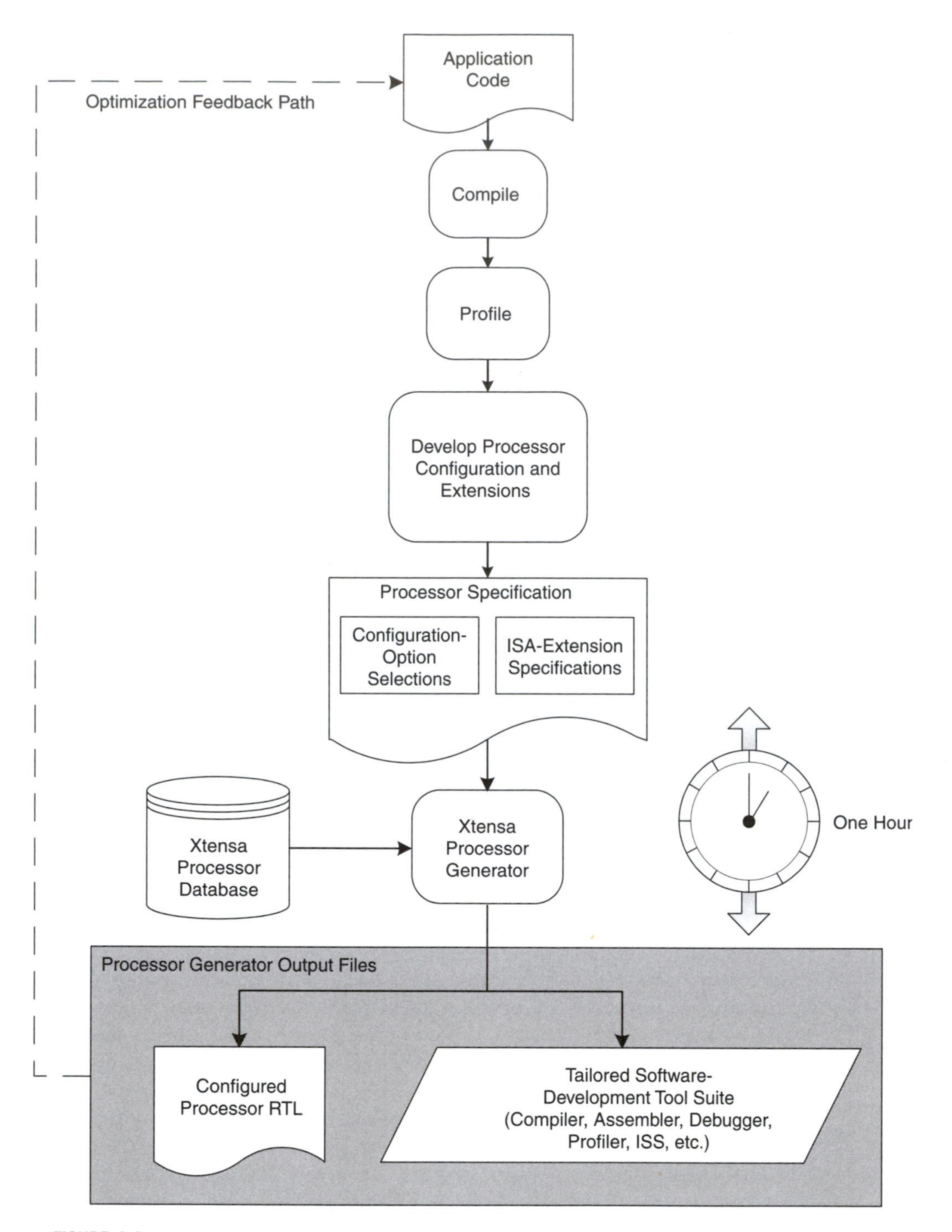

■ **FIGURE 4.1**

A simple tool flow allows SOC designers to quickly tailor an Xtensa microprocessor core for a specific application.

than code profiling for the purpose of improving code through selective recoding of critical routines in assembly language.

However, the process of processor optimization replaces assembly-language programming with processor configuration and extension. In this phase, the developer reconfigures the processor and develops register and instruction extensions that accelerate code performance. This optimization step produces a set of configuration selections and specifications for application-specific extensions to the Xtensa ISA.

Configuration options include selections such as the presence or absence of instruction and data caches, the size of the caches, the width of the processor's external buses, the size of the processor's general-purpose register file, inclusion of a multiplier or multiplier/accumulator, the number of interrupts, the number of load/store units, etc. Figure 4.2 shows the processor block diagram for the Xtensa family. Nearly every part of the processor is configurable. This configurability allows the processor to be sized appropriately for every application on the SOC.

Application-specific extensions for the Xtensa processor are described in the TIE (Tensilica Instruction Extension) language. The TIE language can describe new instructions, new registers and register files of any size, and new ports of any width (limited to 1024 bits per port) into and out of the processor. TIE's syntax closely resembles Verilog with one major difference. Because the processor's structure is completely under the control of the Xtensa Processor Generator (XPG), TIE specifications describe only function, not structure.

The XPG uses the functional descriptions written in TIE to add hardware components to the base Xtensa ISA. These hardware components reside in the Xtensa processor database that is part of the XPG. Because the XPG automatically generates and verifies the HDL (hardware description language) description of the new processor core (base processor ISA + configuration options + extensions), the resulting HDL processor description is guaranteed to be correct by construction. This guarantee is only possible because TIE creates a large, safe sandbox in which to develop and tune application-specific instruction-set processors (ASIPs).

The XPG simultaneously produces two sets of files. First, it generates the RTL description of the extended processor along with a SystemC model of the extended processor (for system-level simulation) and synthesis scripts to ease the flow of the processor description through logic synthesis. Second, the XPG generates a software-development tool suite including the compiler, assembler, debugger, profiler, instruction-set simulator (ISS), etc. All of the tools in the generated tool suite have been augmented so that they understand the extensions made to the base Xtensa ISA.

Processor generation by the XPG runs very quickly. It takes about an hour. After the XPG completes its work, the generated compiler, profiler, and ISS can be used to compile and run the target code to determine how much performance improvement the new processor extensions deliver.

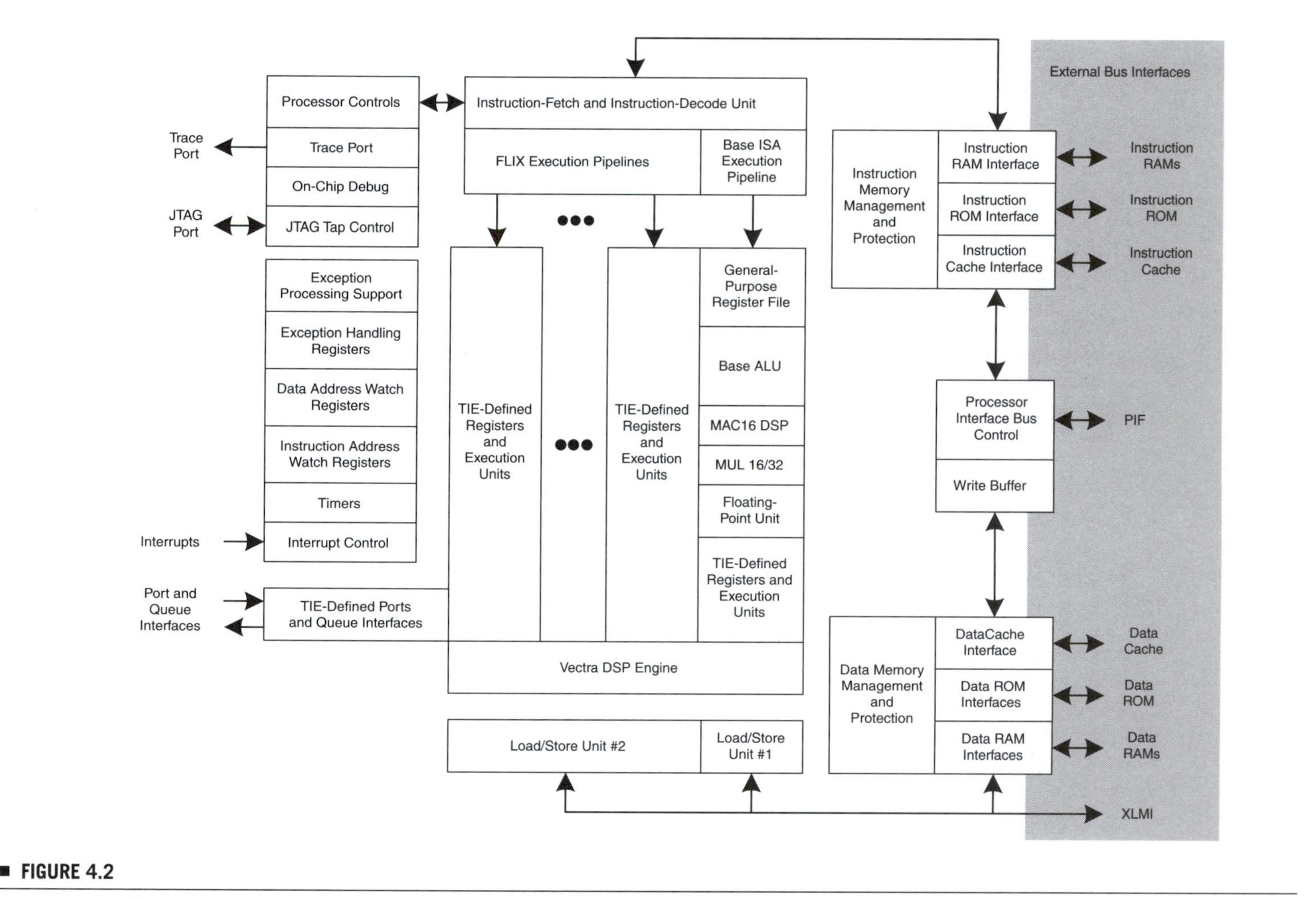

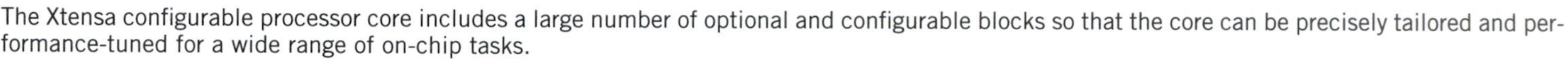

■ **FIGURE 4.2**

The Xtensa configurable processor core includes a large number of optional and configurable blocks so that the core can be precisely tailored and performance-tuned for a wide range of on-chip tasks.

If the performance goals have been met, no further processor enhancement is needed. If more performance is needed, it takes only an hour or so to build another processor. As a result of the XPG's rapid turnaround time, several candidate processor architectures can be built and profiled in a normal working day. This radically different approach to architectural and design-space exploration greatly accelerates SOC design.

4.2 XTENSA PROCESSOR BLOCK DIAGRAM

Figure 4.2 shows a block diagram of the Xtensa processor with all of its optional and configurable components. If the Xtensa processor core was a fixed-ISA machine, every instance of the processor would need to contain all of the elements appearing in Figure 4.2, which would result in a large processor core. Such a large core would be quite powerful but it would not be a precise fit for any particular set of application tasks. Instead, the Xtensa processor core is configurable, which means that the SOC design team can develop processor cores based on the Xtensa ISA that are sized to fit the performance profiles of the various processor "sockets" in the overall system design.

The Xtensa block diagram shows many blocks common to all RISC processors including an instruction-fetch/instruction-decode unit, a load/store unit, a register file, and an ALU. Optional and configurable elements include interfaces to local instruction and data memories; interfaces to instruction and data caches; the processor's main interface bus (called the processor interface, PIF); predefined functional execution units such as a multiplier, multiplier/accumulator, a floating-point unit (FPU), and a vector DSP unit; a second load/store unit; exceptions, interrupts, and timers; memory-management and memory-protection blocks; and trace, and debugging hardware. Table 4.1 describes some of the configurable hardware elements in the Xtensa processor core design.

TABLE 4.1 ■ Configurable elements in the Xtensa processor core

Configurable element	Description and configurability
MAC16	16-bit multiplier/accumulator
MUL32	32-bit multiplier
FPU	Floating-point unit
Vectra	Vector DSP unit
Zero-overhead loop	Optional hardware to accelerate loop performance
Load/store unit	Optional second load/store unit supported
16-bit instructions	16-bit instructions in the base ISA are optional
Clock gating	Adding clock gates reduces dynamic power
General-purpose register file	32 or 64 entries
Pipeline depth	Five or seven stages
Byte ordering	Big- or little-endian byte ordering

(continued)

TABLE 4.1 ■ (*continued*)

Configurable element	Description and configurability
Memory management	Optional Linux-compatible MMU
Memory protection	Optional region-protection unit
Exceptions and interrupts	As many as 32 external interrupts, six priority levels, NMI
PIF	Optional main interface bus, configurable bus width
PIF write-buffer depth	Configurable depth for the PIF write buffer
XLMI	Fast, single-cycle local bus
Cache memory	Optional caches with 1-, 2-, 3-, and 4-way associativity
Cache write policy	Write-through or write-back
Local data RAMs	Optional, multiple local data RAMs supported
Local data ROM	Optional, one local data ROM supported
Local instruction RAMs	Optional, multiple local instruction RAMs supported
Local instruction ROM	Optional, one local instruction ROM supported
Multiple-processor support	Optional synchronization instructions

In addition to the configurable elements in the Xtensa core design, designers can add registers, register files, instructions, and data ports using the TIE language to describe the desired processor extensions. Later sections in this chapter discuss TIE-based processor extensions in more detail.

4.3 PRE-CONFIGURED PROCESSOR CORES

Alternatively, there could be a family of pre-configured cores based on the Xtensa ISA. Each of these pre-configured cores could be tailored for task sets commonly encountered in SOC design. In fact, there is such a family of pre-configured cores: Tensilica's Diamond Standard series of processor cores. The Diamond Standard cores have each been configured and extended for specific applications including control processing, Linux operation, general-purpose digital-signal processing, and audio processing.

The Diamond Standard processor cores are easy, drop-in processors tailored to common SOC tasks. They also demonstrate the types of extensions made possible by the Xtensa configurable ISA, the XPG, and the TIE language. Chapters 7–12 discuss the pre-configured Diamond cores in more detail.

4.4 BASICS OF TIE

The TIE language takes processor configurability to a much higher level. Using TIE's Verilog-like descriptions, a hardware or software SOC designer can enhance the basic Xtensa architecture in three significant ways:

1. New registers and register files that precisely match native application data types.
2. New instructions that operate on the existing registers and general-purpose register file and on any new registers and register files.
3. New, high-speed I/O ports directly into and out of the processor's execution units.

The ability to create new instructions, registers, and register files that precisely meet the native needs of an algorithm's operations and data types can greatly reduce the number of processor cycles required to execute an application. This cycle-count reduction can be exploited by getting more work done at a given clock frequency or by reducing the required clock frequency to accomplish a fixed amount of work during a fixed time period. In other words, processor enhancements can either extend the processor's performance range or reduce its power dissipation.

Although this book is not intended as a TIE tutorial, brief discussions of the language's capabilities will help you to understand how the pre-configured Diamond Standard series cores, which are based on the Xtensa processor architecture, were developed.

4.5 TIE INSTRUCTIONS

The byte-swapping function described in the previous chapter provides an excellent demonstration of the ability of new processor instructions to reduce cycle count. Figure 4.3 illustrates the byte-swap function, which converts a 32-bit word between big- and little-endian formats.

The nine instructions from the base Xtensa ISA that are required to perform this operation are:

slli a9, a14, 24	Form intermediate result bits 24–31 in register a9
slli a8, a14, 8	Shift 32-bit word left by 8 bits, save in register a8
srli a10, a14, 8	Shift 32-bit word right by 8 bits, save in register a10
and a10, a10, a11	Form intermediate result bits 9–15 in register a10
and a8, a8, a13	Form intermediate result bits 16–23 in register a8
or a8, a8, a9	Form intermediate result bits 16–31, save in register a8

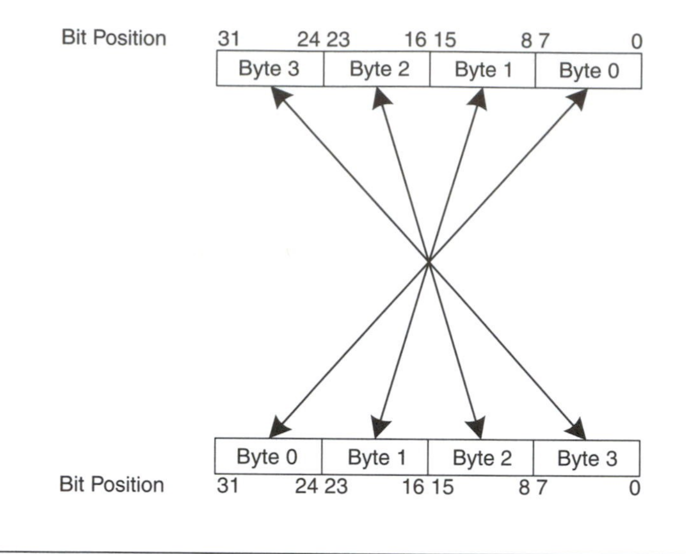

■ **FIGURE 4.3**

Conversion from one endian format to the other for a 32-bit word.

extui a9, a14, 24, 8	Extract result bits 0–7, save in register a9
or a10, a10, a9	Form result bits 0–15, save in register a10
or a10, a10, a8	Form final result in register a10

Through TIE, it's possible to define a BYTESWAP instruction that takes a 32-bit word from one of the processor's general-purpose register-file entries, converts the word from one endian format to the other, and stores the result in another register-file entry. The TIE description to create this new instruction is remarkably short:

```
operation BYTESWAP {out AR outR, in AR inpR} { }
{
wire [31:0] reg_swapped = {inpR[7:0],inpR[15:8],inpR[23:16],inpR[31:24]};
     assign outR = reg_swapped;
}
```

The operation section of a TIE description provides the name, format, and behavior of a TIE instruction. It is the simplest and most compact way to describe a new instruction. In some instances, it is the only section required to completely specify an instruction.

The interface to the BYTESWAP instruction is defined by the information contained in the first set of curly braces of the operation section:

```
operation BYTESWAP {out AR outR, in AR inpR} { }
```

Within the first set of braces, the argument *outR* specifies the destination entry in the AR register file for the result of the instruction. Argument *inpR* specifies the AR register-file entry that provides the source operand for the instruction. The second set of braces in the operation statement can be used to specify additional internal states for this *operation* extension but this feature isn't used in this example.

The behavior of the BYTESWAP instruction is defined within the next set of curly braces:

```
{
wire [31:0] reg_swapped = {inpR[7:0],inpR[15:8],inpR[23:16],inpR[31:24]};
     assign outR = reg_swapped;
}
```

The first line in this group defines how the new machine instruction should compute the byte-swapped 32-bit value and assigns the result of the operation to a temporary variable named *reg_swapped*. Note that the values of the intermediate wires, states, and registers are visible in the tailored Xtensa debugger for a processor incorporating this instruction using the debugger's *info tie_wires* command. This feature greatly facilitates the debugging of TIE instructions in a software environment. The second line above assigns the byte-swapped value to the output argument *outR*.

This 2-line definition is all that's required to add an instruction to the Xtensa processor core. From this single instruction description, the TIE Compiler within the Xtensa Processor Generator builds the necessary execution-unit hardware, adds it to the processor's RTL description, and adds constructs in the software-development tool suite so that the new BYTESWAP instruction can be used as an intrinsic in a C or C++ program.

The BYTESWAP instruction is an example of a fused instruction. The operations of nine *dependent* instructions (each instruction in the sequence depends on results from previous instructions) have been fused into one. In this example, the circuitry required to implement the function is extremely simple. The execution unit for this instruction adds little more than some additional wires to scramble byte lanes, yet this new instruction speeds endian conversion by a factor of 9x. This example demonstrates that a small addition to a processor's hardware can yield large performance gains.

In addition, the new BYTESWAP instruction doesn't need the intermediate-result registers that are used in the 9-instruction byte-swap routine. Some additional gates are included in the processor's instruction decoder to add the new instruction to the processor's instruction set, but these few gates do not make the processor core noticeably larger than the base processor.

In general, most new instructions described in TIE use more complex operations than BYTESWAP. In addition to the *wire* statement used in the above example, the TIE language includes a large number of operators and built-in functions that are used to describe new instructions. These operators and function modules appear in Tables 4.2 and 4.3,

TABLE 4.2 ■ TIE operators

<table>
<tr><th>Operator type</th><th>Operator symbol</th><th>Operation</th></tr>
<tr><td>Arithmetic</td><td>+</td><td>Add</td></tr>
<tr><td></td><td>−</td><td>Subtract</td></tr>
<tr><td></td><td>*</td><td>Multiply</td></tr>
<tr><td>Logical</td><td>!</td><td>Logical negation</td></tr>
<tr><td></td><td>&&</td><td>Logical and</td></tr>
<tr><td></td><td>||</td><td>Logical or</td></tr>
<tr><td>Relational</td><td>></td><td>Greater than</td></tr>
<tr><td></td><td><</td><td>Less than</td></tr>
<tr><td></td><td>>=</td><td>Greater than or equal</td></tr>
<tr><td></td><td><=</td><td>Less than or equal</td></tr>
<tr><td></td><td>==</td><td>Equal</td></tr>
<tr><td></td><td>!=</td><td>Not equal</td></tr>
<tr><td>Bitwise</td><td>~</td><td>Bitwise negation</td></tr>
<tr><td></td><td>&</td><td>Bitwise and</td></tr>
<tr><td></td><td>|</td><td>Bitwise or</td></tr>
<tr><td></td><td>^</td><td>Bitwise ex-or</td></tr>
<tr><td></td><td>^~ or ~^</td><td>Bitwise ex-nor</td></tr>
<tr><td>Reduction</td><td>&</td><td>Reduction and</td></tr>
<tr><td></td><td>~&</td><td>Reduction nand</td></tr>
<tr><td></td><td>|</td><td>Reduction or</td></tr>
<tr><td></td><td>~|</td><td>Reduction nor</td></tr>
<tr><td></td><td>^</td><td>Reduction ex-or</td></tr>
<tr><td></td><td>^~ or ~^</td><td>Reduction ex-nor</td></tr>
<tr><td>Shift</td><td><<</td><td>Left shift</td></tr>
<tr><td></td><td>>></td><td>Right shift</td></tr>
<tr><td>Concatenation</td><td>{ }</td><td>Concatenation</td></tr>
<tr><td>Replication</td><td>{ { } }</td><td>Replication</td></tr>
<tr><td>Conditional</td><td>?:</td><td>Conditional</td></tr>
<tr><td>Built-in modules</td><td><module-name> (...)</td><td>See Table 4.3</td></tr>
</table>

respectively. Nearly any sort of data manipulation can be performed using the *wire* statement in conjunction with the other TIE operators and built-in functions.

4.6 IMPROVING APPLICATION PERFORMANCE USING TIE

As demonstrated above, TIE extensions can improve the execution speed of an application running on an Xtensa processor by enabling the creation of instructions that each perform the work of multiple general-purpose instructions. Several different techniques can be used to combine

TABLE 4.3 ■ Built-in TIE function modules

Format	Description	Result definition
TIEadd(a, b, c_{in})	Add with carry-in	$a + b + c_{in}$
TIEaddn(a_0, a_1, ... a_{n-1})	N-number addition	$a_0 + a_1 + \cdots + a_{n-1}$
TIEcmp(a, b, sign)	Signed and unsigned comparison	$\{a < b, a <= b, a == b, a >= b, a > b\}$
TIEcsa(a, b, c)	Carry-save adder	$\{a \& b \mid a \& c \mid b \& c, a \wedge b \wedge c\}$
TIEmac(a, b, c, sign, negate)	Multiply-accumulate	negate? $c - a * b : c + a * b$ where sign specifies how a and b are extended in the same way as for TIEmul
TIEmul(a, b, sign)	Signed and unsigned multiplication	$\{\{m\{a[n-1] \& sign\}\}, a\} * \{\{n\{b[m-1] \& sign\}\}, b\}$ where n is size of a and m is size of b
TIEmulpp(a, b, sign, negate)	Partial-product multiply	negate? $- a * b : a * b$
TIEmux(s, d_0, d_1, ..., d_{n-1})	n-way multiplexer	$s == 0 ? d0 : s == 1 ? d_1 : \ldots : s == n-2 ? d_{n-2} : d_{n-1}$
TIEpsel(s_0, d_0, s_1, d_1, ..., s_{n-1}, d_{n-1})	n-way priority selector	$s_0 ? d_0 : s_1 ? d_1 : \ldots : s_{n-1} ? d_{n-1} : 0$
TIEsel(s_0, d_0, s_1, d_1, ..., s_{n-1}, d_{n-1})	n-way 1-hot selector	$(size\{S_0\} \& D_0) \mid (size\{S_1\} \& D_1) \mid \ldots (size\{S_{n-1}\} \& D_{n-1})$ where size is the maximum width of $D_0 \ldots D_{n-1}$

multiple general-purpose operations into one instruction. Three common techniques available through TIE are:

1. Fusion
2. SIMD/vector transformation
3. FLIX

To illustrate these three techniques, consider a simple example where profiling indicates that most of an application's execution time is spent computing the average of two arrays in the following loop:

```
unsigned short *a, *b, *c;
...
for (i=0; i<n; i++)
        c[i] = (a[i] + b[i]) >> 1;
```

The piece of C code above adds two short data items together and shifts the sum right by one bit in each loop iteration. Two base Xtensa instructions are required for the computation, not counting the instructions required for loading and storing the data. These two operations can be fused into a single TIE instruction:

```
operation AVERAGE{out AR res, in AR input0, in AR input1} {} {
     wire [16:0] tmp = input0[15:0] + input1[15:0];
     assign res = tmp[16:1];
}
```

This fused TIE instruction, named *AVERAGE*, takes two input values (*input0* and *input1*) from entries in the general-purpose AR register file, computes the output value (*res*), and then saves the result in another AR register-file entry. The semantics of the instruction, an add feeding a shift, are described using the above TIE code. A C or C++ program uses the new *AVERAGE* instruction as follows:

```
#include <xtensa/tie/average.h>
unsigned short *a, *b, *c;
...
for (i=0; i<n; i++)
          c[i] = AVERAGE(a[i], b[i]);
```

Assembly code can also directly use the *AVERAGE* instruction. In fact, the entire software tool chain recognizes *AVERAGE* as a valid instruction for processors built using this TIE extension.

Fused instructions are not necessarily as expensive in hardware as the sum of their constituent parts. Often they are significantly cheaper because they operate on restricted data sets. In the example above, the addition operation performed inside the *AVERAGE* instruction is a 16-bit addition and requires only a 16-bit adder, while the base Xtensa ISA implements 32-bit additions and uses a 32-bit adder. Because the sum in the *AVERAGE* instruction is always right-shifted by 1 bit, the shift operation doesn't require a general-purpose shifter—it is essentially free of hardware cost because a fixed 1-bit shift operation can be performed by simply selecting the appropriate bits from the result of the addition operation. Consequently, the *AVERAGE* instruction described above requires very few gates and easily executes in one cycle.

However, TIE instructions need not execute in a single cycle. The TIE *schedule* construct allows the creation of TIE instructions with computations that span multiple clock cycles. Such instructions can be fully pipelined (multiple instances of a multi-cycle instruction can be issued back-to-back) or they can be iterative. Fully pipelined instructions achieve higher performance because they can be issued back-to-back, but they may also require extra implementation hardware to store intermediate results in the instruction pipeline.

An iterative instruction spanning multiple clock cycles uses the same set of hardware, over two or more clock cycles. This design approach saves hardware but attempts by the running software to issue back-to-back iterative multi-cycle instructions during adjacent instruction cycles will stall the processor until the first dispatch of the multi-cycle instruction completes its computation.

Instruction fusion is not the only way to form new instructions for Xtensa processors. Two other techniques are SIMD (single instruction, multiple data) and FLIX (flexible-length instruction extensions), the Xtensa version of VLIW (very-long instruction word). SIMD instructions gang multiple, parallel execution units that perform the same operation on multiple operands simultaneously. This sort of instruction is particularly useful for stream processing in applications such as audio and video. Although multiple operations occur simultaneously, they are *dependent* operations. VLIW instructions bundle multiple *independent* operations into one machine instruction.

In the fused-instruction example shown above, one TIE instruction combines an add and a shift operation, which cuts the number of instruction cycles for the overall operation in half. Other types of instruction combinations can also improve performance. The C program in the above example performs the same computation on a new data instance during each loop iteration. SIMD instructions (also called vector instructions) perform multiple loop iterations simultaneously by performing parallel computations on different data sets during the execution of one instruction.

TIE instructions can combine fusion and SIMD techniques. Consider, for example, a case where a TIE instruction computes four *AVERAGE* operations in one instruction:

```
regfile VEC 64 8 v

operation VAVERAGE{out VEC res, in VEC input0, in VEC input1} {} {
    wire [67:0] tmp = {input0[63:48] + input1[63:48],
                       input0[47:32] + input1[47:32],
                       input0[31:16] + input1[31:16],
                       input0[15:0] + input1{15:0]};
    assign res = {tmp[67:52], tmp[50:35], tmp[33:18], tmp[16:1]};
}
```

Computing four 16-bit averages simultaneously requires that each data vector be 64 bits wide (containing four 16-bit scalar quantities). However, the general-purpose AR register file in the Xtensa processor is only 32 bits wide. Therefore, the first line in the SIMD TIE example above creates a new register file, called *VEC*, with eight 64-bit register-file entries that hold 64-bit data vectors for the new SIMD instruction. This new instruction, *VAVERAGE*, takes two 64-bit operands (each containing four 16-bit scalar quantities) from the *VEC* register file, computes four simultaneous averages, and saves the 64-bit vector result in a *VEC* register-file entry. To use the instruction in C/C++, simply modify the original example as follows:

```
#include <xtensa/tie/average.h>
VEC *a, *b, *c;
...
for (i=0; i<n; i+=4) {
        c[i] = VAVERAGE(a[i], b[i]);
```

The C/C++ compiler generated for a processor built with this TIE description automatically recognizes a new 64-bit C/C++ data type called *VEC*, which corresponds to the 64-bit entries in the new register file. In addition to the *VAVERAGE* instruction, the Xtensa Processor Generator automatically creates new load and store instructions to move 64-bit vectors between the *VEC* register file and memory. The XCC compiler uses these instructions to load and store the 64-bit vectors of type *VEC*.

Compared to the fused instruction *AVERAGE*, the SIMD vector-fused instruction *VAVERAGE* requires significantly more hardware (in the form of four 16-bit adders) because it performs four 16-bit additions in parallel. The four 1-bit shifts do not require any additional gates. The performance improvement gained by combining vectorization and fusion is significantly larger than the performance improvement from fusion alone.

The addition of SIMD instructions to an Xtensa processor nicely dovetails with Tensilica's XCC C/C++ compiler, which has the ability to unroll and vectorize the inner loops of application programs. The loop acceleration achieved through vectorization is usually on the order of the number of SIMD units within the enhanced instruction. Thus a 2-operation SIMD instruction approximately doubles loop performance and an 8-operation SIMD instruction speeds up loop execution by about 8x.

FLIX instructions are multi-operation instructions like fused and SIMD instructions. They allow a processor to perform multiple, simultaneous, *independent* operations by encoding the multiple operations into a wide instruction word, in contrast with the dependent multiple operations of fused and SIMD instructions. Each operation in a FLIX instruction is independent of the others and the XCC C/C++ compiler for Xtensa processors can bundle these independent operations into a FLIX-format instruction as needed to accelerate code. While TIE-defined fused and SIMD instructions are 24 bits wide, FLIX instructions are either 32 or 64 bits wide, to provide enough instruction-word bits to fully describe the multiple independent operations.

Xtensa instructions of different sizes (base instructions, single-operation TIE instructions, and multi-operation FLIX instructions) can be freely intermixed. Xtensa processors have no mode settings for instruction size. The instruction size is encoded into the instruction word itself. Xtensa processors will identify, decode, and execute any mix of 16-, 24-, and 32- or

64-bit instructions in the incoming instruction stream. The 32-bit or 64-bit FLIX instruction are divided into slots, with independent operations placed in either, all, or some of the slots. FLIX slots need not be equally sized, which is why this feature is called FLIX. Any combination of the operations allowed in each slot can occupy a single FLIX instruction word.

Consider again the *AVERAGE* example. Using base Xtensa instructions, the inner loop contains the *ADD* and *SRAI* instructions to perform the actual computation. Two *L16I* load instructions and one *S16I* store instruction move the data as needed and three *ADDI* instructions update the address pointers used by the loads and stores. A 64-bit FLIX instruction format with one slot for the load and store instructions, one slot for the computation instructions, and one slot for address-update instructions can greatly accelerate this code as follows:

```
format flix3 64 {slot0, slot1, slot2}

slot_opcodes slot0 {L16I, S16I}
slot_opcodes slot1 {ADDI}
slot_opcodes slot2 {ADD, SRAI}
```

The first declaration creates a 64-bit instruction and defines an instruction format with three opcode slots. The last three lines of code list base ISA instructions that are to be available in each opcode slot defined for this FLIX configuration. Note that all the instructions specified are predefined, core processor instructions, so their definition need not be provided in the TIE code. The Xtensa Processor Generator already knows about all base Xtensa instructions.

For this example, the C/C++ program need not be changed at all. The generated C/C++ compiler for a processor built using these FLIX extensions will compile the original source code while exploiting the FLIX extensions automatically. The generated assembly code for this processor implementation would look like this:

```
loop:
{addi a9,a9,4;      add a12,a10,a8;     l16i a8,a9,0       }
{addi a11,a11,4;    srai  a12,a12,1;    l16i a10,a11,0     }
{addi a13,a13,4;    nop;                s16i  a12,a13,0    }
```

A computation that requires eight cycles per iteration on a base Xtensa processor now requires just three cycles per iteration, which is nearly a 3× performance increase. It took only five lines of relatively simple TIE code to specify a complex FLIX configuration format with three instruction slots using existing Xtensa instructions.

Instruction fusion, SIMD, and FLIX techniques can be combined to further reduce cycle count. Tensilica's XPRES compiler uses all three techniques to optimize processors after analyzing C and C++ code for optimization opportunities.

4.7 TIE REGISTERS AND REGISTER FILES

New instructions alone can help improve a processor's performance for specific applications, but instruction extension is aided immensely when processor state can also be added. The first, perhaps most familiar form of processor state is the register file. All Xtensa processors contain a 32-bit, general-purpose register file that has either 32 or 64 entries, depending on processor configuration. However, the world of applications is hardly limited to 32-bit data types, as demonstrated in the SIMD vectorized example shown in the previous section.

It's not difficult to use 32-bit register-file entries to store smaller data types but some applications may employ data types wider than 32-bits. The conventional way to deal with wide data types is to split them into 32-bit chunks and then store the individual chunks in 32-bit register-file entries. Multiple-precision math algorithms routinely perform operations on split operands. Such operations are not cycle efficient, but they fit the traditional processor architecture because they match the processor's native data type (a 32-bit word) to the application's preferred data type.

A far more efficient way to deal with data types larger or smaller than 32 bits is to create register files (or individual registers) to store operands in the proper format and to create instructions that operate directly on the application's native data types.

A TIE description contains sections for creating specialized registers and register files. For example, the statement:

```
regfile VR 128 16
```

creates a new 128-bit-wide, 16-entry register file called *VR*. TIE operation statements can then define new instructions that operate directly on this wide register file. A C data-type (*ctype*) section of a TIE description can then syntactically bond the new register files to corresponding C data types.

Applications often have varying needs with respect to register width. Video and audio applications often employ 24-bit data types, for example. A processor designed to run an application such as DES encryption that stores 56-bit encryption keys as two 28-bit subkeys can make use of 56- and 28-bit registers with instructions that directly operate on those registers. TIE can create such registers because TIE registers and register files are not limited to power-of-two widths. Processors extended in this manner are far more efficient in executing targeted applications that employ unusually sized data types.

4.8 TIE PORTS

A processor's I/O capabilities are often overshadowed by discussions of ISAs and pipeline architectures. However, I/O bandwidth has a substantial influence on a processor's overall performance. Most processor core designs are limited to one main bus and a few local-memory buses. As discussed in previous chapters, buses immediately set processors apart from other sorts of hardware blocks. A bus conducts only one transaction per clock cycle so processors with only one bus to communicate with the rest of a system are similarly limited. The TIE language provides powerful ways to significantly boost the Xtensa processor's I/O bandwidth through the addition of ports and queue interfaces.

Direct processor-to-processor port connections reduce cost and latency for communication between two processors or between a processor and another hardware block. Direct port connections allow data to move directly from one processor's registers to the registers and execution units of another processor. A simple example of a direct port connection appears in Figure 4.4.

A conventional approach to moving data from processor #1 to processor #2 would involve an external communications register. Processor #1 would produce data, normally through a computation. That data normally would be the result of an instruction or instruction sequence and the data would reside in one of the processor's registers as a result of the computation. Processor #1 would then write this data to the external communications register using a store instruction. The store instruction would initiate a bus cycle, which normally consumes several clock cycles for conventional RISC processor cores. If the data being generated is larger than 32 bits, then several internal register-file entries and external communication registers will be involved in this output operation.

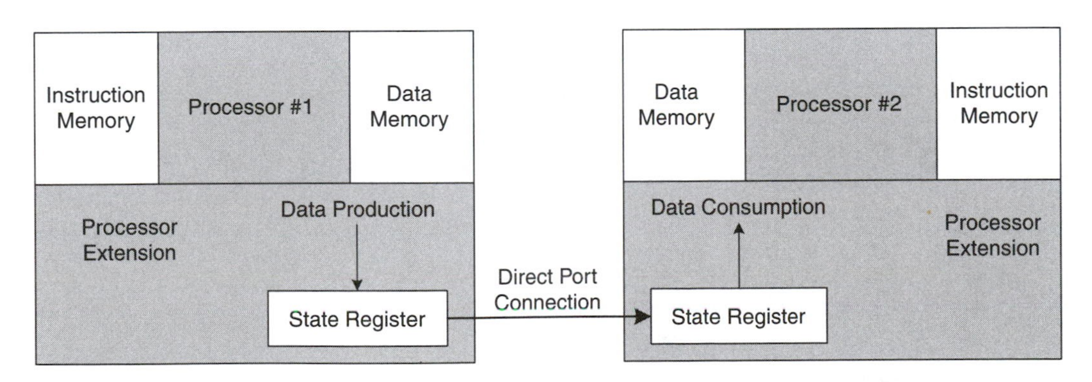

FIGURE 4.4

A direct processor-to-processor port connection.

When processor #1 has completed the data-generation phase and stored the data in an external communications register, processor #2 can access the data and use it. Processor #2 will initiate one or more read cycles to retrieve the data from the external communications register using one or more load instructions that will initiate one or more read cycles on processor #2's main bus. The load instructions will place the data into one or more entries in processor #2's main register file where the data can then be processed. This extended sequence of stores and loads has become so natural to designers of processor-based systems that the huge amounts of overhead involved no longer appear to the designers' eye. Invisible or not, that overhead is present and it erodes the system's performance.

Direct communication ports incorporated into the two processors using TIE descriptions can significantly shorten the sequence of events required to transfer data between processors. A TIE instruction can create the data. A TIE state register in processor #1 can serve as the destination for data produced in processor #1 by the TIE instruction. When the TIE instruction places the instruction result in State Register #1, the value in the register becomes available on the attached communication-port pins. Note that no additional store instruction is needed to place data on these port pins.

The act of placing a result in State Register #1 automatically places the data on the associated communication-port pins. That value is immediately available as an input value to operations in Processor #2 through State Register #2. Any TIE instruction can use State Register #2 as an input operand so no explicit instruction to read the register is needed. TIE port connections can be arbitrarily wide, allowing large and non-power-of-two-sized operands to be transferred easily and quickly between processor #1 and processor #2 and this same mechanism can be used to transfer data from a processor to a block of hardware or from a hardware block into a processor.

Note that the above data-transfer sequence ignores the need for a handshake mechanism between processors. Generally, some sort of polled or interrupt-driven mechanism is required to sequence data flow between the processors. Interrupts and polling routines take time, so simple port interfaces do not deliver the ultimate in transfer speed, although they are faster than bus-based interconnection schemes. For the highest transfer rates, SOC designers should use FIFO queues.

4.9 TIE QUEUE INTERFACES

The highest-bandwidth mechanism for task-to-task communication is hardware implementation of data queues. One data queue can sustain data rates as high as one transfer every cycle or more than 10 Gbytes per second for wide operands (tens of bytes per operand at a clock rate of

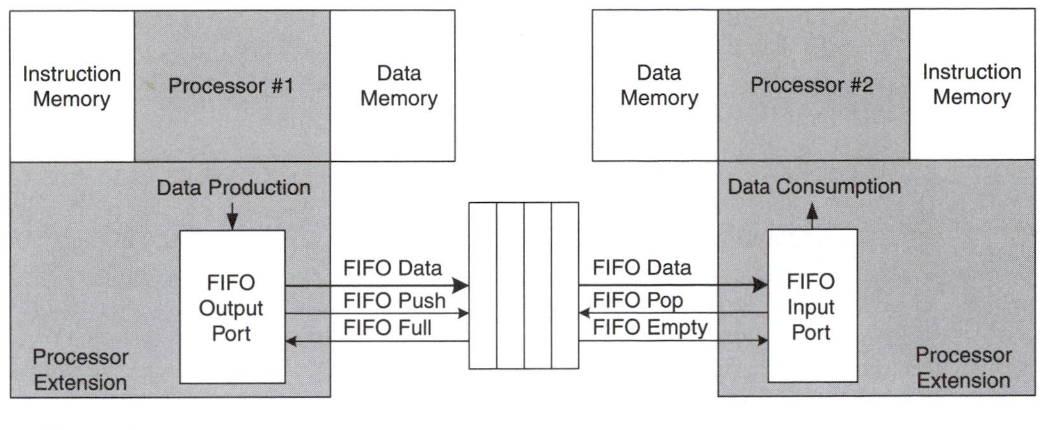

■ **FIGURE 4.5**

Processors linked by a FIFO queue.

hundreds of MHz) because queue widths need not be tied to a processor's bus width or general-register width. The handshake between producer and consumer is implicit in the queue interfaces.

When the data producer has created data, it pushes that data into the tail of the queue, assuming the queue is not full. If the queue is full, the producer stalls. When the data consumer is ready for new data, it pops it from the head of the queue, assuming the queue is not empty. If the queue is empty, the consumer stalls. This self-regulating data-transfer mechanism ensures maximum I/O bandwidth between data producer and consumer.

TIE allows direct implementation of queue interfaces as source and destination operands for instructions. An instruction can specify a queue as one of the destinations for result values or use an incoming queue value as one source. This form of queue interface, shown in Figure 4.5, allows a new data value to be created or used each cycle on each queue interface. Hardware-handshake lines (FIFO push, FIFO full, FIFO pop, and FIFO empty) ensure that data flows at the maximum allowable rate with no software intervention. If the producer or consumer is not ready, the chain stalls. However, the overall process of data production, data transfer, and data consumption always proceeds at the chain's maximum rate.

TIE queues can be configured to provide non-blocking push and pop operations, where the producer can explicitly check for a full queue before attempting a push and the consumer can explicit check for an empty queue before attempting a pop. This mechanism allows the producer or consumer task to move on to other work in lieu of stalling.

A TIE processor extension can perform multiple queue operations per cycle, perhaps combining inputs from two input queues with local data and sending result values to one or more output queues. The high aggregate bandwidth and low control overhead of queues allows application-specific processors to be used for applications with very high data rates where processors with conventional bus or memory interfaces are not appropriate because they lack the ability to handle the required high data rates.

Queues decouple the performance of one task from another. If the rate of data production and data consumption are quite uniform, the queue can be shallow. If either production or consumption rates are highly variable or bursty, a deep queue can mask the data-rate mismatch and ensure throughput at the average rate of producer and consumer, rather than at the minimum rate of the producer or the minimum rate of the consumer. Sizing the queues is an important design optimization driven by good system-level simulation. If the queue is too shallow, the processor at one end of the communication channel may stall when the other processor slows for some reason. If the queue is too deep, the silicon cost will be excessive.

Queue interfaces are added to an Xtensa processor through the following TIE syntax:

```
queue <queue-name> <width> in|out
```

The name of the queue, its width, and direction are defined with the above TIE syntax. An Xtensa processor can have more than 300 queue interfaces and each queue interface can be as wide as 1024 bits. These limits are set well beyond the routing limits of current silicon technology so that the processor core's architecture is not the limiting factor in the design of a system. The designer can set the real limit based on system requirements and EDA flow. Using queues, designers can trade off fast and narrow processor interfaces with slower and wider interfaces to achieve bandwidth and performance goals.

Figure 4.6 shows how TIE queue interfaces are easily connected to simple Designware FIFOs. TIE queue push and pop requests are gated by the FIFO empty and full status signals to comply with the Designware FIFO specification.

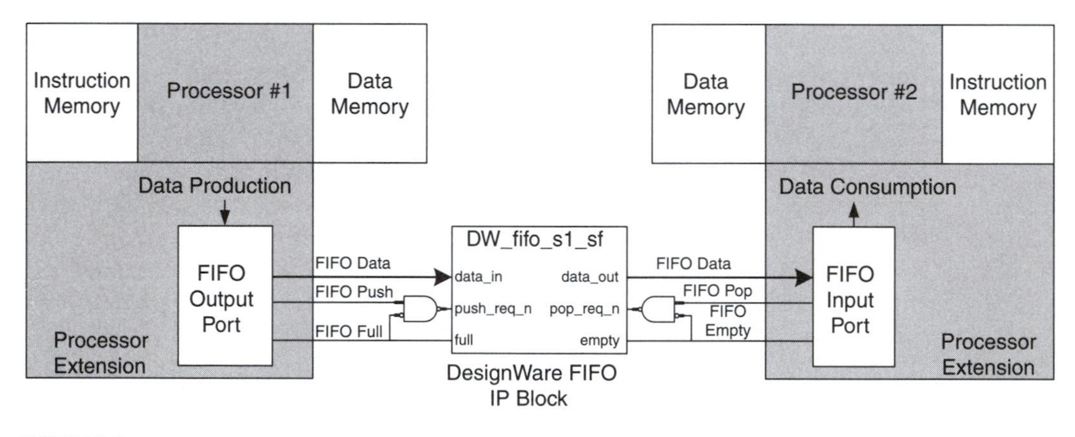

FIGURE 4.6

Designware synchronous FIFO used with TIE queues.

TIE queue interfaces serve directly as input and output operands of TIE instructions, just like a register operand, a register-file entry, or a memory interface. The Xtensa processor includes 2-entry buffering for every TIE queue interface defined. The area consumed by a queue's 2-entry buffer is substantially smaller than a load/store unit, which can have large combinational blocks for alignment, rotation, and sign extension of data, as well as cache-line buffers, write buffers, and complicated state machines. The processor area consumed by TIE queue interface ports is relatively small and is under the designer's direct control.

The FIFO buffering incorporated into the processor for TIE queues serves three distinct purposes. First, the buffering provides a registered and synchronous interface to the external FIFO. Secondly, for output queues, the buffer provides two FIFO entries that buffer the processor from stalls that occur when the attached external FIFO is full. Thirdly, the buffering is necessary to hide the processor's speculative execution from the external FIFO.

4.10 COMBINING INSTRUCTION EXTENSIONS WITH QUEUES

The availability of queue interfaces tied directly to a processor's execution units permits the use of Xtensa processors in an application domain previously reserved for hand-coded RTL logic blocks: flow-through processing. By combining input-and output-queue interfaces with designer-defined execution units, it's possible to create a firmware-controlled processing block within a processor that can read values from input queues, perform a computation on those values, and output the result of that computation with a pipelined throughput of one clock per complete input-compute-output cycle.

Figure 4.7 illustrates a simple design of such a system with two 256-bit input queues, one 256-bit output queue, and a 256-bit adder/multiplexer execution unit. Although this processor extension runs under firmware control, its operation bypasses the processor's memory busses and load/store unit to achieve hardware-like processing speeds.

Even though there is a substantial amount of hardware in this processor extension, its definition consumes only four lines of TIE code:

```
queue InData1 256 in
queue InData2 256 in
queue OutData 256 out
operation QADD {} {in InData1, in InData2, in SumCtrl, out OutData}
        {assign OutData = SumCtrl ? (InData1 = InData2) : InData1;}
```

The first three lines of this code define the 256-bit input and output queues and the fourth line defines a new processor instruction called *QADD* that

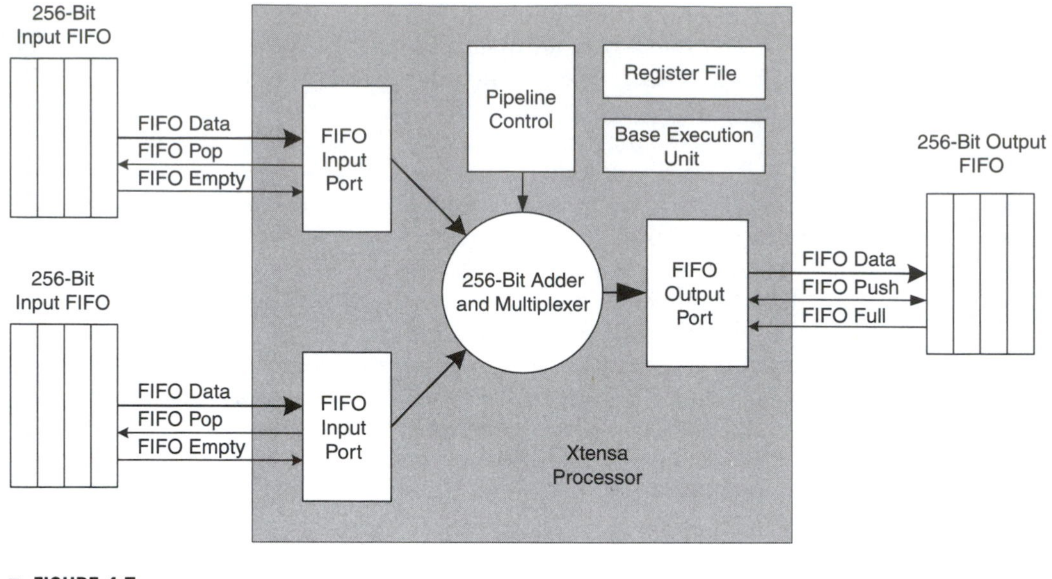

■ **FIGURE 4.7**

Combining queues with execution unit to add flow-through processing to a configurable processor core.

takes data from the input queues, performs 256-bit operations (additions or multiplexing), and passes the results to the 256-bit output queue. *QADD* instructions are pipelined so the effective instruction cycle time is one clock.

No conventional processor could possibly perform the composite *QADD* operation in one clock cycle. It would require many, many clock cycles just to retrieve the two 256-bit source operands, several more clock cycles to perform the 256-bit addition, and even more clock cycles to output the 256-bit result. This example demonstrates the ability of TIE extensions to replicate performance results of HDL-defined hardware while retaining firmware programmability.

4.11 DIAMOND STANDARD SERIES PROCESSOR CORES—DEALING WITH COMPLEXITY

TIE expands the processing capabilities of microprocessor cores far beyond that of fixed-ISA processors. The combination of Xtensa processor configurability and TIE-based extensibility creates a huge, multidimensional design space, as illustrated by Figure 4.8.

Figure 4.8 is divided into the Xtensa technology's four main configurability and extensibility quadrants: configurable features, instruction extensions, processor-state (register and register-file) extensions, and I/O extensions (ports and queue interfaces). Arrows within each quadrant

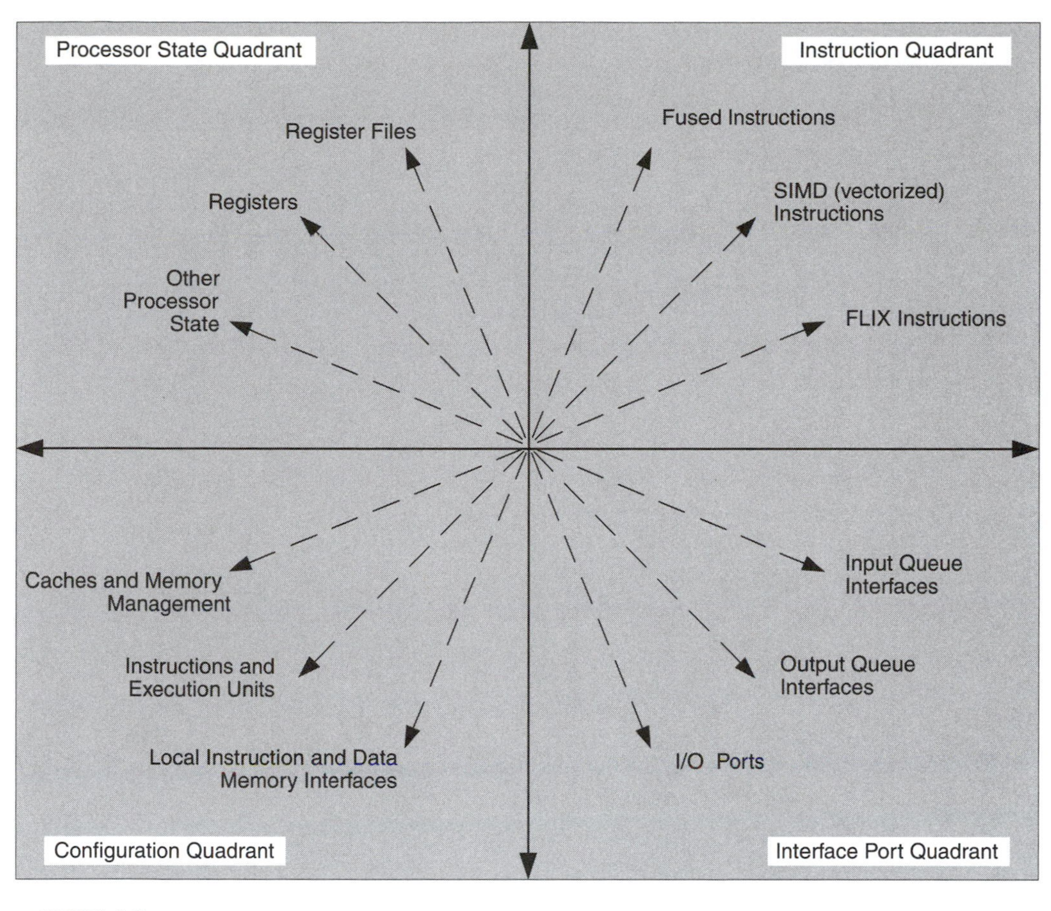

■ **FIGURE 4.8**

Processor configurability and TIE-based extensibility create a huge, multidimensional design space.

show a few of each quadrant's many design-space dimensions, but by no means all of the dimensions. The total number of degrees of freedom is quite large, which makes the combination of the Xtensa architecture and the TIE language a truly powerful design tool. With all of this capability and flexibility comes a dilemma: how to best use all of that raw capability to address SOC design problems.

Because many SOCs share a common set of design problems, Tensilica has used the configurable and extensible features of the Xtensa processor family to create a series of pre-configured cores that address common problem sets encountered in many SOCs. These pre-configured cores are called the Diamond Standard Series and are discussed in detail in Chapters 6–12. Each of the Diamond cores takes the Xtensa architecture in a different direction. Consequently, these cores effectively demonstrate the capabilities of Tensilica's configurable, extensible processor technology.

BIBLIOGRAPHY

Leibson, S. and Kim, J., "Configurable Processors: A New Era in Chip Design," *IEEE Computer*, July 2005, p 51–59.

Rowen, C. and Leibson, S., *Engineering the Complex SOC*, Prentice-Hall, 2004.

Xtensa 6 Microprocessor Data Book, Tensilica, Inc, October 2005.

Xtensa LX Microprocessor Data Book, Tensilica, Inc, August 2004.

CHAPTER 5

MPSOC SYSTEM ARCHITECTURES AND DESIGN TOOLS

In pioneer days they used oxen for heavy pulling,
and when one ox couldn't budge a log,
they didn't try to grow a larger ox.
We shouldn't be trying for bigger computers,
but for more systems of computers.
—Admiral Grace Hopper

If you were plowing a field,
which would you rather use:
Two strong oxen or 1024 chickens?
—Seymour Cray

All SOC architectures in the 21st century will consist of many interconnected hardware blocks. There's simply no other possible design approach that can reconcile the complexity of mega-gate SOCs with the mental abilities and capacities of human designers. Many SOC blocks will contain memories, some will consist of custom-designed logic, some will be predesigned IP, and some will be microprocessor cores. These blocks will intercommunicate over a variety of buses, networks, and point-to-point connections. This chapter ties a number of concepts discussed in previous chapters into an integrated approach to multiple-processor SOC (MPSOC) design.

As previous chapters have already discussed, custom-designed, hand-built logic blocks are usually not programmable and are therefore not flexible. Fixed-ISA microprocessor cores are programmable and flexible but they sometimes lack the processing capability and I/O bandwidth needed to perform a task. Configurable processors share the flexibility and programmability of fixed-ISA microprocessor cores while enabling substantial task-specific computational performance and I/O bandwidth improvements through their extreme hardware flexibility and adaptability. Therefore, you should expect to see a rise in the number of programmable fixed-ISA and configurable microprocessor cores used in each SOC design as design teams strive to develop the most flexible and adaptable silicon possible to deal with the real-world vagaries of shifting standards, rapidly changing markets, and unforeseen competitive forces.

As SOC complexity inevitably grows and the number of SOC blocks increases, the efficient interconnection of these blocks becomes ever more important. Conventional interconnection schemes (namely buses) will experience increasingly severe data congestion as SOC complexity spirals ever upwards. Consequently, SOC designers must adopt new ways of organizing blocks within their designs and must employ new, more efficient interconnection schemes between blocks.

These new interconnection schemes must be efficient—they must move large amounts of data quickly, with little expended energy—and they must be economical. Increasingly faster and more complex buses and bus hierarchies do not meet these criteria for entire, chip-level systems although buses are still quite useful within localized SOC subsystems.

5.1 SOC ARCHITECTURAL EVOLUTION

SOC architectures have evolved in lock step with silicon-fabrication technology. Figure 5.1 shows a simple, single-bus SOC architecture that incorporates one processor core, memory, and some peripheral devices. The system architecture in Figure 5.1 is the stereotypical microprocessor-based system design that's remained unchanged since the first commercial microprocessor became available in 1971.

As the on-chip SOC microprocessor core clock rates climbed from tens to hundreds of MHz, bus bandwidth started to become a serious problem. It no longer made sense to place fast processors and memories on the same bus with slow peripheral devices. The solution to this problem was

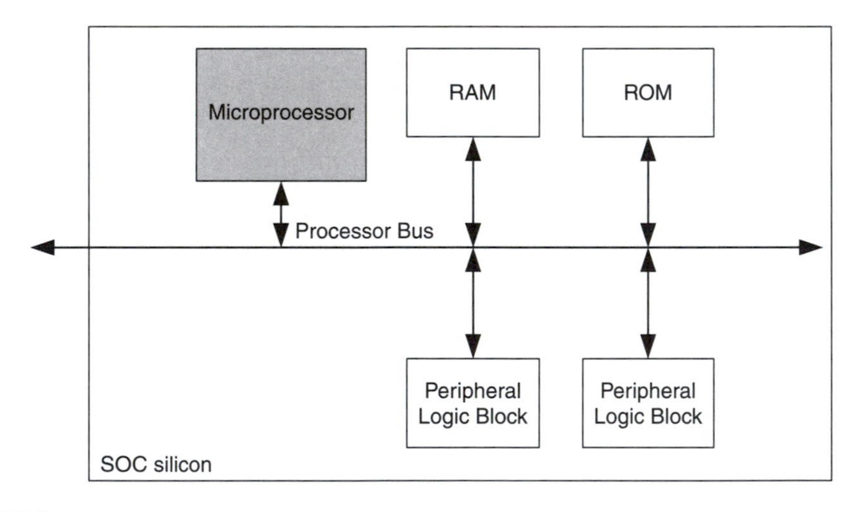

■ **FIGURE 5.1**

The simple, single-bus, microprocessor-based system architecture has remained unchanged since 1971.

to develop split-bus architectures that give slow devices a separate bus bridged to the faster bus. Such an architecture appears in Figure 5.2.

However, as SOC systems became more complex, the amount of software these systems ran and the amount of data processed grew. Consequently, the attached memories started to grow to accommodate the larger programs and bigger data structures. Larger memories are slower due to increased interconnect capacitance in the memory array. Eventually, the speed mismatch between the processor's clock rate and the memory-access times became large enough so that the processor could no longer access main memory in one clock cycle.

This growing problem forced another architectural change to SOC systems. Processor designers started adding fast memory-cache interfaces to their processor core designs to ease this problem. An SOC architecture using a cached processor core appears in Figure 5.3.

As SOC performance levels continued to climb, single-processor architectures proved increasingly inadequate. Many SOC architects started to add a second processor to the system design, as shown in Figure 5.4. The additional processor, often a DSP (digital signal processor), offloaded some of the on-chip tasks from the main processor. This system-design approach allowed both of the on-chip processors to run at slower clock rates, which kept power dissipation low and allowed the use of less expensive IC-fabrication processes.

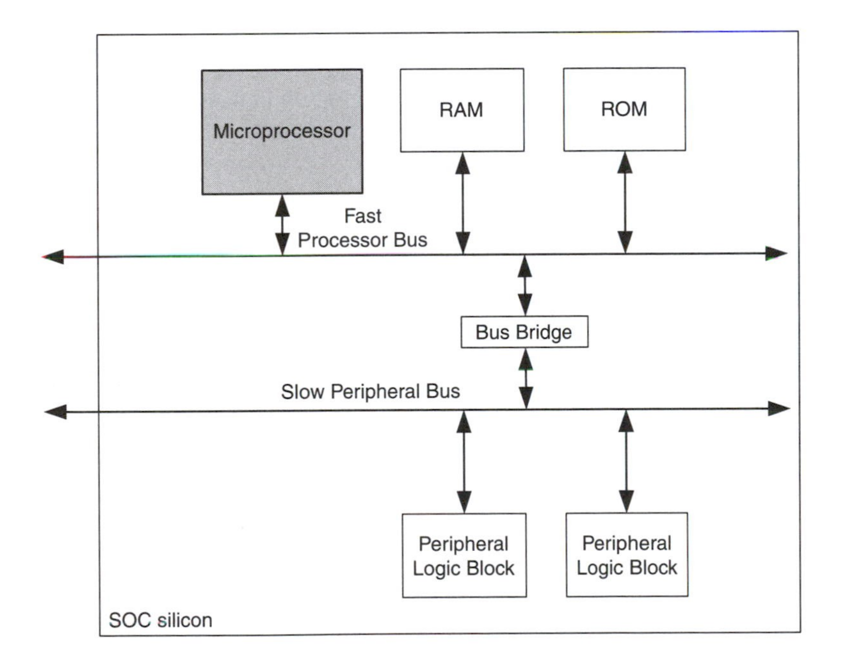

■ **FIGURE 5.2**

Splitting the microprocessor bus into fast and slow buses isolates fast SOC blocks such as processors and memories from slow devices such as peripheral blocks.

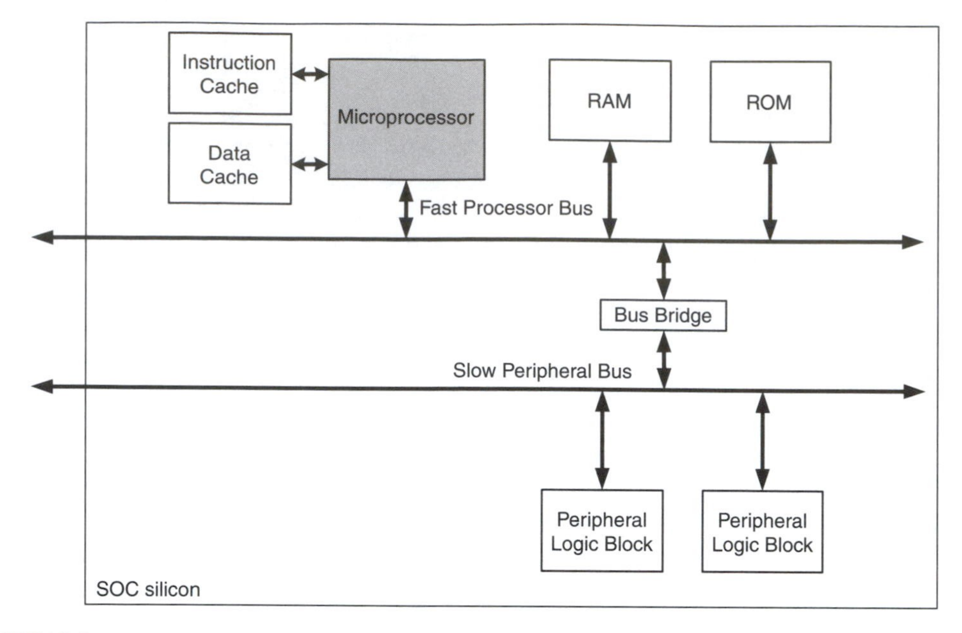

■ **FIGURE 5.3**

Memory caches allow fast processor cores to achieve good performance even when the attached main memory is relatively slow.

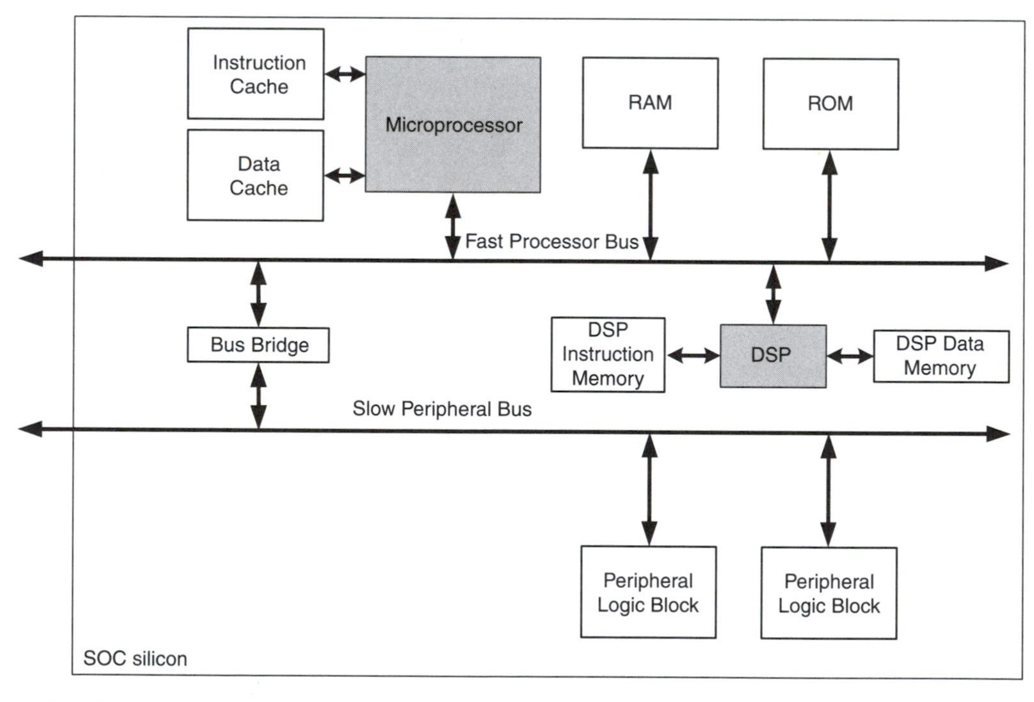

■ **FIGURE 5.4**

To reach even higher performance levels, SOC architectures adopted a second processor—often a DSP.

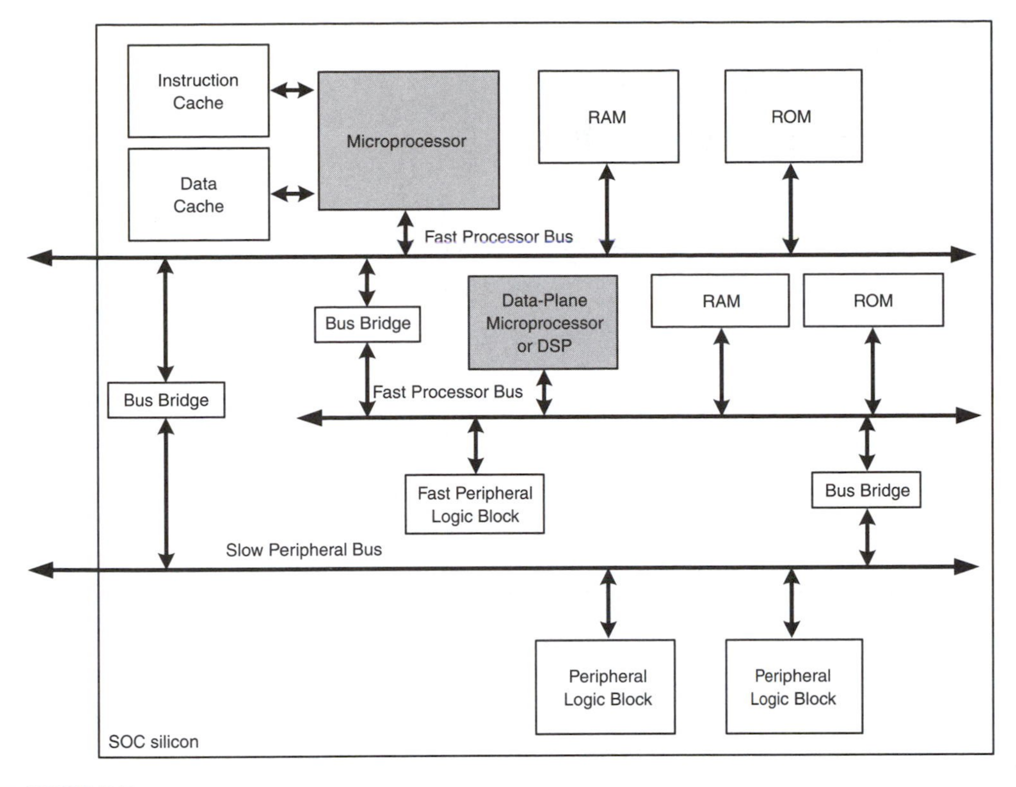

■ **FIGURE 5.5**

SOC architectures added processors and buses to achieve even higher performance levels.

Initially, the second (and sometimes a third) processor shared the fast processor bus with the main processor. However, the fast processor bus quickly became a system bottleneck and auxiliary processors got their own buses, as shown in Figure 5.5.

Sometimes, programmable microprocessors and DSPs cannot reach the desired performance levels required by the SOC. For example, video compression algorithms place immense processing demands on the hardware. In such cases, system designers often resort to hardware accelerators either built as custom-designed logic blocks or purchased as pre-built, pre-verified IP. Figure 5.6 shows such a multiprocessor system with a hardware accelerator block attached to the high-speed bus of the auxiliary processor (often called the "data-plane" processor to distinguish it from the main or "control-plane" processor).

5.2 THE CONSEQUENCES OF ARCHITECTURAL EVOLUTION

One of the many consequences of this years-long SOC architectural evolution is a huge proliferation of processor-core varieties, each optimized for

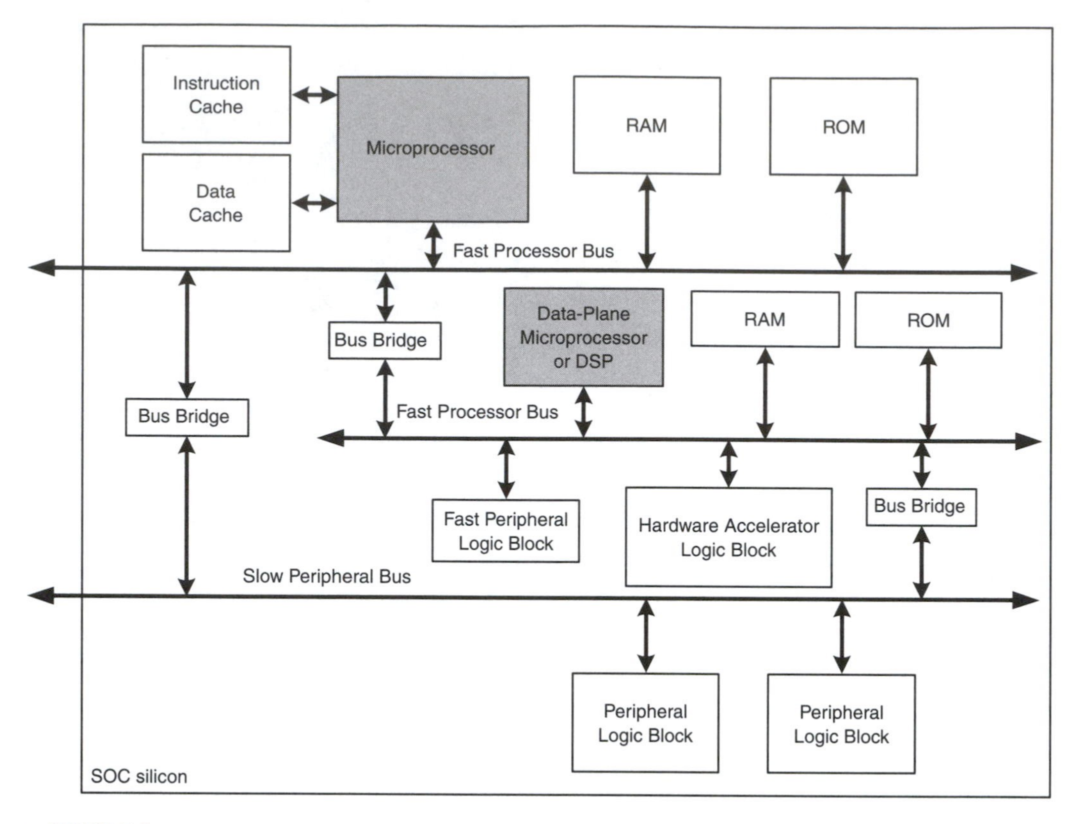

■ **FIGURE 5.6**

Some SOCs need require the extra performance provided by hardware accelerators.

some particular evolutionary stage in system design. Another consequence is that system architects worldwide use a highly varied mixture of all the system architectures shown above, and even more variations not shown in these five figures but discussed later in this chapter.

Successful processor core architectures for SOC designs must have a large number of memory and I/O ports to support all of the connection schemes shown in Figures 5.1–5.6. Tensilica's Xtensa and Diamond Standard Series processor cores do, as shown in Figure 5.7.

5.3 MEMORY INTERFACES

Xtensa and Diamond Standard Series microprocessor cores have a number of local-memory interfaces that allow fast access to instructions and data over a wide range of different architectural configurations. Several

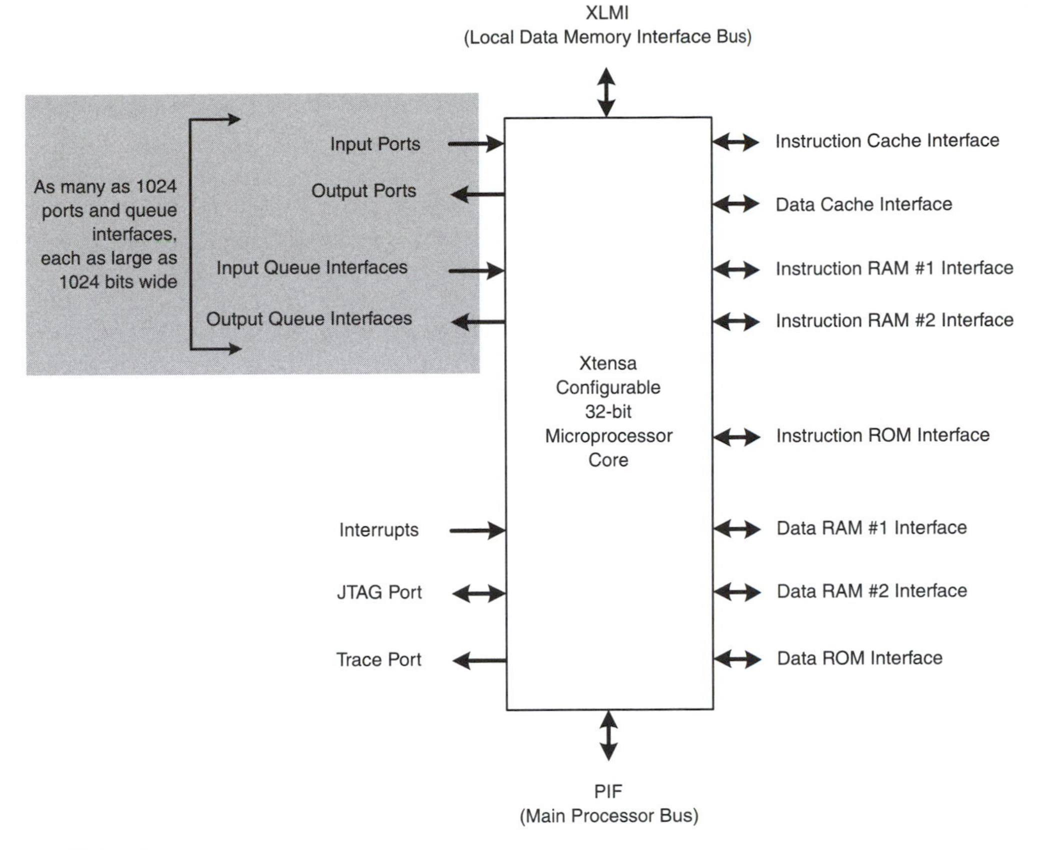

■ **FIGURE 5.7**

The configurable Xtensa microprocessor core has a large number of memory interfaces and I/O ports that support all of the system interconnection schemes shown in Figures 5.1–5.5 and a few more discussed later in this chapter.

memory-specific interface types are supported:

- Instruction and data cache memory
- Local RAM
- Local ROM
- XLMI (high-speed local data bus).

5.4 MEMORY CACHES

Xtensa and some Diamond Standard Series microprocessor cores use separate data and instruction memory caches to significantly improve processor performance. The data caches can be configured as a write-through or write-back cache. Cache locking allows cache lines to be fetched and locked down in the cache to avert cache misses during the execution of critical code.

The Xtensa ISA includes instructions that allow application code to manage and test the memory caches. These instructions allow the code to invalidate cache lines, directly read and write the cache arrays, and lock cache lines. Each cache way actually consists of two differently sized memory arrays: one array for data and one for the cache tag.

5.5 LOCAL ROM AND LOCAL RAM INTERFACES, THE XLMI PORT, AND THE PIF

Xtensa and some Diamond Standard Series microprocessor cores have dedicated RAM and ROM ports for local instruction and data memories. In addition, the Xtensa ISA includes an optional, configurable XLMI data port for high-speed connection to local memories and other device types. Local memories do not incur the same silicon overhead as a memory-cache way, because simple local memories (as opposed to caches) do not have tag arrays.

Each memory interface and the XLMI port have a busy signal that can be used to indicate that the associated memory is being used by some logic external to the processor (or that the memory is otherwise unavailable) and that the processor cannot immediately execute a fetch, load, or store transaction with the selected memory. The memory-busy signals can be used to, among other things, share a memory between multiple processors, or to allow external logic to load the local memory. *Note*: ROM ports do not permit write operations; RAM and XLMI ports do.

The Xtensa and Diamond Standard Series microprocessor cores have a main-bus interface called the PIF (processor interface) that supports load, store, and instruction-fetch transactions initiated by the processor as well as inbound transactions initiated by a device external to the processor (including other processors). Inbound-PIF read and write operations can be directed at a processor's local memories so that the data stored in those memories can be read or modified by external devices.

5.6 THE PIF

The configurable PIF is the main-bus interface for Xtensa and Diamond Standard Series microprocessor cores. Xtensa PIFs can be configured to be 32, 64, or 128 bits wide. Pre-configured Diamond Standard Series processor cores have pre-configured PIF widths and the width depends on the Diamond core selected—higher performance Diamond cores have wider PIFs.

All Xtensa and Diamond core PIFs support single-data transactions (transactions that are less than or equal to the size of the data buses) as well as block transactions where several data-bus widths of data are input or output using multiple PIF-transaction cycles. The PIF employs a split-transaction protocol that accommodates multiple outstanding transaction requests. The PIF-transaction protocol is easily adaptable to a variety of on-chip inter-processor communication schemes.

Like transactions on all microprocessor main buses, PIF transactions occur over several clock cycles. Such is the nature of buses that support ready/busy handshaking, split transactions, and multiple bus masters. Xtensa and Diamond processor cores have faster buses, the XLMI port and the local-memory interfaces, which do not support split transactions or multiple masters, but even these buses require multiple clock cycles to effect a transaction.

Because buses are shared resources, bus protocols must give target slave devices sufficient time to decode the target address presented on the bus at the beginning of the transaction cycle. Consequently, bus transactions are not the fastest possible I/O transaction protocols because bus-transaction speeds do not approach the data rates of direct connections that are one of the hallmarks of RTL hardware design. For applications that require even more I/O bandwidth, Xtensa and Diamond processor cores offer direct ports and queue interfaces.

5.7 PORTS AND QUEUE INTERFACES

One of the most innovative I/O features of Xtensa and Diamond Standard Series microprocessor cores is the availability of TIE-defined ports and queue interfaces that connect the processors' execution units directly to external devices using point-to-point (non-bused) connections. Ports are wires that either make specific internal processor state visible to external logic blocks (and other processors) or bring external signals directly into the processor where it becomes readable processor state. Xtensa processor cores have configurable ports and some pre-configured Diamond Standard Series cores have pre-configured input and output ports.

Similarly, Xtensa and Diamond queue interfaces are designed to connect directly to FIFO memories. Output queue interfaces connect to a FIFO's inputs and drive data into the attached FIFO. Input queue interfaces connect to a FIFO's outputs and accept data from the attached FIFO.

TIE queues allow Xtensa LX processors to directly interface to external queues that have standard push- and pop-type interfaces with full-and-empty hardware flow control. Xtensa port and queue interfaces can be configured to be from 1 to 1024 bits wide and an Xtensa processor can have as many as 1024 ports. Each queue interface actually consumes three TIE ports, so Xtensa processors are "limited" to "only" 340 or so queue interfaces, each up to 1024 bits wide.

Pre-configured Diamond cores have a predetermined number of ports and queues that varies with each core. The raw amount of I/O bandwidth represented by these ports and queue interfaces is unprecedented in microprocessor-based design.

TIE-defined instructions read data from input ports, write data to output ports, push data from the processor onto external queues, and pop data from external queues into processor state registers or register files. Suitably configured Xtensa processors can use the information from input queues directly in computations, bypassing an explicit "load from queue" instruction. Similarly, suitably configured Xtensa processors can store the information from a computation directly to an output queue as part of the computation, bypassing an explicit "store to queue" instruction. Some Xtensa and Diamond cores can perform simultaneous input and output operations over ports and queue interfaces.

An input queue interface connects to a FIFO output and brings data directly into the processor. Single-cycle instructions drive the ports and the queue interfaces, so port and queue transactions are much faster than PIF transactions. In addition, making connections to ports and queue interfaces requires less circuitry than attaching devices using a bus because no address decoding is needed for unshared, point-to-point connections. Addressing is implicit in point-to-point connections.

5.8 SOC CONNECTION TOPOLOGIES

The large number and variety of interfaces on Xtensa and Diamond Standard Series processor cores creates a very wide architectural-design space. The PIF, local memory, and XLMI buses support all of the system topologies shown in Figures 5.1–5.5. In addition, the large number of buses, ports, and queues available on Xtensa and Diamond processor cores allows these processors to be employed in many novel system architectures that provide much higher I/O bandwidth than was previously possible with conventional microprocessor-based designs.

5.9 SHARED-MEMORY TOPOLOGIES

A variety of system topologies employ shared memories to pass large amounts of data between processors in the system. With their many buses and memory interfaces, Xtensa and Diamond processors support a variety of such topologies. Figure 5.8 shows two processors sharing memory over a common multimaster bus. This is a simple and often-used

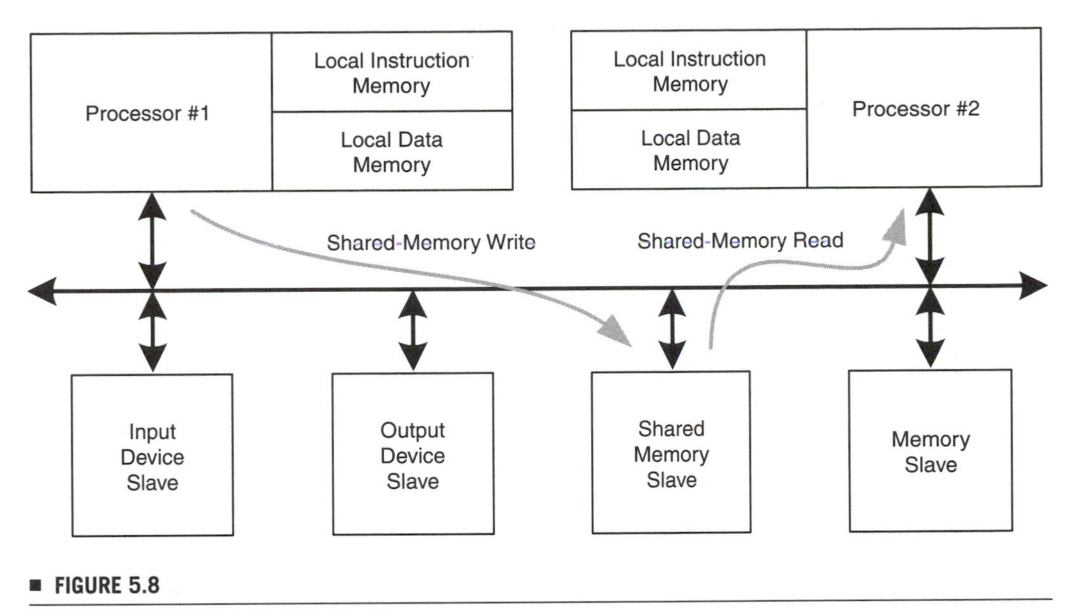

■ **FIGURE 5.8**

Two processors can share memory over a bus.

architectural approach. Xtensa and Diamond processor cores can share memory in this manner over their main PIF buses because the PIF is designed to be a multimaster bus.

Sharing memory over a processors' main buses is a simple interprocessor connection scheme but this approach can exhibit severe performance problems under high traffic loads because the single bus must handle all the traffic from both processors. In addition, the bus must also devote cycles to arbitrating access between the processors. In addition, each processor must, in turn, acquire control of the bus, access the memory, and then signal to the other processor that a memory transaction has been completed.

A method for sharing memory while lessening the bus overhead appears in Figure 5.9. Using this approach, each processor connects to the shared memory through a separate local bus and a hardware-memory arbiter controls access to the shared memory. The shared-memory arbiter can be a separate logic block, which then causes the shared-memory block to emulate a dual-ported memory, or the shared-memory block can be designed as a true dual-ported memory, in which case the memory incorporates the dual-port arbitration logic.

Because the local buses that connect the processors to the shared memory are not themselves shared resources in Figure 5.9, more bandwidth is available to the system and performance improves to the limit set by the shared memory's bandwidth. Xtensa and Diamond processors can connect to shared, dual-ported memories over their XLMI and local RAM buses.

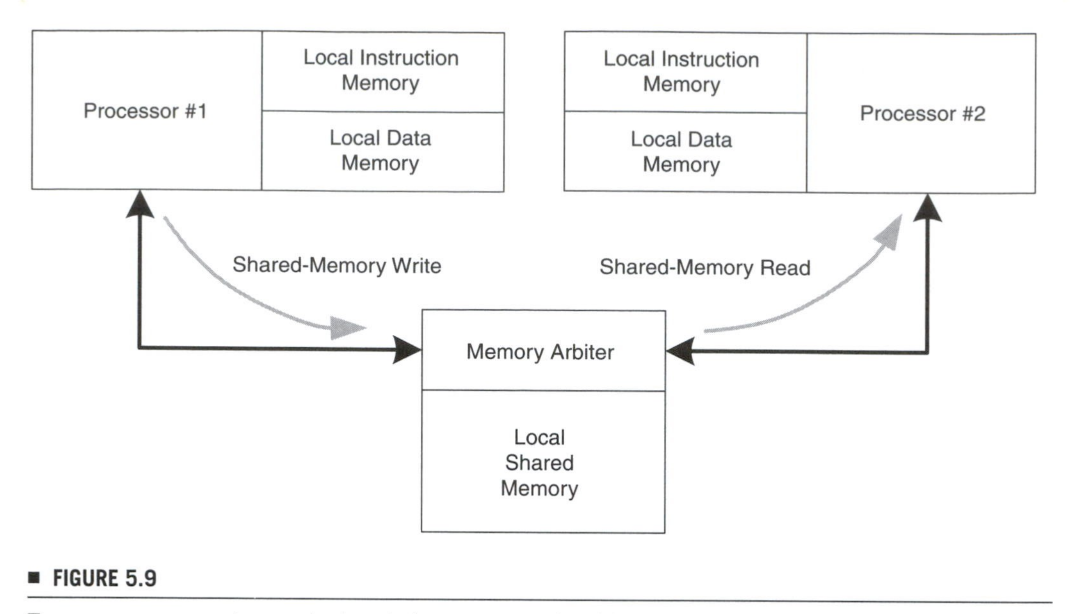

■ **FIGURE 5.9**

Two processors can share a dual-ported memory over local buses.

In addition, Xtensa and Diamond processor cores can directly access each other's local memories over a common PIF bus using the processors' inbound-PIF feature, as shown in Figure 5.10. The PIF supports reads and writes from external agents. Although this method for sharing memory is physically the same as the shared-memory scheme shown in Figure 5.8, it delivers better performance because only half of the shared-memory transactions are conducted over the PIF. The other half of the shared-memory transactions occur over the processors' unshared local-memory buses, so PIF bandwidth requirements are reduced. Further, transactions can occur over each processor's local memory bus simultaneously without creating a resource conflict. This additional concurrency further improves performance for systems that incorporate multiple processors.

5.10 DIRECT PORT-CONNECTED TOPOLOGIES

Buses cannot cleanly support the type of flow-through processing systems commonly implemented using custom-designed logic, shown in Figure 5.11, because buses only permit one I/O transaction at a time. Flow-through processing blocks implemented in custom hardware usually perform input and output operations simultaneously in a pipelined fashion to improve throughput.

The ability to perform simultaneous input and output operations along with internal data computations is one of the key performance advantages

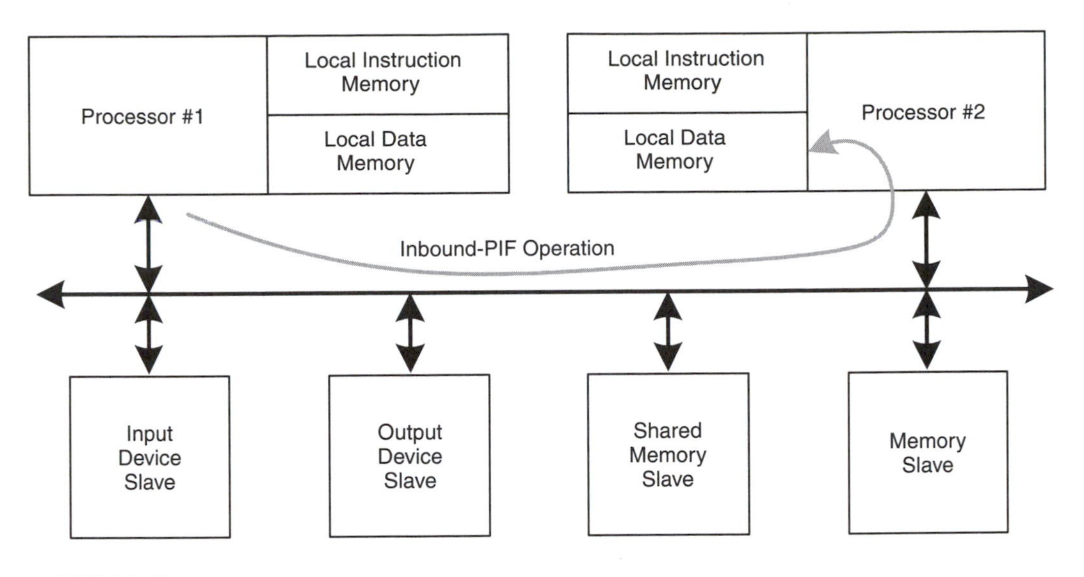

■ **FIGURE 5.10**

Two Xtensa or Diamond processor cores can write to and read from each other's local memory over a shared PIF bus using inbound-PIF operations.

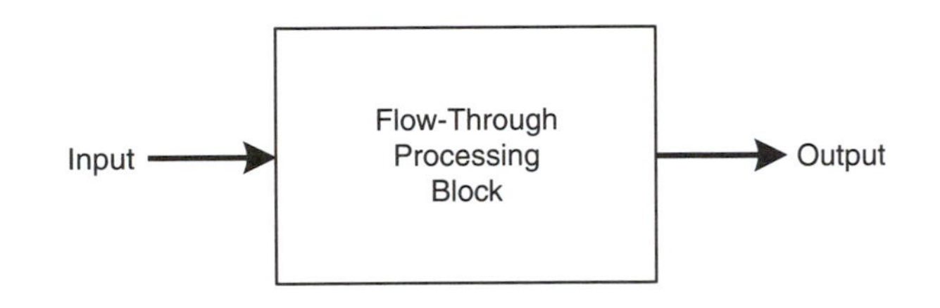

■ **FIGURE 5.11**

A flow-through processing block takes in data, processes it, and outputs the result. All three operations can occur simultaneously. Custom-designed hardware blocks often work in this manner but most processors with their bus-based I/O must emulate flow-through operations because their buses only permit one I/O transaction at a time.

of flow-through processing. Many system designs that construct dataflow architectures from multiple processors—architectures often employed in signal- and image-processing applications for example—can benefit from the additional I/O performance provided by true flow-through processing.

Processor cores that perform all I/O transactions over buses can only emulate the operation of flow-through functional units because they can only operate one of their buses at a time and a bus can only conduct one transaction (input or output) at a time. However, using bus-based processors to emulate flow-through functional units has become so ingrained into the modern system-design lexicon that the overhead of using inherently half-duplex buses to mimic flow-through operation has become all but invisible.

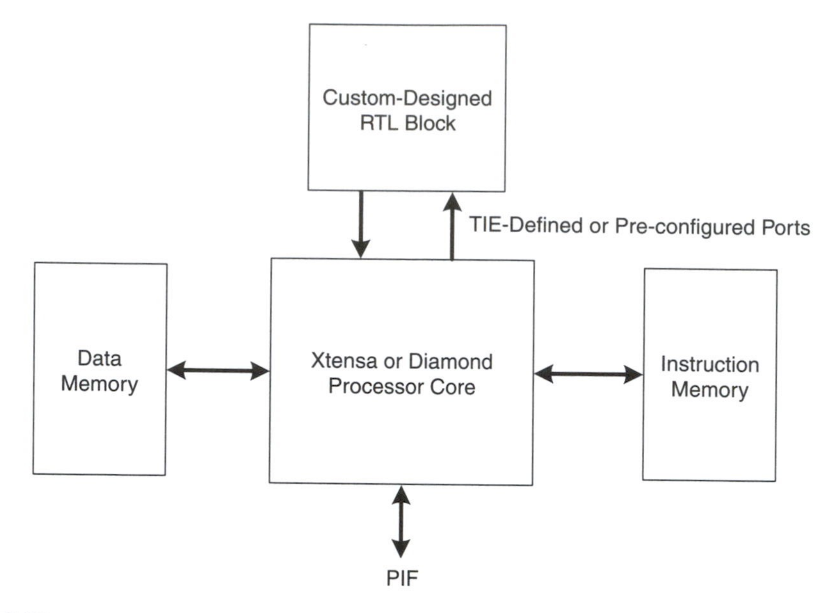

■ **FIGURE 5.12**

Ports allow an Xtensa or Diamond processor core to directly interact with a custom-designed or pre-existing logic block.

Ports and queue interfaces on Xtensa and Diamond processor cores directly support flow-through processing. Figure 5.12 shows how ports can tie an Xtensa or Diamond processor core directly to a custom-designed or pre-existing logic block and Figure 5.13 shows how the RTL block fits into the processor's pipeline stages.

Ports and queue interfaces can also directly tie two processors together, as shown in Figure 5.14. Because these connections exploit the descriptive abilities of TIE, the inter-processor connections are not limited to the widths of their buses. Each port or queue connection can be as wide as 1024 bits and each processor can have hundreds of such connections.

If Xtensa processors were individual chips used on circuit boards, the high I/O pin counts resulting from heavy use of TIE ports and queue interfaces would be unthinkable or merely economically unfeasible. However, Xtensa processors are IP cores implemented in nanometer-fabrication technology so the notion of hundreds of wires connecting two adjacent processors is no longer unthinkable or even unreasonable if you clear your mind of the old-fashioned design constraints previously placed on system architectures that are no longer applicable in the 21st century.

In fact, such high pin counts may actually be very desirable because they potentially reduce processor clock rate. If hundreds of bits are transferred between processors or between a processor and a hardware block during each clock period, there's much less incentive to quest for high I/O bandwidth by running the processor at higher clock rates. System

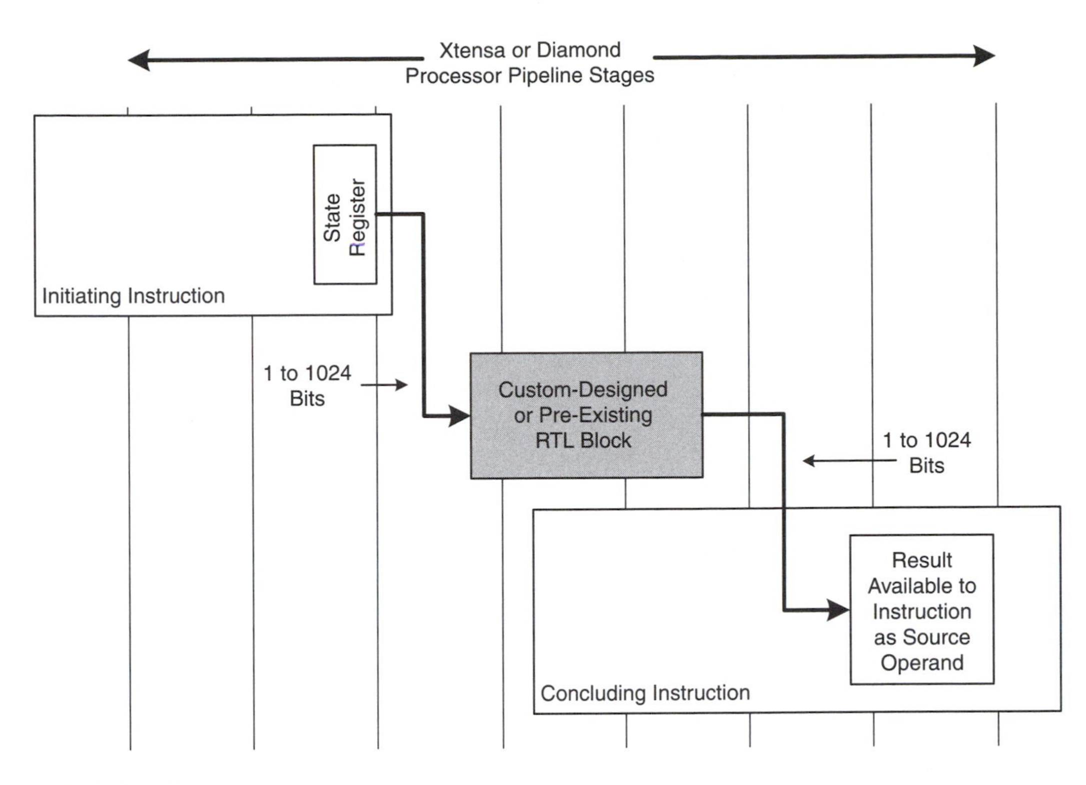

■ **FIGURE 5.13**

One instruction drives the processor's output-port pins that control the external RTL block and another instruction reads the result returned from the RTL block on the processor's input-port pins.

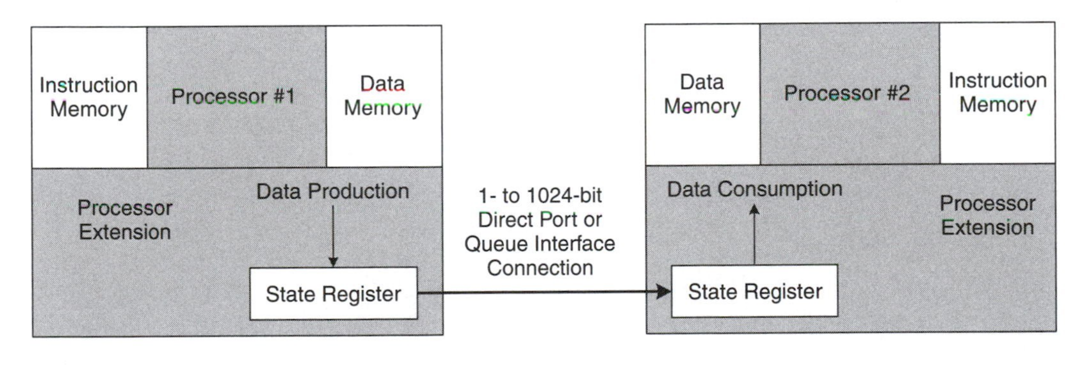

■ **FIGURE 5.14**

Ports and queue interfaces can directly connect two Xtensa or Diamond processor cores.

architectures that employ wide I/O buses between adjacent processors and other logic blocks may become highly desirable for these reasons.

However, connecting widely separated processors and logic blocks with wide buses on an SOC is highly undesirable because stringing wide buses across long distances on a chip can create severe routing congestion and high capacitive loading, which will both degrade signal integrity. Consequently, the use of wide I/O buses for on-chip inter-processor communications must be supported with good chip floorplanning and block placement to minimize the length of the bus runs.

5.11 QUEUE-BASED SYSTEM TOPOLOGIES

Queue-based system topologies maximize the performance of systems built with multiple processors. Use of queues greatly increases concurrency by creating many unshared connections between and among processors. These multiple unshared connections support multiple transactions per clock on multiple and split queues. Figures 5.14–5.16 illustrate several FIFO-based architectural configurations for multiple-processor systems. Figure 5.15 shows a simple system with one FIFO linking two processors. Figure 5.16 shows a FIFO driving other FIFOs to increase I/O concurrency while accommodating dissimilar processing rates of the receiving processors. Figure 5.17 shows a FIFO with address bits selectively driving other FIFOs. The address bits steer the data to the appropriate receiving FIFO.

As a result of their extremely high connectivity levels, Xtensa and Diamond processors permit the development of many new and interesting system topologies that substantially boost an SOC's processing performance. For example, Figure 5.18 shows the block diagram of a wireless audio/video receiver built from four processors. All four processors perform flow-through tasks and there is FIFO buffering between all of the processors to smooth data flow.

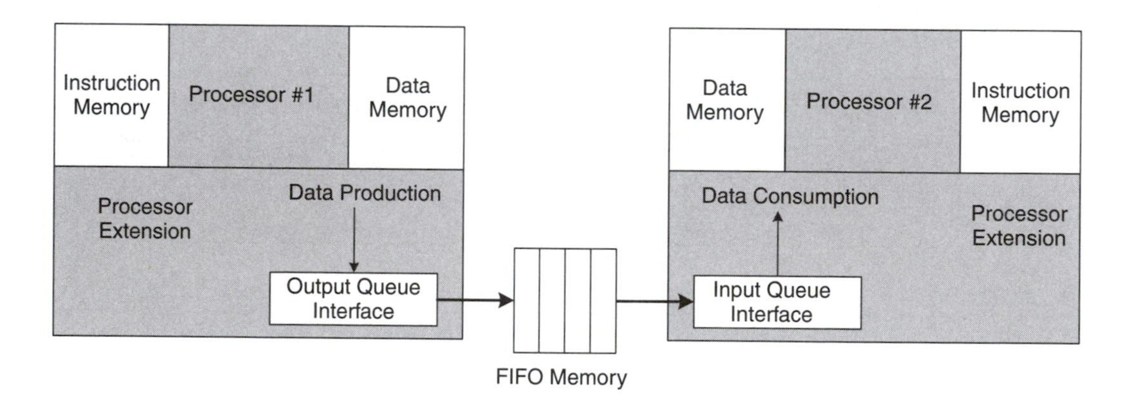

■ **FIGURE 5.15**

A FIFO memory between producing and consuming processors equalizes bursty transfers from and to the processors.

5.12 EXISTING DESIGN TOOLS FOR COMPLEX SOC DESIGNS

With all of the available interconnect schemes for multiple-processor architectures, good system-design tools are essential. A variety of existing SOC construction, simulation and analysis design-tool environments support graphical SOC architecture and platform design using a library of standard components including embedded processors, memories, special hardware blocks, and peripherals all linked with familiar, conventional bus topologies. These existing design tools allow system simulation of processor-based SOC designs using instruction-set simulator (ISS) models for processors and SystemC models (or possibly RTL models if high-level models are not available) for other component types.

Existing design tools can also perform some system-level analysis of design characteristics such as bus loading, bus and resource contention, memory-access activity, and processor loading. These tools are most effective when the system architecture is already known, when the major IP blocks have already been chosen, and when only some micro-architectural tuning and detailed verification are required. However, these tools have

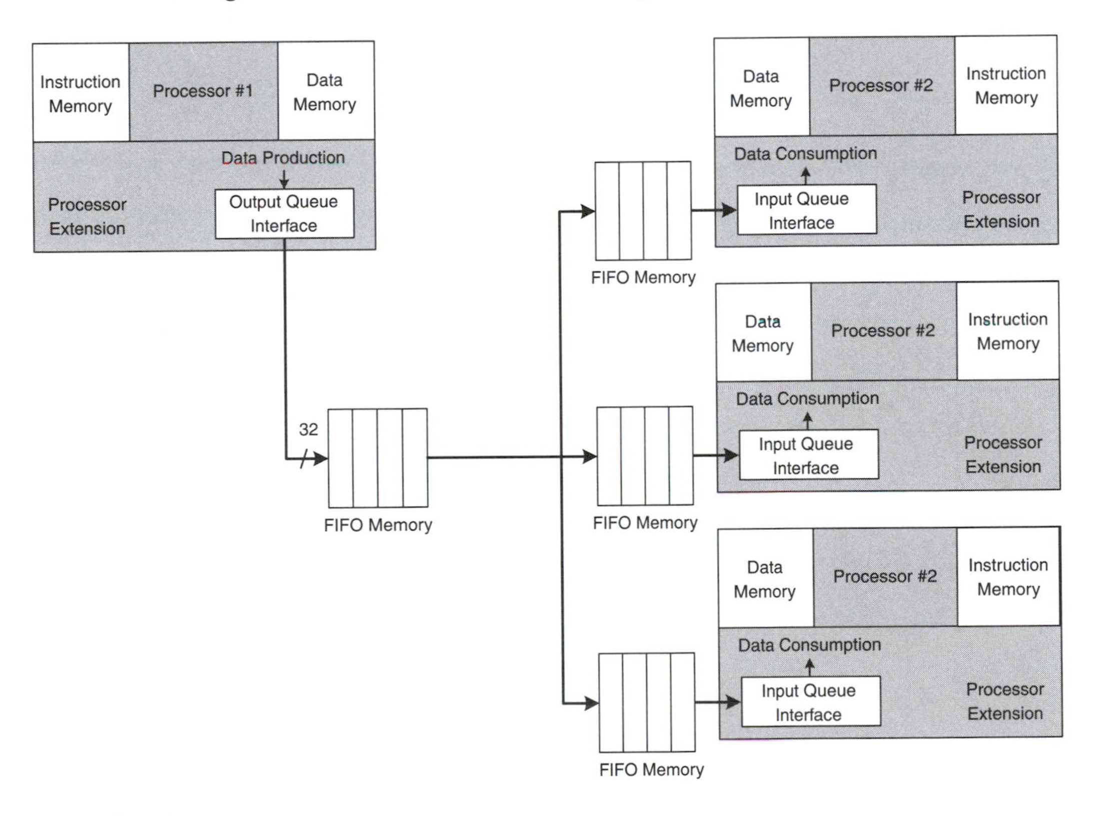

■ **FIGURE 5.16**

One FIFO can drive several others to increase concurrency.

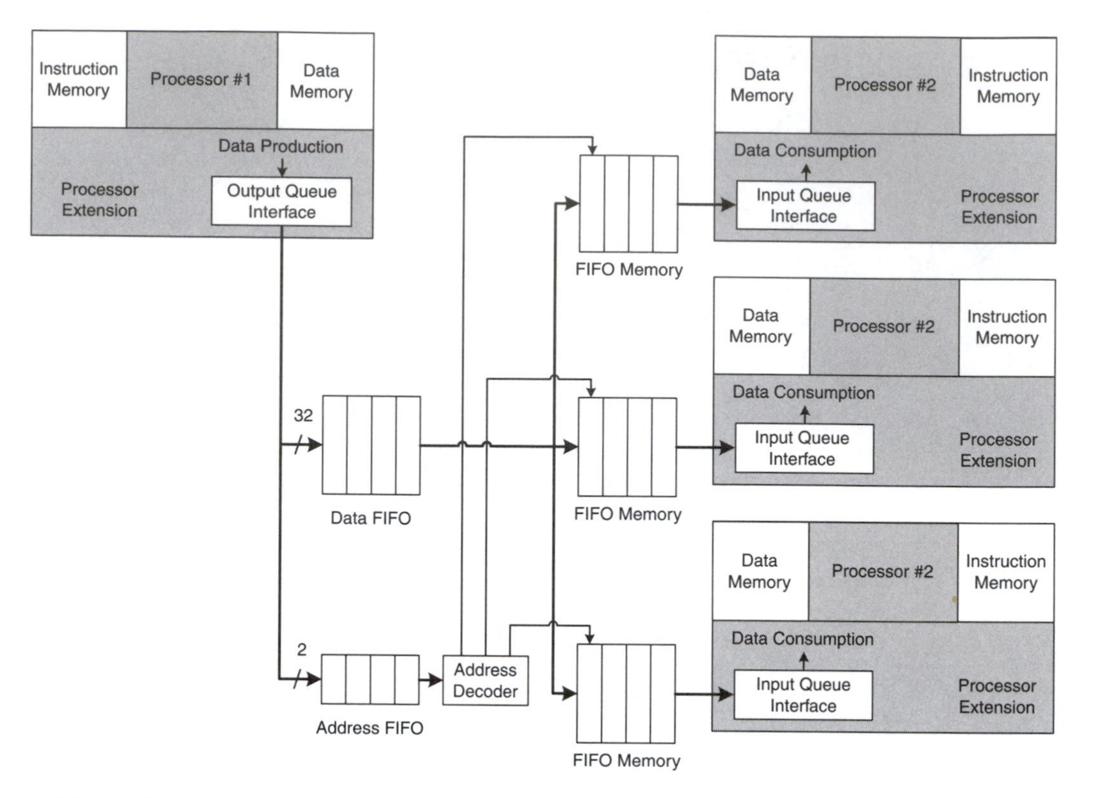

■ **FIGURE 5.17**

One FIFO built from separate address and data FIFOs can automatically and selectively separate out merged data streams containing data intended for different individual FIFOs.

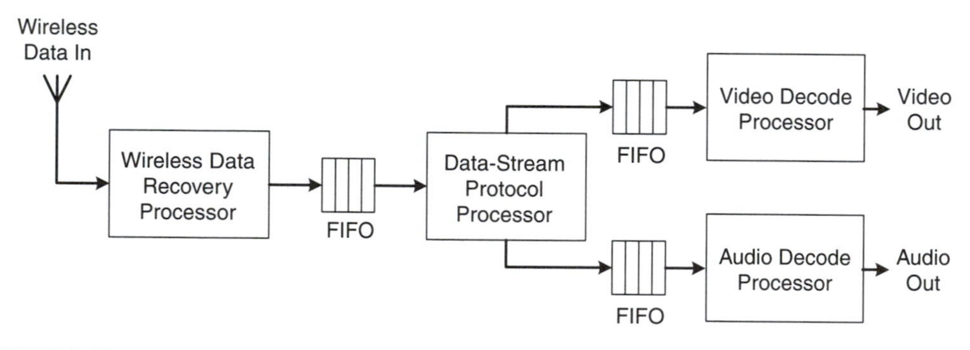

■ **FIGURE 5.18**

Some applications, such as this wireless audio/video receiver, are especially well suited to FIFO-based system architectures.

limited usefulness when the design team wants to explore tradeoffs in task-processor mappings or make informed choices among inter-processor communications alternatives such as buses, ports, and queues.

Virtual system prototyping (VSP) tools can provide some helpful quantitative information about systems designed with more conventional topologies by constructing simulation models of single or multi-processor SOC platforms that execute at speeds at tens of MHz and higher, as opposed to the slower speeds of normal ISSs. VSP tools and models are useful for software developers who want to execute their embedded firmware on models that closely resemble the actual hardware implementation but still provide reasonable simulation performance.

Systems architects who need to test, modify, and characterize the performance of complex applications on a target hardware platform—and who need to run many test scenarios—can also use VSP tools for this purpose as long as the major characteristics of the target hardware platform are fixed. However, none of these existing system-design tools give designers the ability to develop and refine basic SOC architectures, to determine the number and kind of processors needed to achieve system-level performance goals, to design and optimize the on-chip communications architecture (beyond the classical and woefully archaic use of hierarchical buses), to partition their software into multiple tasks mapped onto multiple processors, or to otherwise explore the vast design space that's available.

In 1997, a group of researchers at Delft University of Technology, Philips Research Laboratories, and another group of researchers affiliated with the POLIS Project—a cooperative project conducted by U.C. Berkeley, Cadence Design Systems, Magneti Marelli, Politecnico di Torino, and several other organizations—independently proposed a conceptual scheme for developing and evaluating architectures called the Y-chart, shown in Figure 5.19. This scheme—also called "function-architecture" co-design—was a method of quantitatively evaluating candidate system architectures through system-level simulation of mapped algorithms (functions or application tasks) on candidate architectures.

Function-architecture co-design simulations produce performance results (such as number of computation cycles, amount of inter-block communication traffic, maximum required I/O bandwidth, and memory usage) for each candidate architecture. The quantitative data produced by such simulations would help system architects develop and refine optimal architectures for their systems.

An early function-architecture co-design tool, Cadence's VCC (virtual component co-design) environment, appeared in the late 1990s. This design tool allowed designers to make and verify crucial system-level architectural decisions such as hardware/software system partitioning very early in the system-design cycle. After refining system architectures within VCC, designers could export the hardware/software system design and testbench infrastructure into Mentor Graphics' Seamless Co-Verification Environment

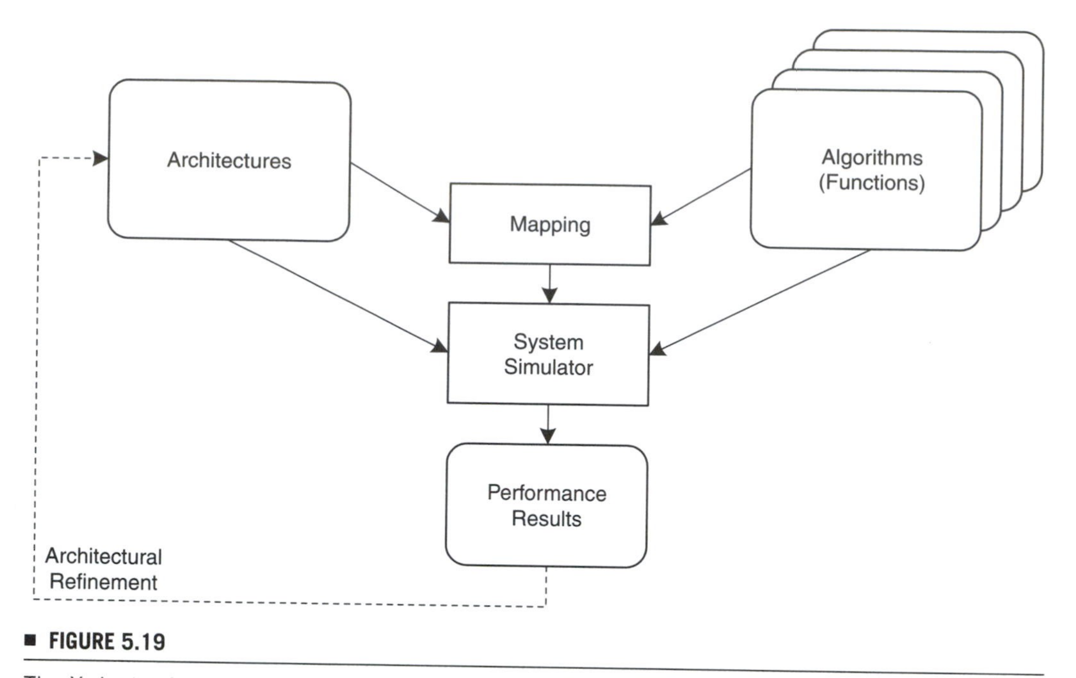

▪ **FIGURE 5.19**

The Y-chart scheme (function-architecture co-design) for system development employs simulation to quantitatively evaluate algorithms mapped to architectures.

(CVE), which allowed them to perform detailed analyses and reconfirm their system-level design decisions. Even though SOC co-design and co-verification provides a development flow that assures architectural-design correctness before IC implementation, Cadence's VCC failed to win a critical mass of designers. Today's commercial electronic system level (ESL) tools do not offer such capabilities.

Cadence's early attempt at function-architecture co-design failed for many reasons. First, the target hardware architectures developed in the mid- to late-1990s often were based on single, fixed-ISA processors plus custom acceleration blocks. The processor had already been chosen for these systems before serious system design began and no amount of simulation would alter that critical selection. Second, poor availability of adequate processor and IP models at the time was a severely limiting problem. Finally, the lack of any standards for system-level modeling posed a real barrier to the creation of such processor and IP models. As a result of these issues, most system architects weren't interested in learning a new, proprietary, non-portable system-modeling language that only offered limited capability.

Consequently, there really has been no way to systematically explore a rich space of system-design alternatives. Without advanced tools for quantitatively exploring system architectures, SOC designers have been artificially constrained to a very limited region of the overall design space even though 21st-century silicon can support vastly superior architectural designs. Existing system-design approaches that employed hand-crafted models, processor ISSs, and HDL simulators usually prove adequate to the relatively simple task of developing the simple single-processor architectures that most SOC-design teams produce.

As a result of this self-supporting circle and a lack of powerful system-simulation tools, most SOC designers continue to rely on tried-and-true architectures that employ relatively simple bus hierarchies. These decades-old system architectures fall far short of exploiting the full capabilities of nanometer silicon and leave untouched a substantial amount of performance that could supply SOC developers with a real competitive advantage in the marketplace.

However, we now find a very different milieu for ESL and IP-based design. The IEEE ratified the SystemC (IEEE 1666) standard on December 12, 2005. SystemC has clearly become the industry's chosen interoperable ESL modeling language for developing complex IP blocks and SoC platforms. System-development tools based on SystemC will inevitably follow ratification of the IEEE 1666 standard.

Methodologies and automated tools for configuring and extending ASIPs (application-specific instruction-set processors) have also begun to emerge. Some of these tools, such as Tensilica's XPRES compiler for Xtensa processors, can be thought of as a non-classical form of behavioral synthesis. Driven by C/C++ code, automated configuration tools like XPRES can produce a processor description in synthesizable Verilog, which effectively results in a kind of "C-to-gates" transformation that produces firmware-programmable hardware. The behavior of firmware-driven, processor-based blocks can be changed far more easily than hard-wired logic and processors are far more capable of executing an expanded task set if necessary.

If SOC designs continue to be limited to single processors with a few custom-designed hardware acceleration blocks, then existing design tools and methodologies will suffice. However, most SOCs already incorporate at least two processors (a RISC processor for control and a DSP) and state-of-the-art SOCs now harness the concurrent processing power of six, ten, or more processor cores. Some use many more. Existing SOC architectural-design methods and tools are distinctly lacking in support for the emerging MPSOC-design style.

5.13 MPSOC ARCHITECTURAL-DESIGN TOOLS

There are several key system-level questions involved in designing an MPSOC based on configurable processors:

- How many processors are needed?
- How should these processors be configured and extended?

- Should the multiple processors be homogeneous or heterogeneous?
- How should the processors communicate? Standard buses? Network on chip (NoC)? Point to point?
- What is the right concurrency model? Pipelined? Dataflow? Multithreading?
- How can software developers extract concurrency from application software?
- How should applications be partitioned to best meet design goals?
- How can the design space encompassing multiple configurable processors, new inter-processor communications architectures, and memory choices best be explored?
- How do you scale a design from 10 to 100 to 1000 processors?

ESL tools that help answer such questions are not presently available from the commercial EDA tools industry because:

- The number of SOC architects is relatively small.
- The appropriate ESL tools may be highly specific to the targeted ASIP technology.

Consequently, this type of ESL-design tool is likely to emerge directly from IP providers to bolster the use of specific IP blocks. Such MPSOC-design tools will help designers and architects by providing an integrated design flow using the following steps:

- Start with the applications or algorithm code.
- Decompose this code into multiple concurrent processing tasks.
- Map the tasks to multiple optimized processors linked with an idealized communications network.
- Simulate the system and iterate processor definition and task mapping.
- Analyze the communications network's requirements.
- Design concurrency control and develop a scheduling model.
- Design an optimized communications network (including shared memories, buses, ports, and queues).

- Analyze the quantitative results of these system optimizations and experiment with alternative configurations to further optimize the system's design.
- Iterate on these design-space exploration steps until a balanced MPSOC system architectural definition is reached.
- Proceed with the detailed hardware/software implementation of the system.

These steps constitute an application-driven, top-down, system-design flow well suited to MPSOC development. In particular, where a wholly new system design is clearly required or when previous subsystem designs simply cannot be extended to meet the requirements of new standards in an application domain, starting over with a clean sheet of paper and driving the subsystem's architecture from the application's characteristics (rather than resorting to convenient, familiar, and thoroughly archaic—"tried-and-true"—subsystem architectures) is far more likely to produce an optimal solution.

5.14 PLATFORM DESIGN

These same system-level design techniques can be used to develop reusable platform architectures. In particular, the ability to specify the structure and architecture of a platform subsystem, to simulate and analyze it, and to iterate on the number and types of processor(s) and the structure of the associated memory hierarchy and communications subsystems, are very helpful to SOC designers who have previously developed their system designs based more on gut feel and the mechanical selection of old, familiar system architectures rather than quantitative simulation results that produce more efficient, better optimized designs.

Often, what's missing is the ability to drive platform design with a product's actual application code, which may not be ready at such an early stage in the product's development. Using the top-down design methodology discussed above, the final code isn't essential for architectural development. Instead of the final application code, similar or related code or code kernels drawn from the general application domain (for example, audio or video encoding or decoding algorithms from a previous generation) can help determine many of the architectural requirements and processor optimizations.

Another possibility is to use artificial code sequences that generate idealized traffic patterns and consume processing and communications resources. Such traffic-generation code can be used to characterize the capabilities of a platform architecture. Once such an architecture has been defined using quantitative system-simulation results, systems and software applications specialists can map the applications onto the target platform and further develop the actual system implementation.

5.15 AN MPSOC-DESIGN TOOL

Configurable, extensible processor cores and pre-configured processor cores are the backbone of high-performance MPSOC architecture. The availability of automatic techniques to derive instruction extensions for configurable processor cores allows the final processor-configuration process to be delayed till almost the end of the system-design process. Of course, pre-configured processors used in the MPSOC's design need no additional configuration.

By automatically generating a configuration and extended ISA for the configurable processor cores early in the MPSOC's design, a lower bound on the MPSOC's performance envelope can be established for the initial task mappings onto the MPSOC's processors, which can be quickly redone whenever changing requirements or quantitative simulation results lead to changes in the task mapping.

Manual improvements made to the configurable processors in the final MPSOC architectural design can create additional performance headroom and allow last-minute processor load balancing and power optimization through the strategic reallocation of some tasks. This refinement process can also help reduce the operating clock frequency for one or more on-chip processor cores by getting more work done per clock cycle, which in turn can lead to an operating-voltage reduction for the MPSOC, which reduces the chip's operating power.

Tensilica is developing just such a processor-IP centric-design methodology that is specifically oriented toward Tensilica's Xtensa and Diamond processor cores. This design methodology is based on the function-architecture co-design scheme first proposed in 1997. Early versions of the design tools associated with this design methodology appear in Figures 5.20–5.22.

Figure 5.20 also illustrates a table-driven user interface for capturing an MPSOC's system structure. Although future MPSOC-design tools may provide graphical, diagrammatic ways of capturing system structure, and this may be a desirable capability in the long term, the presently implemented tabular method captures MP system structure remarkably well.

Modeled at the transaction level, processors and other system components have a reasonable and controlled number of high-level interfaces. Stitching these interfaces together by choosing links from drop-down boxes in a table is sufficient for MP architectures comprising a few tens of key system components. Support for hierarchical subsystem structures with continued use of high-level interfaces between subsystems

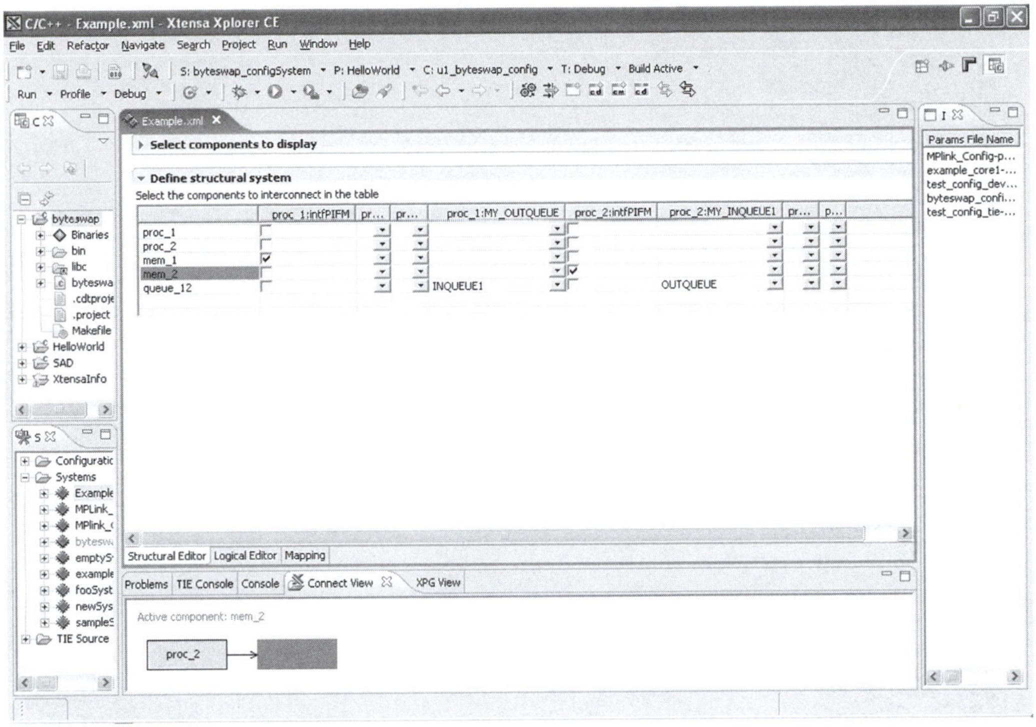

■ **FIGURE 5.20**

One table-driven page can capture an MPSOC's high-level architecture so that quantitative simulation results can be produced.

permits continued use of tabular entry and allows graphical editing of system structure as MP systems grow in complexity.

The captured system structure; the configurable and pre-configured processor models; and the models for other system-level components such as memories, routers, queues, arbiters, and other devices can be used to automatically generate two kinds of system-simulation models, as shown in Figure 5.21.

The first model is a cycle-accurate SystemC model of the subsystem described via the entry table shown in Figure 5.20. This subsystem model has extensive tracing capabilities and can be linked to other SystemC models that represent other portions of the SOC, as long as compatible transaction-level models are used. Alternatively, appropriate wrappers or adapters placed around incompatible models based on different notions of "transaction" can bring compatibility to an incompatible mix of SystemC models.

Such a cycle-accurate, transaction-based system model runs at least two orders of magnitude faster than the equivalent RTL simulation. Faster simulation means better, more thorough simulation because there's more time

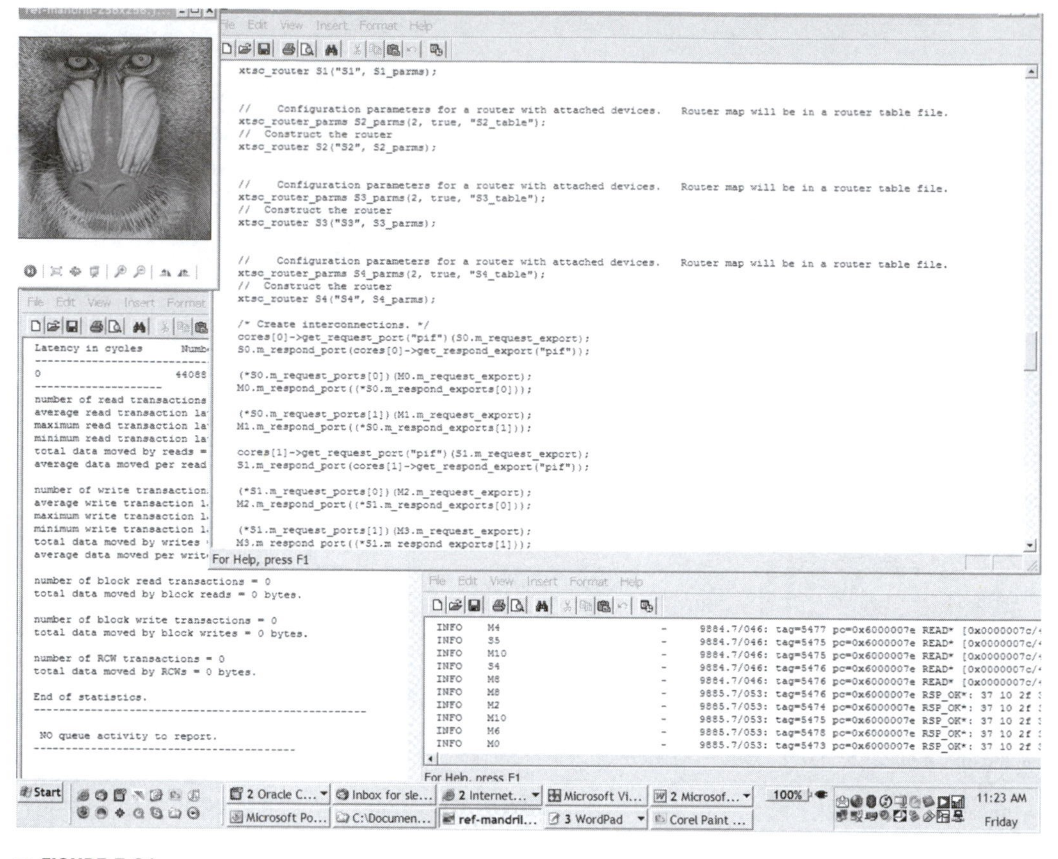

■ **FIGURE 5.21**

Cycle-accurate SystemC models and instruction-accurate fast functional-simulation models can be automatically generated from system high-level system descriptions entered in tabular form.

to try different loading scenarios and operating conditions. Tracing facilities allow system-level transaction performance to be monitored on a statistical basis. From this performance analysis, system designers can derive statistics on overall system throughput and latencies. The tracing facilities also allow detailed transaction-level debug to take place using a visual depiction of the traces.

Figure 5.22 shows a trace file generated in the course of cycle-accurate system simulation. This result display can be used to monitor and debug system-level transactions and to determine the systemic cause for system performance problems. Transaction requests can be examined as they ripple through a hierarchy of devices and their responses analyzed. Stalls, contention, and unusually long transaction-response delays can be visually highlighted as exception conditions.

Tensilica's system-level design methodology provides system-level design capabilities and models some abstract communications mechanisms.

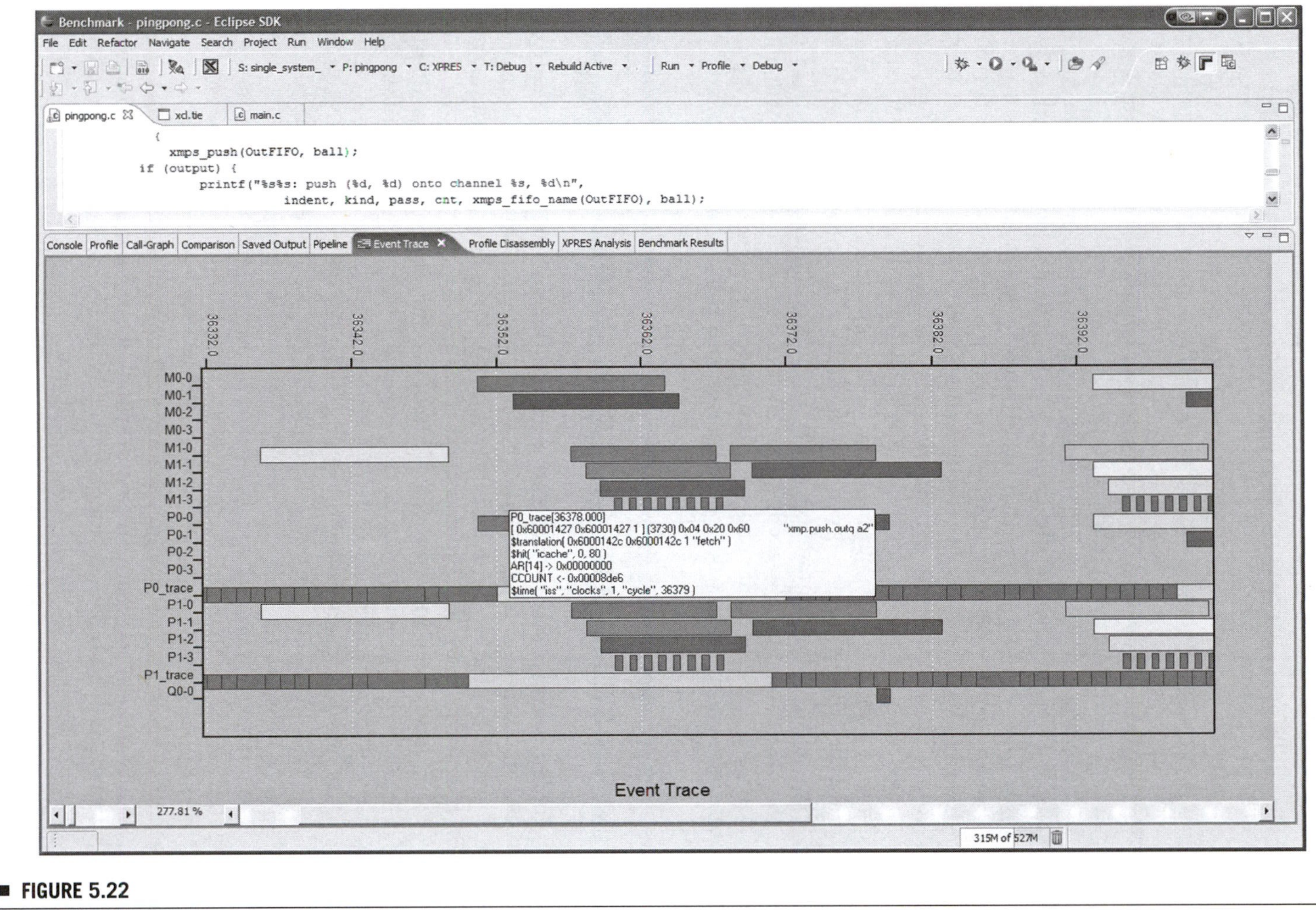

FIGURE 5.22

An event trace created by a system-level design trace facility.

It can map abstract FIFO communication channels into a variety of physical implementations supported by Xtensa and Diamond processor cores including direct hardware queues and shared memories with various locking mechanisms. The methodology can also generate instruction-accurate, fast-functional SystemC models. These models run 10–100× faster than cycle-accurate SystemC simulations for MP systems. Such fast models can only be run under a variety of restrictions but they are particularly useful for software developers, as long as careful attention is paid to the speed-accuracy tradeoff being made.

Fast simulation models require appropriate synchronization mechanisms to be effective. For example, a FIFO queue of fixed depth, which might be used in a cycle-accurate simulation, can stall processor execution for many cycles in an unbalanced system—a processor will stall when it tries to push data into a full FIFO or when it tries to pop data from an empty FIFO. It may be appropriate in such cases to use a FIFO buffer model with a queue depth that's effectively infinite rather than fixed and finite to avoid such stalls in fast simulations, which can communicate with the infinite buffer using direct method calls rather than treating the FIFO as an explicitly modeled device. The behavior of such a modeled system will be functionally accurate for normal operation and thus will allow software development and verification to proceed without getting ensnared in secondary design issues such as optimal queue depths.

5.16 MPSOC SYSTEM-LEVEL SIMULATION EXAMPLE

Table 5.1 shows system-simulation results for a JPEG-encoding subsystem mapped onto a 5-processor MPSOC system. Two of the processors serve as the data source and data sink for the raw and JPEG images. Three processors are linked together in a dataflow style to form the example's processing core. The JPEG-encoding algorithm was divided into three components: color conversion, DCT and quantization, and finally JPEG creation via Huffman encoding. Each processor has access to plentiful local and system-level memory resources. (These resources would be trimmed in a real system to the required sizes, as indicated by quantitative results from more detailed system simulations.) The table shows that the fast, instruction-accurate simulation models are roughly 10–100× faster than the cycle-accurate simulation models. The tradeoff is speed versus accuracy.

Processors in this example system communicate with each other via multiple, direct FIFO queues, as shown in Figure 5.23. These FIFO queues provide fast, high-bandwidth communications between processors without the multi-cycle latency of bus-based communication techniques. Other simulation experiments were run with shared-memory implementations and a mixture of queues and shared memory. All experimental

TABLE 5.1 ■ JPEG-encoding system-simulation results

Image resolution (pixels)	Total cycles required (all processors)	Fast simulation time (sec)	Fast simulation speed (cycles/sec)	Cycle-accurate simulation time (sec)	Cycle-accurate simulation speed (cycles/sec)
32×32	636 K	1.5	370 K	22	29 K
64×64	2.031 M	1.84	1.1 M	70.5	29 K
128×128	21.452 M	4.06	5.8 M	261	82 K
256×256	85.522 M	9.55	9.0 M	1048	82 K

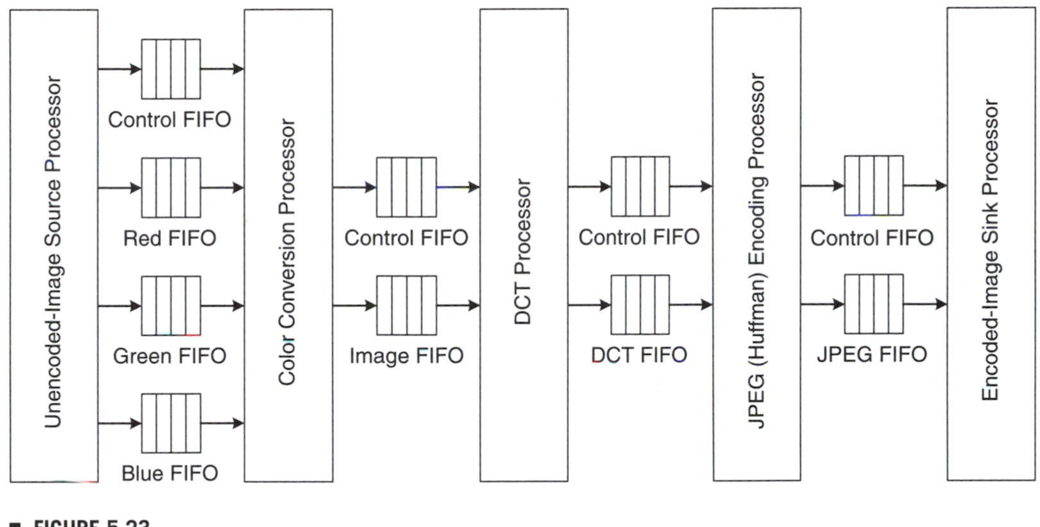

■ **FIGURE 5.23**

The modeled JPEG-encoding system uses five processors in a dataflow configuration linked by multiple hardware FIFO queues.

results listed in Table 5.1 were obtained from simulations running on a Linux workstation containing a Pentium 4 processor running at 3.4 GHz and 1 Gbyte of memory.

Fast-functional and cycle-accurate SystemC simulations were used to validate the software and the system architecture. For a standard picture in 32×32-, 64×64-, 128×128- and 256×256-pixel resolutions, the total number of simulated system cycles on the five processors, the elapsed time for the two simulations, and the corresponding system-simulation rates appear in Table 5.1. The simulated system used enormous hardware queues as communications mechanisms for FIFO channels—20K items deep. Significant processor stalling due to full queues did not occur for the 32×32- and 64×64-pixel resolutions. Table 5.1 illustrates the difference between fast-functional and cycle-accurate system simulations (the cycle-accurate simulation involved 75 system-level devices including processors, local memories, system bus interfaces, routers, system memories, and hardware queue models).

5.17 SOC DESIGN IN THE 21st CENTURY

SOC complexity continues to grow in lockstep with Moore's law. As the number of SOC blocks continues to increase, efficient interconnection of all system blocks and system-level modeling of the resulting complex systems become ever more important. Conventional on-chip interconnection schemes (namely buses and bus hierarchies) derived from board-level system designs of the 1970s and 1980s are increasingly unattractive for global on-chip interconnection because they incur severe routing liabilities and have significant speed limitations due to growing on-chip capacitance. In addition, available design tools from EDA vendors lack support for developing complex systems using multiple processors, and that situation is unlikely to change for a variety of reasons discussed earlier.

Xtensa and Diamond processor cores have multiple on-chip buses, ports, and queues that can support both conventional and newer, more efficient system architectures. The MPSOC-design tool complements these processors by making it possible to perform high-level, SystemC simulations of subsystems and entire SOCs constructed from multiple processor cores and other large logic blocks. The Xtensa and Diamond processor cores and MPSOC allow SOC architectural design to step up to the next level of complexity while leveraging all that's familiar and well understood in the realm of system-level and SOC design.

BIBLIOGRAPHY

Balarin, F., Giusto, P.D., Jurecska, A., Passerone, C., Sentovich, E., Tabbara, B., Chiodo, M., Hsieh, H., Lavagno, L., Sangiovanni-Vincentelli, A., Suzuki, K., *Hardware–Software Co-Design of Embedded Systems: The POLIS Approach*, Kluwer Academic Publishers, 1997.

Chang, H., Cooke, L., Hunt, M., Martin, G., McNelly, A., Todd, L., *Surviving the SOC Revolution: A Guide to Platform-Based Design*, Kluwer Academic Publishers, 1999.

Hellestrand, G., "The Engineering of Supersystems," *IEEE Computer*, January 2005, pp. 103–105.

Kienhuis, B. (ed.), Deprettere, Kees Vissers, Pieter van der Wolf, "An Approach for Quantitative Analysis of Application-Specific Dataflow Architectures," *ASAP '97: IEEE International Conference on Application-Specific Systems, Architectures and Processors*, July 1997, pp. 338–349.

Leupers, R., "Code Generation for Embedded Processors," *13th International Symposium on System Synthesis (ISSS'00)*, 20–22 September 2000, pp. 173–179.

Maestro, J.A., Mozos, D. and Hermida, R., "The Heterogeneous Structure Problem in Hardware/Software Codesign: A Macroscopic Approach," *DATE '99: Proceedings of the conference on Design, automation and test in Europe*, 1999, pp. 766–767.

Martin, G. and Chang, H., *Winning the SOC Revolution: Experiences in Real Design*, Kluwer Academic Publishers, 2003.

Marwedel, P., "Code generation for core processors," *DAC '97: Proceedings of the 34th annual conference on Design automation*, 1997, pp. 232–237.

Rowen, C. and Leibson, S., *Engineering the Complex SOC*, Prentice-Hall, 2004.

Xtensa LX Microprocessor Data Book, Tensilica, Inc, August, 2004.

Xtensa 6 Microprocessor Data Book, Tensilica, Inc, October, 2005.

CHAPTER 6

Introduction to Diamond Standard Series Processor Cores

Adversity is the diamond dust
Heaven polishes its jewels with.
—Thomas Carlyle

With their configurability and extensibility, Tensilica's 32-bit Xtensa microprocessor cores can perform a very wide array of SOC tasks. However, there are many on-chip tasks that do not require the processing and I/O bandwidths achievable with configurable Xtensa cores. Consequently, Tensilica has used the Xtensa ISA and the Xtensa Processor Generator to create architecturally compatible, pre-configured processor cores for specific task sets frequently encountered in SOC design.

These cores, comprising the Diamond Standard Series of pre-configured cores, offer a wide range of performance options without the need to further configure them. Diamond processor cores all employ the Xtensa ISA, so they are software compatible with the configurable Xtensa cores and use the same software-development tools. They carry on the Xtensa system-design philosophy of moving the bulk of a system's capabilities into firmware and defining those capabilities through application code written in C or C++ as much as possible for maximum flexibility and easier system maintenance. In addition to being extremely useful system components for SOC designs, the Diamond processor cores also demonstrate several of the many interesting dimensions of Xtensa configurable-core technology.

6.1 THE DIAMOND STANDARD SERIES OF 32-BIT PROCESSOR CORES

The Tensilica Diamond Standard Series is a family of 32-bit microprocessor cores based on Tensilica's Xtensa ISA. The Diamond Series ISA is implemented as a set of 24-bit instructions that are targeted at a wide range of embedded applications. The most common instructions also have a 16-bit narrow encoding and the Diamond Series architecture allows

TABLE 6.1 ■ Diamond Standard Series microprocessor cores

Diamond Standard controller cores	
108Mini	Compact, cacheless, optimized for low gate counts.
212GP	Mid-performance with instruction and data caches.
Diamond Standard high-performance CPU cores	
232L	CPU with a Linux-compatible MMU.
570T	Extremely high-performance, 3-issue static superscalar, with additional I/O capabilities.
Diamond Standard high-performance and Audio DSP cores	
330HiFi	Dual-issue audio engine optimized for multiple digital audio formats (codecs available for MP3, AC3, AAC, WMA, etc.).
545CK	High-performance, 3-issue static superscalar DSP with 8 MACs.

modeless switching between 24- and 16-bit instructions. Consequently, the Diamond Series processors achieve one of the highest code densities among all 32-bit RISC processor cores, which reduces the amount of on-chip instruction memory required to execute a given task. The initial set of Diamond cores consists of two general-purpose control processors, two high-performance CPUs, and two DSPs, as shown in Table 6.1.

The two Diamond Standard controller cores (108Mini and 212GP) are pre-configured to provide SOC designers with excellent area, power-consumption, code-density, and application-performance specifications. The Diamond 108Mini processor core is a very small, cacheless control processor. It is the only Diamond processor core without memory-cache interfaces and is intended to be used for running tasks that entirely fit within and execute out of local instruction memory. The Diamond 212GP core is a general-purpose controller with instruction- and data-cache interfaces for larger application tasks.

The Diamond Standard CPU cores (232L and 570T) provide additional capabilities over the controller cores. The Diamond Standard 232L processor core upgrades the 212GP controller core with a Linux-compatible memory-management unit (MMU). The Diamond Standard 570T static superscalar processor core is a high-performance CPU that includes 64-bit VLIW (very long instruction word) instruction bundles in its instruction-word lexicon, which means that it can issue and execute two or three independent instructions each clock cycle.

The XCC C and C++ compiler for the Diamond Standard 570T processor core freely intermixes 64-bit instruction bundles with the processor's base-ISA 16- and 24-bit instructions. The Diamond Standard 570T processor modelessly switches between the variously sized instructions. The Diamond Standard 570T processor also includes two 32-bit FIFO queue interfaces that can be used to provide high-bandwidth communication with other blocks in the SOC.

The Diamond 330HiFi Audio Engine consists of a base RISC core with additional dedicated audio data registers and dual MACs (multiplier/accumulators) that operate on 24×24- or 32×16-bit data to achieve full 24-bit audio precision. The 330HiFi Audio Engine includes specialized instructions defined by Tensilica to improve the execution speed of audio codecs. These instructions include loads and stores to auxiliary audio registers, bit-stream operations, and specialized Huffman-coding operations.

Like the Diamond Standard 570T and 545CK processor cores, the Diamond Standard 330HiFi Audio Engine also executes 64-bit instruction bundles. However, the 330HiFi instruction bundles contain two instruction slots. Like the Diamond Standard 570T and 545CK processor cores, the Diamond Standard 330HiFi Audio Engine modelessly switches between the variously sized instructions. Optional audio codec software for encoding and decoding most popular audio formats such as MP3, AC3, AAC, and WMA is available directly from Tensilica. These audio codecs are pre-verified to execute efficiently on the Diamond Standard 330HiFi processor.

The Diamond 545CK processor core is a high-performance DSP that provides the industry's highest performance for licensable DSP core, according to benchmarks developed by BDTI (Berkeley Design Technology, Inc.). Like the Diamond Standard 570T processor core, the Diamond Standard 545CK DSP can issue 64-bit instruction bundles containing three instruction slots per bundle. The 545CK also modelessly switches between its variously sized instruction formats. The 545CK DSP's two 128-bit load/store units and eight MACs allow the DSP to perform eight 16-bit MACs per clock cycle. The MACs are pipelined, which means that the processor can sustain this computation rate for long periods.

The Diamond 545CK DSP core is well suited to communications, audio, and imaging applications. It employs a highly efficient and easy-to-program vector architecture that couples closely to the software-loop vectorizing features in Tensilica's XCC C/C++ compiler. The 545CK core provides higher data throughput, lower power dissipation, and better DSP performance per watt and per area than any other commercially available DSP core.

6.2 DIAMOND STANDARD SERIES SOFTWARE-DEVELOPMENT TOOLS

Software-development tools provided with the Diamond Standard Series processor cores include:

- A software tool suite that precisely matches each processor's architecture. This tool suite includes XCC (the Xtensa C/C++ compiler), a macro assembler, linker, debugger, diagnostic software, reference test benches, and a basic software library. While XCC's operation resembles that of the GNU C/C++ Compiler (GCC), XCC is an advanced optimizing compiler that provides better execution performance and smaller code size. XCC is a vectorizing compiler with direct support for the vector capabilities built into the Diamond Standard 545CK DSP core. In addition, XCC can bundle multiple independent operations into VLIW instructions for the 570T, 330HiFi, and 545CK multi-issue Diamond Standard Series processors.

- Xtensa Xplorer Diamond Edition, an integrated development environment based on the Eclipse platform. Xtensa Xplorer serves as a cockpit for single- and multiple-processor SOC hardware and software design using Xtensa and Diamond processor cores. Xtensa Xplorer Diamond Edition integrates software development and system analysis tools into one common visual design environment that provides powerful graphical visualization abilities and makes creating processor-based SOC hardware and software much easier.
- A cycle-accurate instruction-set simulator (ISS) for each of the Diamond Standard Series processors.

6.3 DIAMOND STANDARD SERIES FEATURE SUMMARY

All Diamond Standard Series processor cores share a common base of 16- and 24-bit instructions. Some Diamond processor cores add 64-bit, VLIW-style instructions. Tensilica's VLIW capability—called FLIX (flexible-length instruction extensions)—allows some of the Diamond Standard Series processor cores to issue and execute multiple independent operations per instruction. This feature significantly boosts application performance. Various key features of the Diamond Standard Series processor cores appear in Table 6.2.

TABLE 6.2 ■ Significant Diamond Standard Series processor core features

Specialized functional units (not on every Diamond Standard Series processor core)
Multipliers, 16-bit MAC, SIMD, VLIW
Memory management and memory protection
Region-based memory protection (108Mini, 212GP, 570T, 545CK, and 330HiFi)
Full Linux MMU (232L)
Miscellaneous processor attributes
Little-endian byte ordering
5-stage pipeline
Exceptions: non-maskable interrupt (NMI), nine external interrupts, six interrupt priority levels, three 32-bit timer interrupts
32- or 64-entry windowed register file
Write buffer: 4/8/16 entries
Available interfaces
32-, 64-, or 128-bit PIF width to main system memory or to an on-chip system bus (Vera-based tool kit for PIF bridge implementation and verification)

(Continued)

TABLE 6.2 ■ (Continued)

Inbound-PIF requests allow external access to the processor's local-memory buses
AMBA AHB-Lite interface available
Two 32-bit TIE ports for the 108Mini, 212GP, and 570T processor cores
Two 32-bit TIE queues for the 570T, 330HiFi, and 545CK processor cores

On-Chip memory architecture (depends on core)
2- or 4-way set-associative caches (all Diamond cores except 108Mini)
Write-through or write-back cache-write policy (all Diamond cores except 108Mini)
Line-based cache locking set-associative caches (all Diamond cores except 108Mini)
8- or 16-Kbyte instruction cache, 32- or 64-byte cache line (all Diamond cores except 108Mini)
8- or 16-Kbyte data cache, 32- or 64-byte cache line (all Diamond cores except 108Mini)
Designer-selectable number of data RAMs and instruction RAMs
Size of data RAM: 0/0.5/1/2/4/8/16/32/64/128 Kbytes
Size of instruction RAM: 0/0.5/1/2/4/8/16/32/64/128 Kbytes

Processor development and debug capabilities
C/C++ callable ISS
On-chip debug (OCD) capability: Trace and instruction/data breakpoint support (two hardware-assisted instruction breakpoints and two hardware-assisted data breakpoints)
GDB debugger support
ISS and Co-Simulation Model (CSM) support for Mentor Graphics' Seamless CVE (Co-verification environment)

Software-development tools
High-performance Xtensa C/C++ compiler (XCC) and companion GNU tool chain
Eclipse-based Xtensa Xplorer Diamond Edition integrated development environment with graphical visualization capabilities

EDA environment support
Physical synthesis design flow for major EDA supplier tool sets (Cadence, Magma, Mentor Graphics)

Verification support
Comprehensive diagnostics for the core

Clock-cycle-accurate, pipeline-modeled ISS

OSKit overlay for supported real-time operating system
Nucleus from Accelerated Technology, Embedded Systems Division, Mentor Graphics
Linux from MontaVista Software

Bus designer's toolkit (PIF kit) for bus-bridge design help

6.4 DIAMOND STANDARD SERIES PROCESSOR CORE HARDWARE OVERVIEW AND COMPARISON

This section highlights the major hardware blocks comprising the Diamond processor cores including the core register file and execution units, load/store units, issue width, external interfaces, and on-chip trace and debug units. Table 6.3 compares the major hardware features included in each Diamond core. Table 6.4 compares several key physical attributes of the six Diamond processor cores.

The feature sets of each pre-configured Diamond core were set so that these six ISA-compatible processor cores provide the SOC designer with

TABLE 6.3 ■ Diamond Standard Series processor core feature comparison

Hardware features	108Mini	212GP	232L	570T	330HiFi	545CK
Instruction width (bits)	16/24	16/24	16/24	16/24/64	16/24/64	16/24/64
Multiple instruction issue (static superscalar)	No	No	No	3 or 2 issue	2 issue	3 issue
Local-memory bus width (bits)	32	32	32	64	64	128
General-purpose registers	32	32	32	32	32	64
DSP vector registers					8×48-bit and 4×56-bit	16×160-bit
Instruction-cache size (Kbytes)	N/A	8	16	16	8	NA
Instruction-cache associativity	N/A	2-way	4-way	2-way	2-way	N/A
Data-cache size (Kbytes)	N/A	8	16	16	8	N/A
Data-cache associativity	N/A	2-way	4-way	2-way	2-way	N/A
Load/store units	1	1	1	1	1	2
Local instruction RAM, selectable size, max size (Kbytes)	128	128	N/A	128	128	128
Local data RAM, selectable size, max size (Kbytes)	128 (dual)	128	N/A	128	128 (dual)	128 (dual)
Local XLMI interface	No	Yes	No	Yes	No	No
32-bit I/O ports	Yes	Yes	No	Yes	No	No
32-bit I/O queue interfaces	No	No	No	Yes	Yes	Yes
Main bus interface (PIF) width	32	32	32	64	64	128
Zero-overhead looping	No	Yes	Yes	Yes	Yes	Yes
Sign-extend, NSA, MIN/MAX, CLAMPS instructions	Yes	Yes	Yes	Yes	Yes	Yes
Synchronization instructions	Yes	Yes	No	Yes	Yes	Yes
MUL16 instructions	No	Yes	Yes	Yes	No	Yes
16/32-bit MAC16 instructions	No	Yes	Yes	Yes	No	No
MUL32 Instructions	No	No	No	Yes	No	No
Audio instructions	No	No	No	No	Yes	No
Specialized DSP instructions	No	No	No	No	No	Yes
External interrupts	9	9	9	9	9	9
Timer interrupts	3	3	3	3	3	3
Non-maskable interrupt	1	1	1	1	1	1
On-chip debug (OCD)	Yes	Yes	Yes	Yes	Yes	Yes

TABLE 6.4 ■ Diamond Standard Series processor core physical comparison

Diamond core	Maximum frequency (MHz) (0.13G WC)	Dhrys. 2.1 MIPS/MHz	Die area* (mm^2) (0.13G)	Gate count	Instruction width	mW/MHz (0.13G)
108Mini	250	1.2	0.41	47K	16/24 bits	0.09
212GP	266	1.3	0.68	73K	16/24 bits	0.135
232L	233	1.3	0.84	84K	16/24 bits	0.145
570T	250	1.52	1.13	114K	16/24/64 bits	0.20
330HiFi	220	1.3	1.33	142K	16/24/64 bits	0.18
545CK	230	1.3	2.93	310K	16/24/64 bits	0.196

Note: All area, power, and frequency are representative only, and subject to variation based on the process technology, cell library, and design tools used.
*Area is post synthesis, post layout, assuming 85% routing efficiency.

a wide range of core size and performance, and a broad selection of features. The cores' physical attributes range widely, tracking the wide range in capabilities among the cores.

6.5 DIAMOND-CORE LOCAL-MEMORY INTERFACES

Each Diamond processor core has a number of local-memory interfaces that provide fast access to instructions and data. There are three types of local-memory interfaces including the cache interfaces, local RAM interfaces, and the XLMI port. All three local-memory interface types are read/write interfaces. (*Note*: The Diamond 232L core has no local-memory interfaces or XLMI port. However, it does have four instruction-cache and four data-cache interfaces needed to support 4-way instruction- and data-cache memories.)

Diamond processor cores, like all pipelined RISC processors, perform speculative read operations on the local-memory buses. Data obtained from these read operations may or may not be used, depending on circumstances in the processor's pipeline. For example, a branch instruction or interrupt can cause the processor to discard data that it has read speculatively. Consequently, devices connected to local-memory buses must not have any associated read side effects.

Simple RAMs do not have read side effects. FIFO memories and most I/O devices do have read side effects. For example, reading the status register of an I/O device often has an irreversible effect on the I/O device and reading a word from the head of a FIFO permanently removes that word from the FIFO.

Attaching devices with read side effects to the local-memory buses of Diamond processor cores will result in unpredictable and incorrect system behavior. Speculative reads only occur on the Diamond cores' local-memory and XLMI buses. Speculative reads do not occur on the Diamond processor cores' main PIF bus.

All Diamond processor cores except the 108Mini processor core have interfaces for separate data and instruction caches to significantly improve processor performance for large programs that will not fit in local instruction memory. The 108Mini core is designed to run programs primarily out of its local instruction memory. Diamond-core data caches can be programmed to employ either a write-through or write-back cache policy. Cache locking allows cache lines to be fetched from main memory and locked down in the cache to avoid cache misses in critical code.

The Diamond ISA includes instructions that allow the application code to manage and test the cache memories. These instructions allow the code to invalidate cache lines, directly read and write the cache arrays, and lock lines in the cache.

Diamond processors have dedicated interface ports for local instruction memories and local data memories. In addition, two of the Diamond cores (the 212GP and 570T processor cores) have a fast, special-purpose interface called the XLMI port. The base addresses for the local-memory and XLMI ports on Diamond processor cores are pre-configured and the address spaces reserved for memories (or other blocks) attached to these ports do not overlap.

The XLMI port is unique among the Diamond cores' local buses. Although speculative read transactions do occur on the XLMI port, the port includes signals that indicate when the oldest outstanding load retires (which signals that the oldest outstanding load is no longer speculative) and when the processor has flushed all outstanding loads (data from all outstanding loads has been discarded). These interface-port signals allow SOC designers to design circuitry that can accommodate speculative operations. If only simple RAM or ROM is attached to the XLMI port, these signals need not be used.

Devices with read side effects can be attached to the XLMI bus as long as they adhere to the "load-retired" and "load-flushed" signaling protocols. Attaching devices with read side effects to a Diamond processor core's XLMI bus without conforming to the interface's "load-retired" and "load-flushed" protocols will result in incorrect and unpredictable system behavior.

The simple way to avoid dealing with speculative read operations is to attach devices that have read side effects only to the processor core's PIF (processor interface) bus, which incorporates logic that prevents speculative reads from appearing on the bus. However, the tradeoff is that it takes extra clock cycles for load and store transactions to appear on the PIF—partly to allow the speculative nature of the read operations to be resolved one way or the other—so PIF transactions consume more clock cycles than XLMI transactions.

Each RAM and XLMI interface port has a busy signal that can be used to indicate that the attached memory is not available to the processor for a load or store transaction. For example, this signal can be used to share a single RAM between multiple processors or to allow external logic to store data in the local memory while holding off accesses from the local processor. The interface busy signals are not used to stretch the read or write transaction. Asserted busy signals indicate rejection of the transaction. The processor may elect to rerun the cycle later or, if the transaction is a speculative operation that's later flushed, the processor may never rerun the transaction.

Note that inbound-PIF operations can be used to make local memories into de-facto dual-port memories by allowing external devices and other processors attached to the PIF to access a processor's local RAM through the PIF. In this mode, the processor itself acts as the memory arbiter. This feature allows external devices to load a Diamond processor's instruction RAM with a program and to use a processor's local RAM as a dual-ported memory with very low hardware overhead.

6.6 THE PIF MAIN BUS

The PIF is the Diamond processor's main bus. The PIF has separate data buses for input and output transactions, and has an inbound-PIF request that allows external devices to access the Diamond processor core's local memories. Figure 6.1 shows a single-data PIF read cycle and Figure 6.2 shows a single-data PIF write cycle. The PIF transaction protocol allows single-data transactions (less than or equal to the size of the data buses) as well as block transactions where several data-bus widths of data are input or output using multiple bus cycles. The PIF protocol employs split transactions and supports multiple outstanding requests. Speculative reads do not appear on the PIF.

Along with the processor's operating clock rate, the PIF's bus width governs its I/O bandwidth. Wider PIF implementations have larger bandwidths. The 108Mini, 212GP, and 232L cores have 32-bit PIFs; the 330Hifi and 570T cores have 64-bit PIFs; and the 545CK core has a 128-bit PIF. Widths for the Diamond cores' local-memory interfaces are the same as the width of each core's PIF.

Appropriate bus bridges for translating between the PIF and AMBA AHB-Lite bus are supplied with all of the Diamond processor cores. These bridges allow the use of existing AMBA-AHB peripheral devices in SOCs designed with Diamond processor cores.

6.7 DIAMOND-CORE PORTS AND QUEUES

In addition to the various local buses described above, most of the Diamond processor cores have a pair of 32-bit I/O ports, a pair of 32-bit I/O

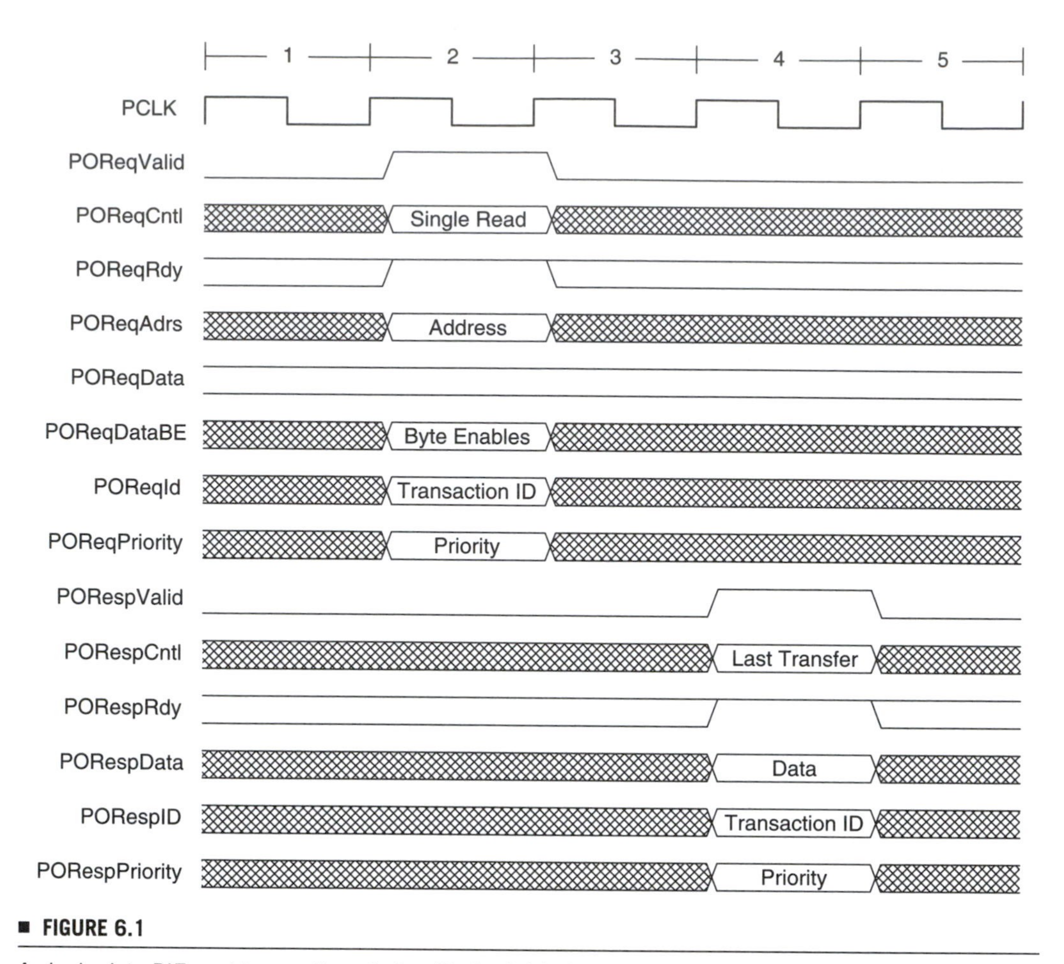

■ **FIGURE 6.1**

A single-data PIF read transaction starts with the initiation of a read request by the PIF master followed by response from the PIF slave that delivers the requested data.

queue interfaces, or both the I/O-port and queue-interface pairs. (Detailed port and queue-interface discussions appear in Chapter 4.) The Diamond processor cores' ports and queue interfaces allow fast I/O transactions to occur without the use of a multi-cycle bus transaction—that's why port and queue I/O is so fast.

The 108Mini and 212GP processor cores each have a 32-bit input and a 32-bit output port. One processor instruction can read the input port and another instruction places data on the output port. Both the 330HiFi and 545CK processor cores have a 32-bit input-queue interface and a 32-bit output-queue interface. One processor instruction pops a data value from the head of a FIFO memory attached to the input-queue interface and another instruction pushes a data value into the tail of a FIFO memory attached to the output-queue interface. The 570T processor core has a pair of 32-bit I/O ports and a pair of 32-bit queue interfaces.

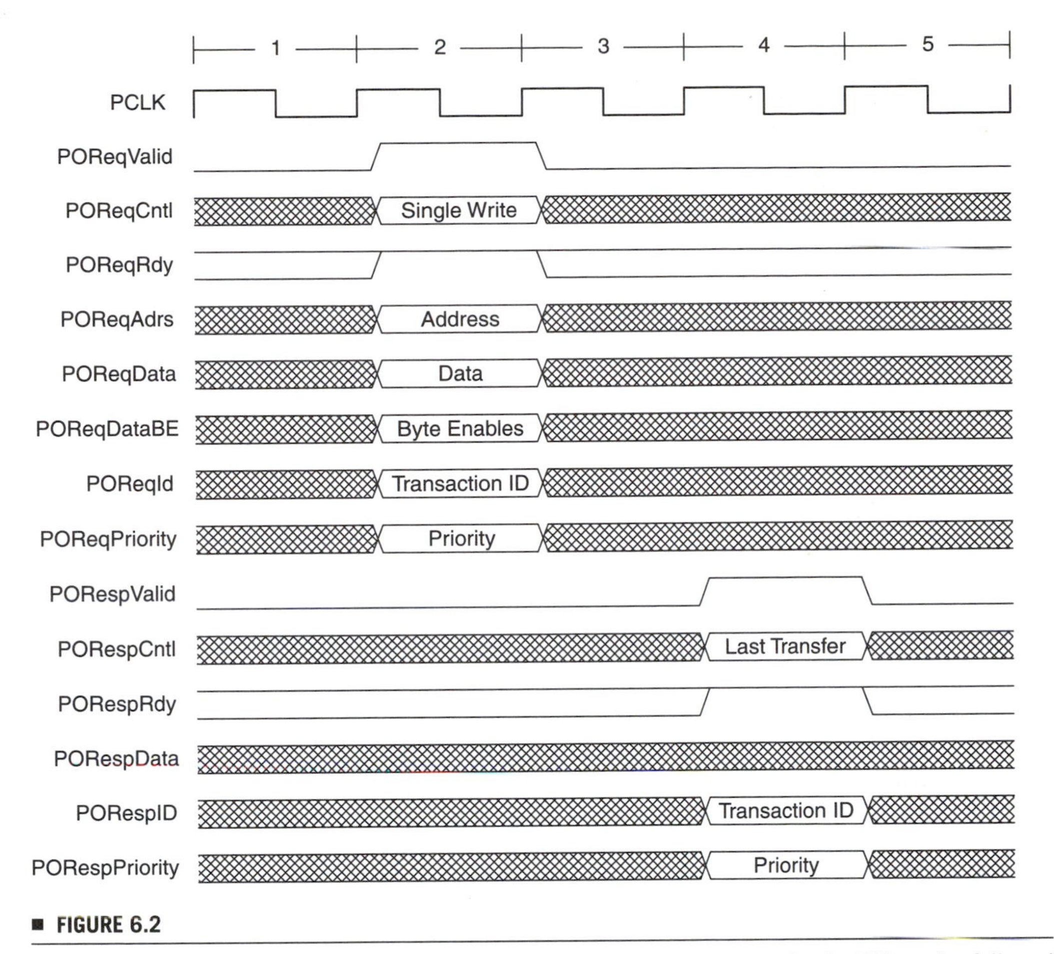

■ **FIGURE 6.2**

A single-data PIF write transaction starts with the initiation of a write request by the PIF master followed by response (acknowledgement) from the PIF slave.

All queue push and pop operations are blocking operations. If a Diamond core attempts to pop information from an empty input FIFO, the processor will stall until data is available from the FIFO. Similarly, if a Diamond core attempts to push data into a FIFO that's already full, the processor will stall until there's room in the FIFO for the operation to complete. It's possible to prevent such blocking behavior from occurring by having the application program sample the input FIFO's empty and the output FIFO's full status signals before attempting the respective pop and push operations. Instructions to check these status signals are available in Diamond processor cores that have queue interfaces.

Diamond ports and queue interfaces open up new architectural design possibilities. Input and output ports replicate the abilities of input and output registers attached to external processor buses. However, the integral ports and queue interfaces have certain advantages over their bus-attached

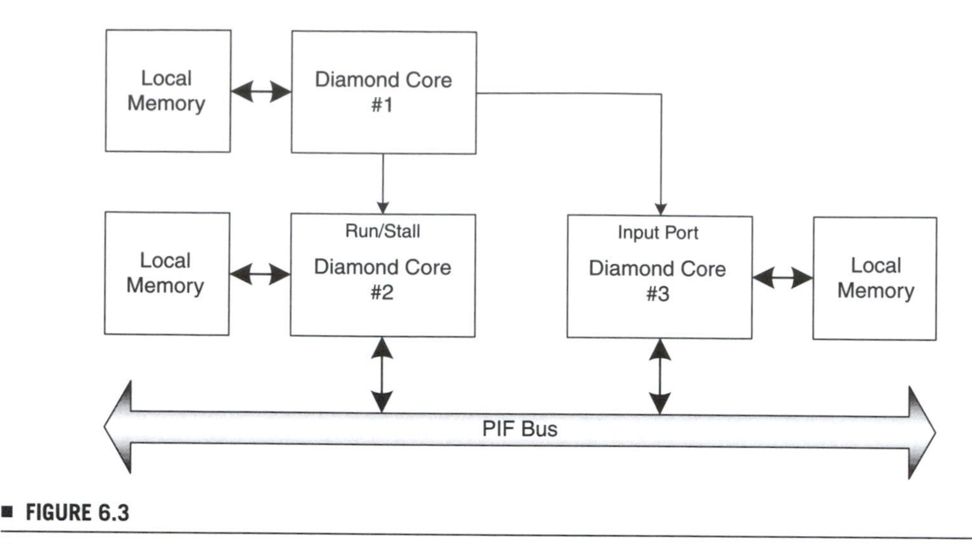

▪ **FIGURE 6.3**

Processor #1 uses two output-port pins to control the Run/Stall input on processor #2 and to signal an input-port pin on processor #3, which alerts processor #3 to the availability of the PIF bus for inbound-PIF access to processor #2's local memory.

counterparts. In particular, they are faster than ports based on bus-attached registers because a Diamond processor accesses these ports without initiating a multi-clock bus cycle. Ports and queue interfaces employ implicit addressing (the port read and write instructions implicitly operate on the appropriate port or queue registers) so explicit addresses need not be set up in a register prior to the execution of a conventional load or store instruction to access the port or queue register.

Figure 6.3 illustrates the use of ports with a 3-processor system. Diamond processor core #1 uses two port pins to respectively drive the Run/Stall input of processor core #2 and an input-port pin on processor core #3. When processor #2 is stalled by the appropriate state on its Run/ Stall input pin, processor #3 is able to read and write processor #2's local memory over the PIF bus using the inbound-PIF feature. This configuration allows processor #1 to halt processor #2 and then signal processor #3 to initiate a data transfer or program load under processor #3's control.

For this mechanism to work, processor #3 must periodically poll its input pin, looking for a command from processor #1. If the input port were implemented conventionally using a latch and address decoder on a bus, the polling operation could generate bus traffic. However, polling an input-port pin generates no external bus traffic.

Figure 6.4 shows two Diamond processor cores connected by both a PIF bus and a FIFO. Because the two processors are both connected to the PIF, they very likely have the same clock. This entire subsystem is synchronous and so is the FIFO that connects the two processors. Processor #1 can send data to processor #2 either to its local memory over the PIF

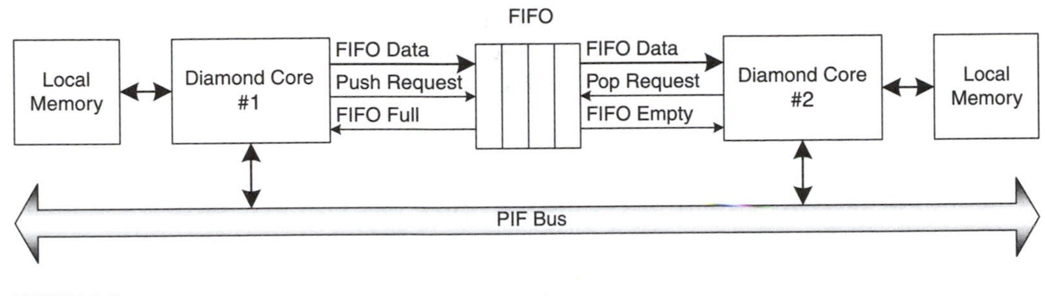

FIGURE 6.4

Processor #1 can communicate with processor #2 over the PIF bus, which is a shared resource, or through the FIFO, which is a private resource.

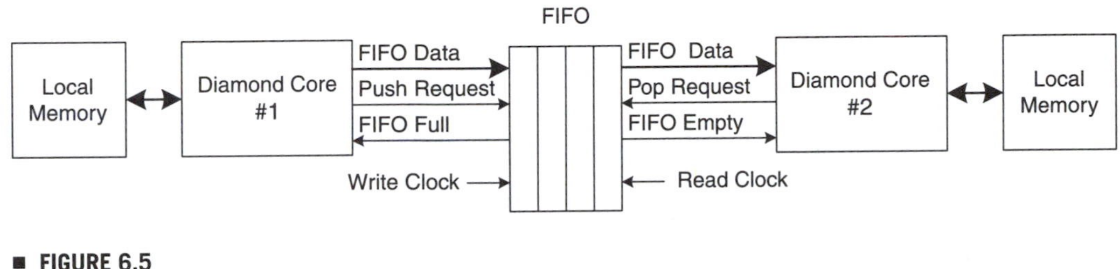

■ **FIGURE 6.5**

Asynchronous FIFOs allow interconnected Diamond processor cores to run at different clock rates.

using processor #2's inbound-PIF capability or using the FIFO attached to the processors' queue interfaces.

Traffic over the PIF must contend with other PIF traffic that will include instruction and data traffic for both processors and traffic from other devices attached to the PIF. By contrast, the inter-processor FIFO connection is private. It is not shared. Therefore, the connection between the FIFO and processor #1 is always available to processor #1 and the connection between the FIFO and processor #2 is always available to processor #2. Using this FIFO link, a Diamond processor core can sustain one transfer per clock through the FIFO link. As a consequence, the FIFO configuration can potentially support much higher data rates than PIF-based communications.

In Figure 6.4 the two processors used a common clock. Asynchronous FIFOs allow interconnected Diamond processor cores to run at different clock rates as shown in Figure 6.5.

Asynchronous FIFOs can be very handy in breaking up large clock domains on an SOC. Creating smaller synchronous clock domains on the chip makes it easier to achieve timing closure in large, complex SOC designs. Further, removing the need to route one clock around an entire chip design can reduce the power dissipation associated with large, high-frequency clock trees. Separate clock domains also allow different blocks within the SOC to operate at different frequencies. Each block runs only at the frequency required to achieve its assigned task, which can further reduce on-chip power dissipation.

6.8 DIAMOND STANDARD SERIES CORE INSTRUCTIONS

All Diamond Standard Series processor cores implement the base Xtensa ISA instructions listed in Chapter 3. These core instructions make the Diamond Standard Series of processor cores software compatible across a very wide performance range. Each of the Diamond processor cores also implements additional instructions that are tailored for each core's intended application space.

6.9 ZERO-OVERHEAD LOOP INSTRUCTIONS

All of the Diamond cores except the 108Mini implement a set of zero-overhead loop instructions. Loops are a fundamental programming structure and are usually implemented with a processor's decrement-test-and-branch instructions. Like all instructions, these instructions must be fetched and executed. These operations take time and create memory cycles and bus traffic, which can increase power dissipation. In addition, branch instructions inevitably create pipeline bubbles.

All of these considerations generate loop overhead. During these overhead cycles, the processor performs no useful work. The zero-overhead loop instruction uses three additional 32-bit registers to set up and keep track of loop conditions so that no housekeeping instructions for the loop are executed within the loop. These zero-overhead loop registers become part of the special-register group in the processor. Table 6.5 lists the three zero-overhead loop registers and Table 6.6 lists the three new instructions that implement the zero-overhead loops.

The XCC compiler automatically uses the zero-overhead loop registers and instructions, if present, to accelerate program execution. Once the loop is set up, no instructions within the loop are needed for loop housekeeping. The processor hardware automatically manages the loop using

TABLE 6.5 ■ Zero-overhead loop registers

Register mnemonic	Register name	Special register number
LBEG	Loop Begin	0
LEND	Loop End	1
LCOUNT	Loop Count	2

logic circuits that perform the loop computations and comparisons in the background. This mechanism accelerates the execution of all loops and greatly accelerates the execution speed of small loops.

It's possible to terminate zero-overhead loops prematurely by directly branching to the address stored in the *LEND* register. It's also possible to short-circuit one loop iteration and conditionally end the loop by branching to a *nop* instruction placed just before the address contained in the *LEND* register.

6.10 MISCELLANEOUS INSTRUCTIONS

All Diamond processor cores implement the miscellaneous instructions listed in Table 6.7. These instructions are useful in a wide range of applications.

TABLE 6.6 ■ Zero-overhead loop instructions

Instruction mnemonic	Instruction definition
LOOP	Set up the zero-overhead loop by initializing the *LBEG*, *LEND*, and *LCOUNT* registers.
LOOPGTZ	Set up the zero-overhead loop by initializing the *LBEG*, *LEND*, and *LCOUNT* registers. Skip the loop if *LCOUNT* is not positive.
LOOPNEZ	Set up the zero-overhead loop by initializing the *LBEG*, *LEND*, and *LCOUNT* registers. Skip the loop if *LCOUNT* is zero.

TABLE 6.7 ■ Miscellaneous Diamond processor core instructions

Instruction mnemonic	Instruction definition
NSA	Calculates the left-shift amount needed to normalize a 32-bit two's-complement number.
NSAU	Calculates the left-shift amount needed to normalize a 32-bit, unsigned number.
MIN	Selects the minimum value operand from two input operands stored in two's-complement format.
MAX	Selects the maximum value operand from two input operands stored in two's-complement format.
MINU	Selects the minimum value operand from two input operands stored in unsigned format.
MAXU	Selects the maximum value operand from two input operands stored in unsigned format.
CLAMPS	Signed clamping function, used for saturating arithmetic.
SEXT	32-bit sign extension.

TABLE 6.8 ■ MP synchronization instructions

Instruction mnemonic	Instruction definition
L32AI	Load with non-speculative characteristics. Forces all subsequent loads and stores to occur after completion of the L32AI instruction.
S32RI	Store with non-speculative characteristics. Forces all previous stores to complete before initiation of S32RI-initiated store operation.
S32C1I	Atomic conditional store used for updating synchronization variables in memory.

TABLE 6.9 ■ 16-bit multiply instructions

Instruction mnemonic	Instruction definition
MUL16S	16-bit, two's-complement multiply with 32-bit result.
MUL16U	16-bit, unsigned multiply with 32-bit result.

6.11 SYNCHRONIZATION INSTRUCTIONS

All Diamond processor cores except the 232L implement the multiprocessor synchronization instructions listed in Table 6.8. These instructions can be used to implement communications structures such as semaphores between multiple tasks and multiple processors.

6.12 16-BIT MULTIPLY AND MULTIPLY/ACCUMULATE INSTRUCTIONS

The 212GP, 232L, and 570T Diamond processor cores implement the 16-bit multiply and 16/32-bit multiply/accumulate instructions listed in Tables 6.9 and 6.10, respectively. The 16-bit multiply instructions use the general-purpose register file entries as sources and destinations.

The 16-bit multiply/accumulate (MAC16) group adds a 40-bit accumulator and four 32-bit multiplication data registers (*MR[0]–MR[4]*) to the processor's special register set and load, multiply, multiply/accumulate, and multiply/subtract instructions to the instruction set. The MAC16 instruction group includes multiply and multiply/accumulate instructions, and loads and stores for the *MR* registers. The multiply and multiply/accumulate instructions can take source operands from the processor's general-purpose register file or from the *MR* data registers. Results from these instructions are placed in a 40-bit accumulator addressed as two special registers. The XCC C/C++ compiler will automatically generate code that uses the MUL16 and MAC16 instructions for multiplication functions with *short* source operands.

TABLE 6.10 ■ 16-bit multiply/accumulate instructions

Instruction mnemonic	Instruction definition
LDDEC	Load MAC16 data register with auto decrement.
LDINC	Load MAC16 data register with auto increment.
UMUL.AA.*qq*	Unsigned multiply using source operands from the general-purpose register file.
MUL.AA.*qq*	Signed multiply using source operands from the general-purpose register file.
MUL.AD.*qq*	Signed multiply taking the first source operand from the general-purpose register file and second from an *MR* register.
MUL.DA.*qq*	Signed multiply taking the first source operand from an *MR* register and second from the general-purpose register file.
MUL.DD.*qq*	Signed multiply using source operands from the *MR* registers.
MULA.AA.*qq*	Signed multiply/accumulate using source operands from the general-purpose register file.
MULA.AD.*qq*	Signed multiply/accumulate taking the first source operand from the general-purpose register file and second from an *MR* register.
MULA.DA.*qq*	Signed multiply/accumulate taking the first source operand from an *MR* register and second from the general-purpose register file.
MULA.DD.*qq*	Signed multiply/accumulate using source operands from the *MR* registers.
MULS.AA.*qq*	Signed multiply/subtract using source operands from the general-purpose register file.
MULS.AD.*qq*	Signed multiply/subtract taking the first source operand from the general-purpose register file and second from an *MR* register.
MULS.DA.*qq*	Signed multiply/subtract taking the first source operand from an *MR* register and second from the general-purpose register file.
MULS.DD.*qq*	Signed multiply/subtract using source operands from the *MR* registers.
MULA.AA.*qq*.LDDEC	Signed multiply/accumulate using source operands from the general-purpose register file followed by a load from an *MR* register with autodecrement.
MULA.AD.*qq*.LDINC	Signed multiply/accumulate taking the first source operand from the general-purpose register file and second from an *MR* register followed by a load from an *MR* register with autoincrement.
MULA.DD.*qq*.LDDEC	Signed multiply/accumulate using source operands from the *MR* registers followed by a load from an *MR* register with autodecrement.
MULA.DD.*qq*.LDINC	Signed multiply/accumulate using source operands from the *MR* registers followed by a load from an *MR* register with autoincrement.

Note: The *qq* designation in the above instructions indicates whether the instruction's source operands are taken from the upper (q = H) or lower (q = L) 16 bits of the 32-bit register. The first q designator specifies the upper or lower location for the first source operand and the second q designator specifies the upper or lower location for the second source operand.

6.13 32-BIT MULTIPLY INSTRUCTIONS

The 570T Diamond processor core implements the 32-bit multiply instructions listed in Table 6.11. The 32-bit multiply instructions use the

TABLE 6.11 ■ 32-bit multiply instructions

Instruction mnemonic	Instruction definition
MULL	32-bit multiply with result taken from low 32 bits of the full 64-bit result.
MULSH	32-bit, two's-complement multiply with result taken from the high 32 bits of the 64-bit result.
MULUH	32-bit, unsigned multiply with result taken from the high 32 bits of the 64-bit result.

general-purpose register file entries as sources and destinations. The XCC C/C++ compiler will generate code that uses the MUL32 instructions for multiplication functions with *int* source operands.

6.14 DIAMOND-DEVELOPMENT TOOLS

A full-featured software cross-development environment supports the Diamond Standard Series of processor cores. The Xtensa Xplorer Diamond Edition is an Eclipse-based cockpit for code development using the compiler tool chain, the ISS, and the hardware emulation and development boards. The Xplorer environment also includes a software project manager and a performance-modeling tool.

The basic Xplorer GUI (graphical user interface) incorporates the XCC C/C++ compiler, which generates high-density 16/24-bit instructions for all Diamond processor cores and automatically includes 64-bit instruction bundles in the instruction stream for the 330HiFi, 545CK, and 570T Diamond cores. Proprietary compiler-optimization techniques incorporated into the XCC compiler improve density and performance in generated code for all of the Diamond processor cores. On average, the XCC compiler delivers a 20–40% improvement in performance compared to the widely used GCC compiler. When compared to other RISC processor compilers, the XCC compiler's superior code density helps to reduce system costs (by requiring less external and internal memory) and cuts processing overhead.

The pipeline-accurate ISS for Diamond processor cores can be used for code benchmarking. Used in conjunction with the XCC/GNU tool chain, the ISS produces feedback results that enable faster code development and better system-level architectural tradeoffs.

6.15 OTHER SPECIALIZED DIAMOND STANDARD SERIES PROCESSOR INSTRUCTIONS

Two more specialized groups of Diamond processor core instructions are worth noting. These are the audio instructions included in the 330HiFi processor core's instruction set and the DSP instructions included in the 545CK processor core's instruction set. These specialized instructions are discussed in more detail in the chapters devoted to the 330HiFi and 545CK processor cores.

6.16 CHOOSING A DIAMOND

All of the Diamond Standard Series processor cores are based on a general-purpose, 32-bit RISC ISA, so all of the cores can be assigned general-purpose tasks on an SOC. However, each of the Diamond processor cores has been shaped for specific applications:

- The 108Mini core is a good, general-purpose processor that's been trimmed to minimum size. It's intended for use as a control processor with code running almost exclusively out of the processor core's local instruction memory. This instruction memory can be loaded by the 108Mini itself using load instructions directed to main memory over the processor core's PIF bus or by an external agent using the 108Mini's inbound-PIF feature. The 108Mini core sacrifices the performance-enhancing zero-overhead-loop hardware for reduced core size.
- The 212GP core is a fast, general-purpose processor with instruction and data caches to achieve good overall performance when executing larger programs.
- The 232L processor core is similar to the 212GP core but adds a full-featured MMU for running operating systems that require a full demand-paged MMU with memory protection, such as Linux.
- The 570T processor core is a high-performance CPU with 3-way static superscalar operation. Although only about twice as large as its scalar processor counterparts, the 570T delivers more software execution speed per clock cycle than any processor core in its class. The core also includes two queue interfaces for good performance in flow-through applications.
- The 330HiFi processor core contains a specialized, 24-bit DSP for audio applications. It includes approximately 300 audio-specific instructions for handling and manipulating special audio data types and two queue interfaces for good performance in flow-through applications. A full set of firmware audio codecs (written entirely in C) is available for use with the 330HiFi core.
- The 545CK DSP is a 3-issue static superscalar processor with a SIMD unit containing eight 16-bit MACs. The 545CK DSP core offers the highest performance of any available DSP core, as measured by BDTI

benchmarks. It's targeted at DSP applications that require very fast execution speeds such as signal and image processing while retaining a general-purpose processor core that excels at efficiently executing control code.

All of the Xtensa and Diamond processor cores are accompanied by tailored versions of the XCC C/C++ compiler, which exploits the unique abilities of each processor whenever possible. The close pairing of processor and compiler means that firmware developers can code in C or C++ and only rarely will they need to drop down to assembly-language programming to improve code performance.

BIBLIOGRAPHY

Cummings, C.E., "*Simulation and Synthesis Techniques for Asynchronous FIFO Design*," Sunburst Design, www.sunburst-design.com

Cummings, C.E., "*Simulation and Synthesis Techniques for Asynchronous FIFO Design with Asynchronous Pointer Comparisons*," Sunburst Design, www.sunburst-design.com

Diamond Standard Processors Data Book, Tensilica, Inc., February 2006.

Leibson, S. and Kim, J., "Configurable Processors: A New Era in Chip Design," *IEEE Computer*, July 2005, pp. 51–59.

Rowen, C. and Leibson, S., *Engineering the Complex SOC*, Prentice-Hall, 2004.

CHAPTER 7

The Diamond Standard Series 108Mini Processor Core

Mini Me, you complete me.
—Mike Meyers as Dr. Evil

The Diamond Standard Series 108Mini processor core is the smallest of the Diamond Standard Series processor cores. The processor consumes less than one half mm^2 of silicon and approximately 110 μW/MHz when implemented in a 130 nm, G-type (general-purpose) process technology. Although the 108Mini processor core is physically small, it's still a full-featured, 32-bit RISC processor that can run any program compiled by the Diamond Edition XCC C/C++ compiler and it can run real-time operating systems (RTOSs) such as the Nucleus Plus RTOS from Accelerated Technology.

The Diamond 108Mini is well suited to roles previously assigned to 8- and 16-bit controller cores but it brings many performance benefits of a 32-bit processor to bear on the designated tasks:

- Large 4-Gbyte address space
- 32-bit computations
- Large 32-entry register file
- 5-stage pipelined operation resulting in a 250-MHz maximum clock rate in 130 nm technology.

The Diamond 108Mini processor core has a 32-bit version of the general-purpose PIF bus for global SOC communications but it is intended to be used as a control processor that executes code from local instruction memory and accesses data primarily from local data memories. An optional AMBA AHB bus bridge supplied with the Diamond 108Mini processor core adapts the PIF to peripheral devices designed for the AMBA AHB-Lite bus. The Diamond 108Mini core also incorporates direct input and output ports that accelerate certain types of I/O.

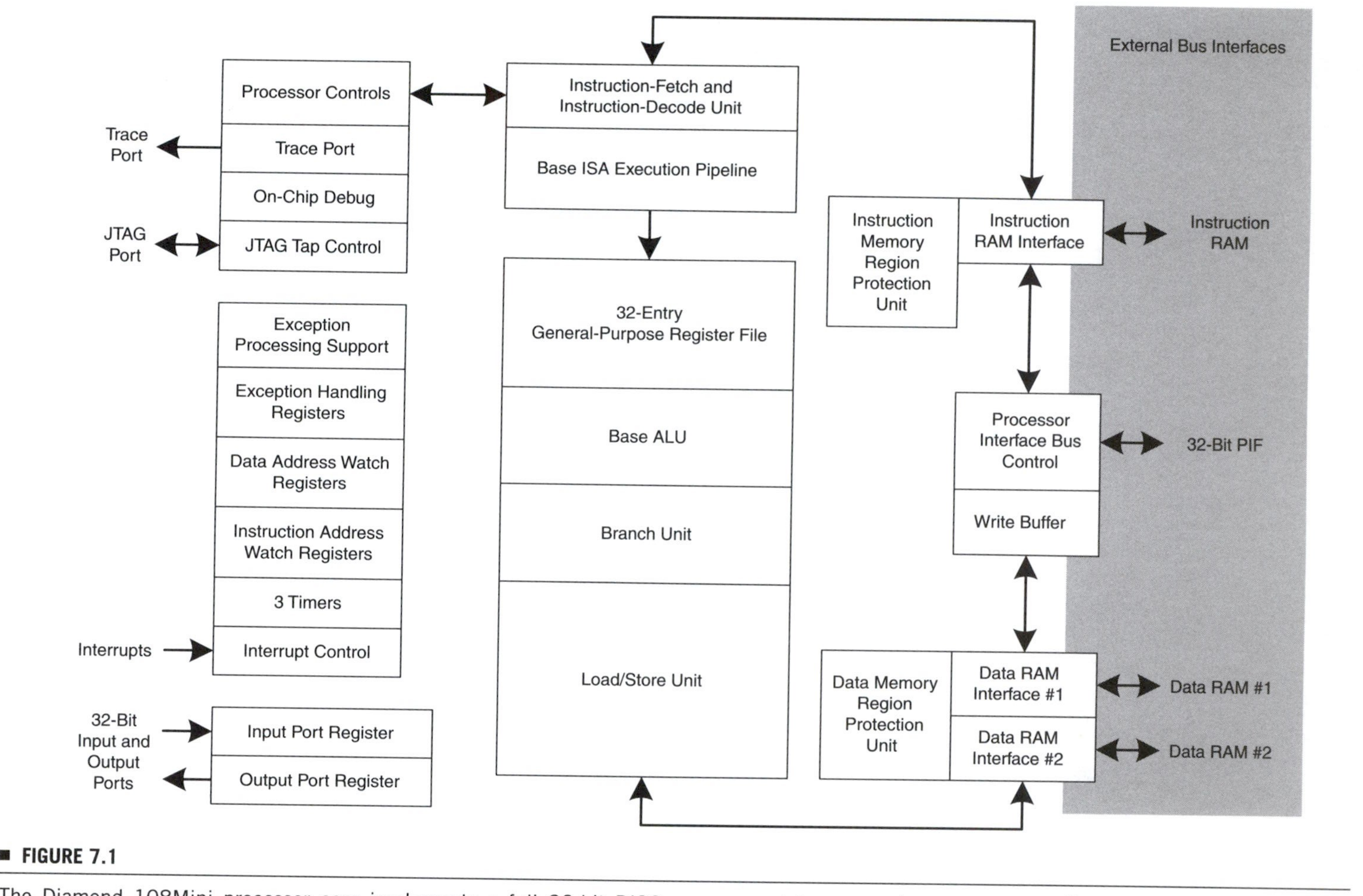

▪ **FIGURE 7.1**

The Diamond 108Mini processor core implements a full 32-bit RISC processor while consuming only 0.41 mm^2 of silicon and approximately 100 μW/MHz when implemented in a 130 nm, G-type process technology.

7.1 THE CONFIGURABLE PROCESSOR AS CONTROLLER

Configurable processor technology is often regarded as a path to creating processors with extremely high performance. The technology can also create processors with extremely small area and the Diamond 108Mini processor is an example of such a core. Figure 7.1 is a block diagram of the Diamond 108Mini processor core.

The Diamond 108Mini processor architecture contains all of the basic elements of the Xtensa ISA (instruction-set architecture). It has a 32-entry general-purpose register file, a 5-stage execution pipeline, 32-bit addressing, and a simple region-protection unit (RPU) that can be used to implement straightforward memory management. The Diamond 108Mini also incorporates the Diamond Series processor core software-debug stack, as shown on the left of Figure 7.1. This debug stack provides external access to the processor's internal, software-visible state through a 5-pin IEEE 1149.1 JTAG TAP (test access port) interface and through a trace port that provides additional program-trace data.

7.2 DIAMOND 108Mini PROCESSOR CORE INTERFACES

The Diamond 108Mini processor core has a number of interfaces for high-bandwidth I/O, as shown in Figure 7.2. These interfaces include a 32-bit implementation of the Xtensa PIF (main) bus and separate memory interfaces for one local instruction-memory block and two local data-memory blocks. Each of the three local-memory interfaces can accommodate memories as large as 128 Kbytes, for a total of 384 Kbytes of local memory.

Table 7.1 lists the address spaces assigned to the Diamond 108Mini processor core's three local-memory blocks. Note that the instruction RAM and first data RAM memory spaces are adjacent so they form one contiguous 256-Kbyte memory block from 0x3FFE0000 to 0x4001FFFF. The address space for the second data memory is just below and contiguous with the address space for the first data-memory block so all three local-memory spaces also form one contiguous 384-Kbyte address space from 0x3FFC0000 to 0x4001FFFF.

The processor core has a 32-bit, 4-Gbyte address space, so additional memory can be attached to the PIF bus if needed. However, the address spaces dedicated to local memory are hardwired to the appropriate local-memory ports and memory transactions to those address spaces will not appear on the PIF bus. All Diamond processor cores including the Diamond 108Mini automatically send memory accesses to non-local address locations out over the PIF bus.

Table 7.2 lists the Diamond 108Mini processor core's assigned reset, NMI (non-maskable interrupt), and other interrupt vectors. Note that at

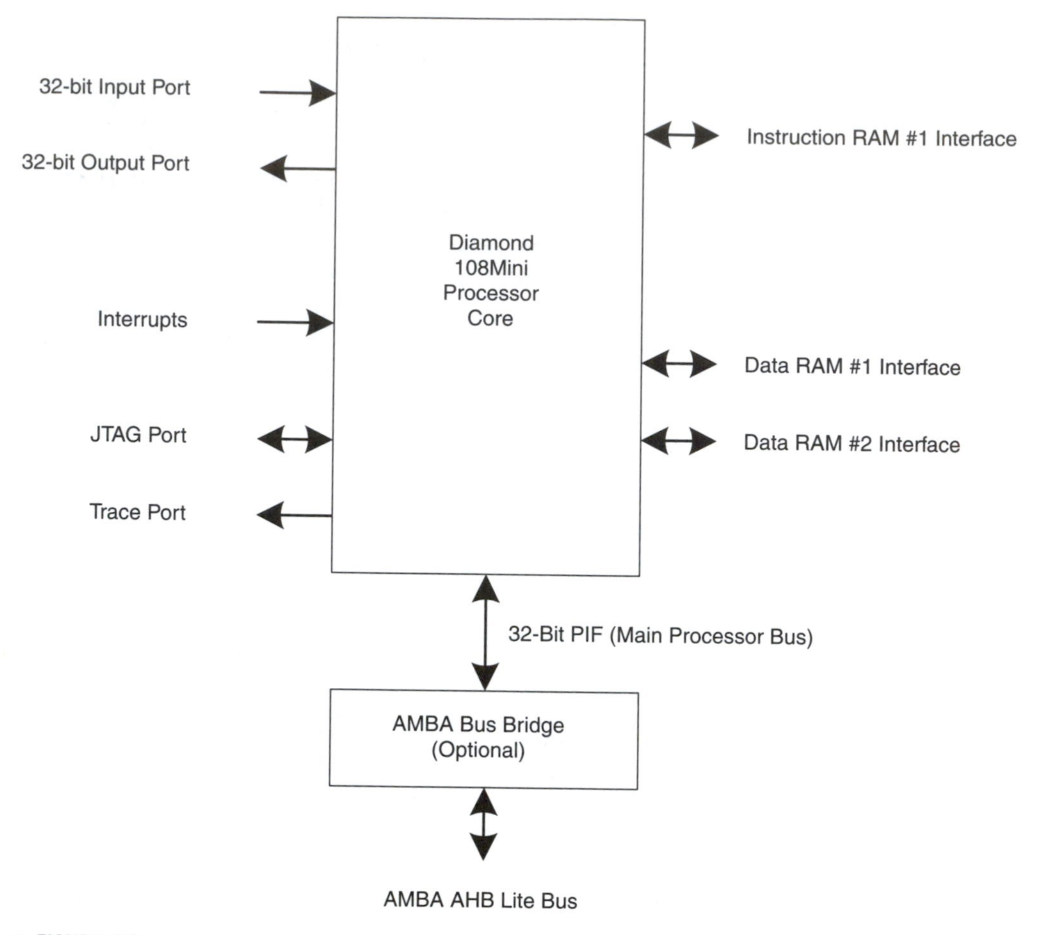

■ **FIGURE 7.2**

The Diamond 108Mini processor core has three 32-bit buses that can be used for attaching local memories and a 32-bit implementation of the Xtensa PIF (main processor interface) bus for attaching additional memory and other system devices.

TABLE 7.1 ■ Diamond 108Mini processor core memory-space assignments

Local memory	Start address	End address
Local instruction RAM	0x40000000	0x4001FFFF
Local data RAM #1	0x3FFE0000	0x3FFFFFFF
Local data RAM #2	0x3FFC0000	0x3FFDFFFF

least 1 Kbyte of RAM must be attached to the Diamond 108Mini's local instruction-memory port to hold the exception vectors.

All of the Diamond 108Mini processor's interrupt and exception vectors are mapped to locations in the local instruction RAM space. The

TABLE 7.2 ■ Diamond 108Mini processor reset, interrupt, and exception vector address mapping

Vector	Address
Reset	0x50000000
Base address for register window underflow/overflow exception vectors	0x40000000
Level 2 high-priority interrupt	0x40000180
Level 3 high-priority interrupt	0x400001C0
Level 4 high-priority interrupt	0x40000200
Level 5 high-priority interrupt	0x40000240
Debug exception	0x40000280
NMI	0x400002C0
Level 1 interrupt (Kernel mode)	0x40000300
Level 1 interrupt (User mode)	0x40000340
Double exception	0x400003C0

processor's timer interrupts and external interrupt pins are pre-assigned to various high-level interrupt vectors. Note that the processor's reset vector is mapped to location 0x50000000, which is not located in the processor's local address space. Consequently, there must be some memory attached to the Diamond 108Mini processor core's PIF bus to hold the processor's initial reset code. Figure 7.3 shows the various important addresses mapped into the Diamond 108Mini processor's address space.

The PIF implementation on the Diamond 108Mini processor core is 32 bits wide. The PIF bus uses a split-transaction protocol and the Diamond 108Mini processor core has a 4-entry write buffer to accommodate as many as four simultaneous outstanding write transactions. This version of the PIF also supports inbound-PIF operations, which means that external devices connected to the Diamond 108Mini processor's PIF bus can access the processor's local memories through the processor's PIF interface. This feature allows "glueless MP" systems to be built from multiple Xtensa and Diamond processor cores. The inbound-PIF portion of the Diamond 108Mini processor core's PIF implementation has an 8-entry request buffer.

The Diamond 108Mini processor core has three internal 32-bit timers in that can be used to generate interrupts at regular intervals. The processor also has nine external, level-triggered interrupt input pins and one edge-triggered, NMI pin.

7.3 THE DIAMOND RPU

Early microprocessors of the 1970s had one unified memory space. As microprocessor usage became more sophisticated, microprocessors started to run multiple tasks and single, unified memory spaces became

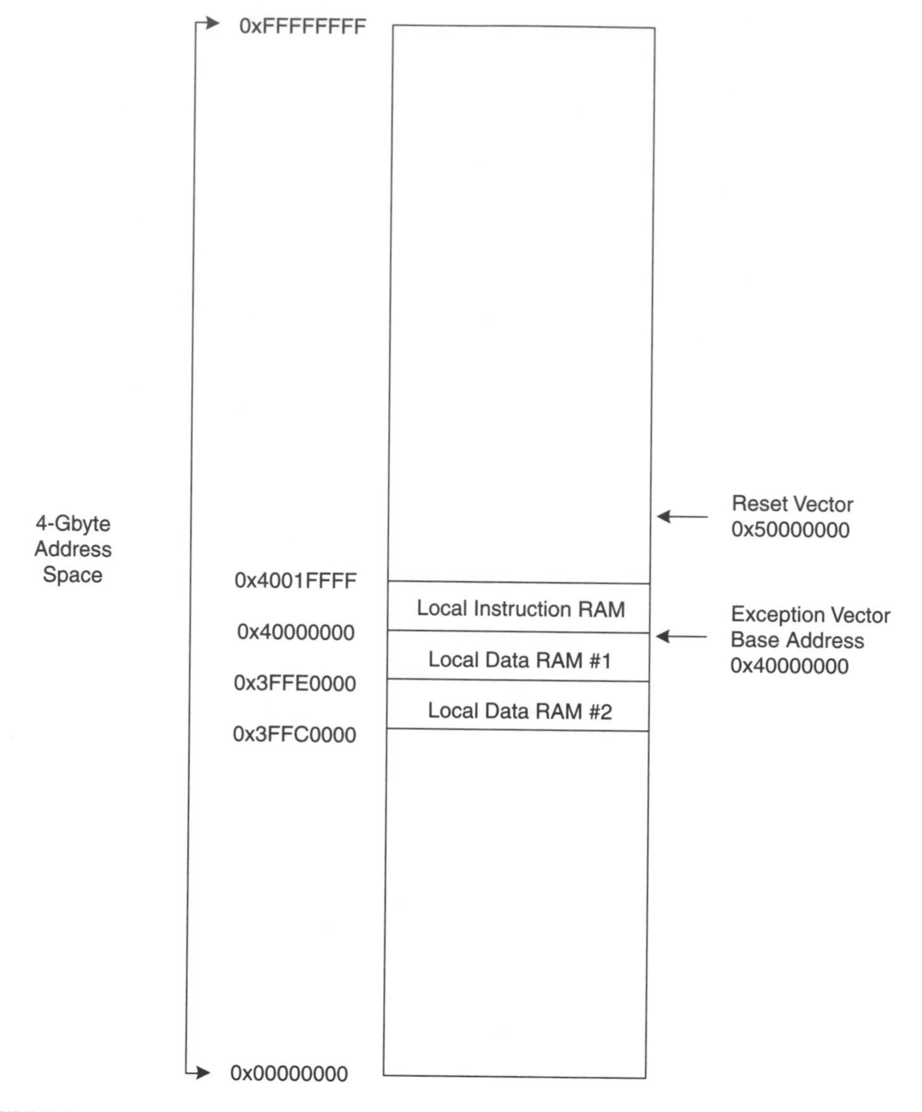

▪ FIGURE 7.3

The Diamond 108Mini processor core's pre-configured address space maps the three local memories into adjacent blocks in the address space.

cumbersome because the unified memory space allowed independent tasks to easily access to memory regions assigned to other tasks. This lack of inter-task memory barriers resulted in spectacular system crashes, which gave rise to the use of the RTOSs to manage the multiple tasks. All of the Diamond Standard Series processor cores except for the 232L core provide hardware support for memory protection through an RPU.

Note: The 232L CPU core incorporates a memory-management unit (MMU) that supports more complex operating systems such as Linux. The access modes used in the TLB contained in the 232L's MMU are similar to those implemented in the TLBs contained in the RPUs of the other Diamond cores.

As Figure 7.4 illustrates, the Diamond RPU divides the processor's 4-Gbyte memory space into eight equally sized, 512-Mbyte regions. The Diamond 108Mini processor core's two local data-memory address spaces fall into memory-protection region 1 and its local instruction-memory address space falls into memory-protection region 2. Thus the RPU can prevent accidental writes to instruction memory through the proper use of its protection mechanisms. The Diamond 108Mini core's non-local address space (assigned to the PIF) falls into all eight memory-protection regions, so the 108Mini controller core's RPU is also useful for managing access to PIF-attached memory and devices.

The Diamond processor sets the memory-protection attributes for each region independently by setting 4-bit access-mode values in separate, 8-entry, instruction and data TLBs. Each TLB has an entry for each of the eight memory-protection regions.

The TLB access modes control both the protection level and the cache behavior for each of the eight memory-protection regions. The access modes appear in Table 7.3 and descriptions of the modes appear in Table 7.4.

Note: Because the Diamond 108Mini processor core has no instruction or data caches, it ignores all of the TLB's cache-related attributes and always acts as though cache-bypass mode is set. For completeness, Tables 7.3 and 7.4 list all of the Diamond RPU TLB's access-mode behaviors for all Diamond cores including the cache-access controls not supported (or needed) by the Diamond 108Mini processor core.

7.4 DIRECT INPUT AND OUTPUT PORTS

Two of the Diamond 108Mini processor core's unique features are its 32-bit direct input and output ports. The input and output wires on these ports directly connect to special software-visible registers in the processor core and can be used for general-purpose signaling, interrupt generation, serial communications, pulse-width modulation, and other sorts of status and control functions on the SOC. Special instructions read and manipulate the contents of these port registers. Table 7.5 lists these port instructions.

The processor's input and output ports can be accessed in one cycle just like simple 32-bit registers attached to one of the processor's local-memory buses, but without the speculation-related read side effects to deal with. (*Note*: All pipelined RISC processors must contend with speculation-related side effects; it's the nature of pipeline-based processor

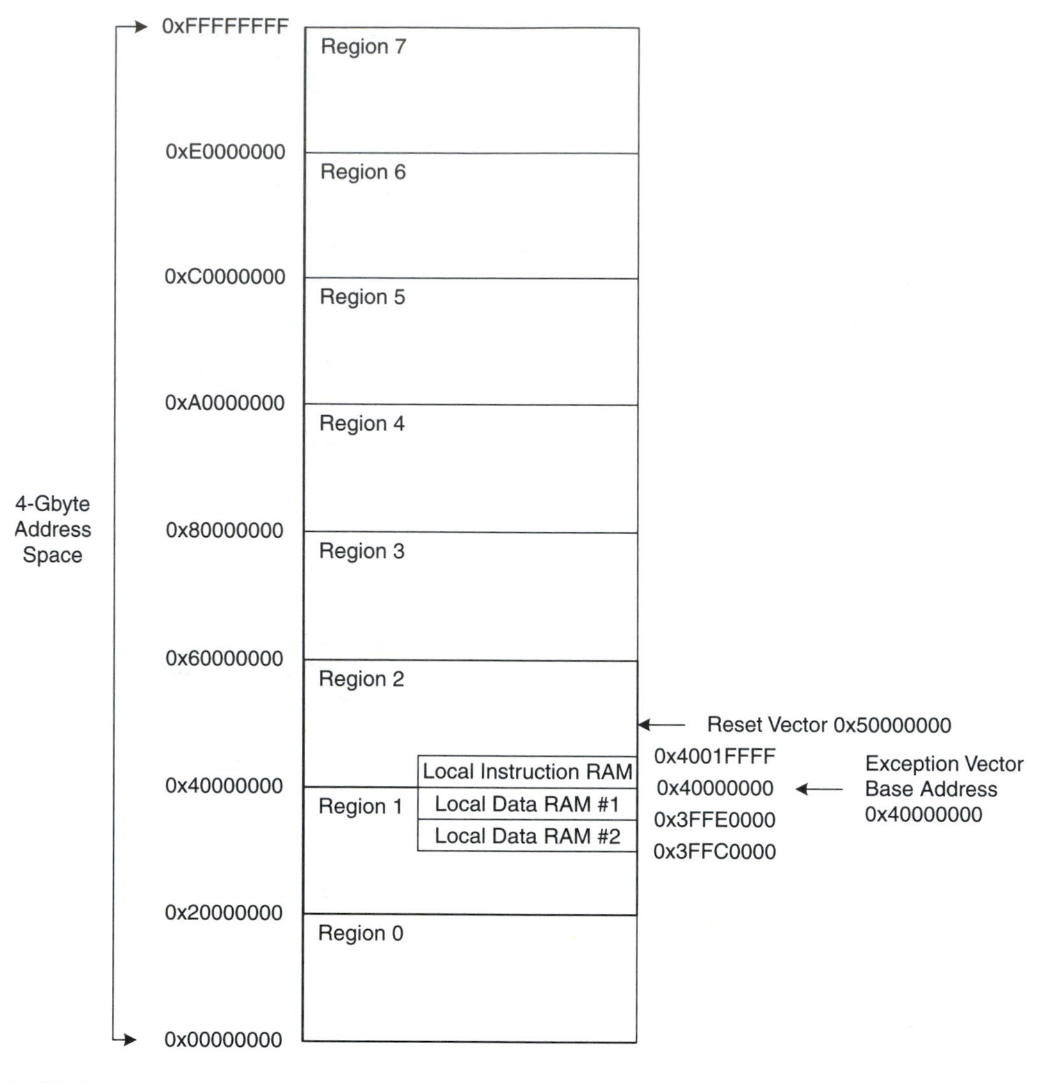

FIGURE 7.4

The Diamond cores' RPU divides the processors' memory space into eight protected regions and the Diamond 108Mini core's local data and instruction memories map into memory-protection regions 1 and 2, respectively so the RPU can prevent accidental writes to instruction memory through the proper use of its protection mechanisms.

architecture.) The special input- and output-port instructions mask the processor's speculative operations so that firmware developers can completely ignore the issue when using these ports.

The input and output ports on the Diamond 108Mini processor have no handshaking lines. Devices driving the processor's input-port wires cannot determine when the processor has read the state of these wires.

TABLE 7.3 ■ Diamond RPU access modes

Access-mode value	Access-mode name	Instruction-fetch behavior	Load behavior	Store behavior
0000	No allocate	Instruction-fetch exception	No allocate	Write-through/ No allocate
0001	Write-through/ No write allocate	Allocate	Allocate	Write-through/ No write allocate
0010	Bypass	Bypass	Bypass	Bypass
0011	Not supported	Undefined	Load exception	Store exception
0100	Write-back mapped region with write-back cache option	Allocate	Allocate	Write-back/ Write allocate
	Write-back mapped region without write-back cache option	Allocate	Allocate	Write-through/ No write allocate
0101–1101	Reserved	Instruction-fetch exception	Load exception	Store exception
1110	Isolate	Instruction-fetch exception	Direct processor access to memory cache	Direct processor access to memory cache
1111	Illegal	Instruction-fetch exception	Load exception	Store exception

TABLE 7.4 ■ Diamond RPU access-mode descriptions

RPU access mode	Access-mode description
No allocate	Do not allocate a cache line for this address. If the address is already cached, fetch or load the cached value. If the address has an allocated cache line but the cache line is not already in the cache, fetch or load the value from main memory and place the value in the cache.
Bypass	Do not use the cache.
Write-back	Write the value to the cache. Update main memory only when the cache line is evicted or when the processor forces the cache line to be written to main memory.
Write-through	Write the value to the cache and to main memory simultaneously.
Isolate	Permits direct read/write access to the cache's data and tag RAM arrays.
Illegal	Any access causes an exception.

TABLE 7.5 ■ Special input- and output-port register instructions

Instruction	Definition
CLRB_EXPSTATE	Clear any bit in the output-port register. Leave other bits unchanged.
SETB_EXPSTATE	Set any one bit of the 32-bit output-port register. Leave other bits unchanged.
READ_INPWIRE	Read the 32-bit value of the input-port register and place result in a general-purpose register-file entry.
RUR_EXPSTATE	Read the 32-bit value of the output-port register and place the result in a general-purpose register-file entry.
WRMASK_EXPSTATE	Write a masked bit pattern to the output-port register.
WUR_EXPSTATE	Write a 32-bit value to the output-port register.

Similarly, there is no output signal to alert devices that the output-port wires have changed state. They simply change state. Consequently, external devices connected to the Diamond 108Mini processor core's direct input and output ports are somewhat uncoupled from the timing of the firmware manipulating those ports. In this sense, the Diamond 108Mini processor core's input and output ports are like the direct input and output lines on a block of custom-designed RTL logic and are intended to be used the same way.

7.5 SYSTEM DESIGN WITH DIAMOND 108Mini PROCESSOR CORES

Figure 7.5 shows a system built with four Diamond 108Mini processor cores. The processor cores can communicate with global memory over the shared 32-bit PIF bus and with each other's local memories using the Diamond processor cores' inbound-PIF feature. A bus arbiter controls access to the PIF bus. Local/Global address-translation blocks attached to each processor's PIF bus perform the critical function of mapping the attached processor's local address space into one unified global address map.

Without these address-translation blocks, the four Diamond 108Mini processors shown in Figure 7.5 could not communicate over the PIF because their local address maps would overlap each other and conflict. For example, if the master Diamond 108Mini processor attempted to write to the local data memory of slave processor #1, it would use a target destination address (say 0x3FFE0000) that would result in the master processor writing to its own local data memory instead of the intended destination in slave processor #1's local data-memory address space. With the address-translation blocks, the master processor writes to an address in the PIF's

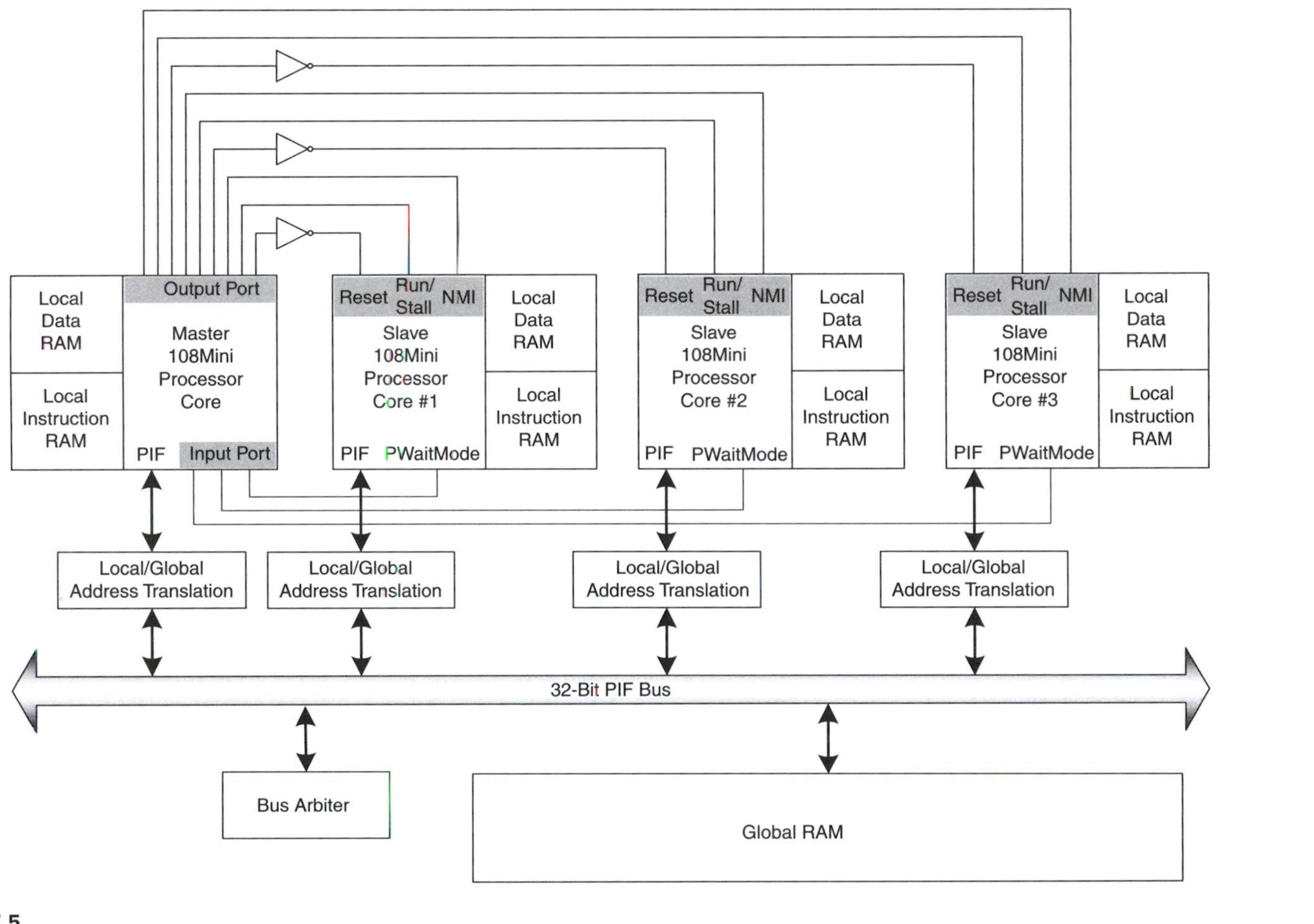

■ **FIGURE 7.5**

This 4-processor system design allows the master Diamond 108Mini processor (shown on the left) to control the operation of the other three processors through their Reset, Run/Stall, and NMI inputs.

global address space and that address is then mapped to the target processor's local address space.

The processor on the left of the figure is configured as the system master and the other three processors are slaves. Three wires from the master processor's direct output port are connected to the Reset inputs of the three slave processors; three more output-port wires are connected to the slave processors' Run/Stall inputs; and yet another three output-port wires are connected to the slave processors' NMI inputs.

This system configuration allows the master processor to independently reset and halt all three slave processors. In fact, the three slave processors are automatically reset when the master processor is reset because the reset-initialized state for each of the pins associated with the master processor's output port is zero. Inverters attached to the master processor's output-port wires used for resetting the slave processors will therefore assert the slave processors' Reset inputs when the master processor is reset. After initializing itself, the master processor can assert the Run/Stall input to each slave processor and then remove the reset signal to each slave processor. While each slave processor is stalled, the master processor can transfer program code from the large global memory (or from its own local memory) to each slave processor's local instruction memory using inbound-PIF write operations—which are permitted even while the slave processors are stalled.

When the master processor releases a slave processor's Run/Stall input, the slave processor will commence program execution starting at its reset-vector address. Because the Diamond 108Mini processor's reset vector is set to 0x50000000, each slave processor's first instruction fetch will be directed to the PIF-connected global memory. That first instruction however can set up a jump to a location in the slave processor's local instruction memory, which will then isolate the slave's operations from the PIF bus.

This scheme allows two of the slave processors to independently execute code from their local memories concurrently while the master processor is programming or reprogramming the third slave processor's instruction memory. Although this design is based on four processors that together consume less than 2 mm^2 on the SOC die, the resulting multiprocessor system is capable of quite sophisticated behavior and very good processing performance.

There are two noteworthy, performance-related reasons why the four Diamond 108Mini processor cores shown in Figure 7.5 should not execute code directly out of the global memory attached to the PIF bus. First, as mentioned above, the local memories provide faster access times to the processor than PIF-connected memory and therefore, operating almost exclusively from the local memories gives the master and slave processors much better performance. Memory transactions on local-memory buses are about five times faster than the same transactions conducted over the PIF bus.

Note that this speed imbalance between the local-memory buses and the PIF bus is not due to poor PIF design. It is because the processor's main bus must hide speculative activity caused by the processor's 5-stage RISC pipeline. At any time, there are as many as five instructions in this pipeline and each of these instructions is at a different stage of completion. If the processor completes the execution of a branch instruction, the processor will discard all of the upstream instructions in its pipeline, including load instructions that may already have generated read operations. The processor fetched these instructions and started to execute them speculatively, before discovering the branch. This speculative, pipeline-oriented operation is one of the reasons behind the RISC processor's good performance. However, extra cycles are required to resolve speculative operations in the processor's pipeline before load or store transactions appear on the PIF bus.

Local-memory buses do not hide speculative activity so they are inherently faster than the PIF bus. Any devices connected to the local-memory buses must be immune to read side effects (the main problem caused by allowing speculative activity to be visible on a processor's external bus). Simple memories—static and dynamic RAMs and ROMs—are immune to read side effects. Peripheral devices and FIFO memories generally aren't immune unless they have been specially designed for speculative operation. System designers who attach devices with read side effects to a RISC processor's local-memory buses will encounter unexpected and incorrect system behavior. Avoid such designs.

All four processors share one 32-bit PIF bus in the system design shown in Figure 7.5, which leads to the second reason for minimizing each processor's PIF activity. Under ideal conditions—and system conditions are rarely ideal—each processor would get a bit less than 25% of the PIF bus' bandwidth (assuming all processors need equal bus bandwidth) when multimaster bus overhead is accounted for. Looked at another way, the PIF bus must operate at a speed that delivers somewhat more than the sum of all four processors' required bus bandwidth. Otherwise, the processors will all starve for instructions and data and overall system performance will suffer.

RISC processors often employ instruction and data caches to avoid such starvation problems. However, cache memories are larger than simple local memories because they require additional RAM bits to hold the cache tags. In addition, integral cache controllers increase a processor's gate count. Consequently, the Diamond 108Mini processor core, which was created to be as small as possible, was designed as a cacheless core with local instruction- and data-memory interfaces to minimize processor gate count.

To get good performance from such a processor core, the processor's code should be stored in local instruction memory and data should be stored in local data memory. Local-memory access is as fast as cache-memory access, so a cacheless processor experiences no performance loss as long it can perform its work using just its local memories. This use model explains why the Diamond 108Mini processor core was configured for two local data memories: so that it could hold larger data sets (as large as 256 Kbytes) completely in its local memory and still deliver good performance.

Note: Other members of the Diamond Standard Series of processor cores include cache-memory controllers to take on even larger tasks.

7.6 LOW-POWER SYSTEM DESIGN AND OPERATION

As gate counts soar into the hundreds of millions per chip, SOC power dissipation becomes an increasingly important design consideration for all systems. There are many ways to reduce system power. Some of these power-reducing methods relate to the processor's design and are in the domain of the processor designer. Other methods are in the system designer's domain.

A processor designer can reduce a processor's power dissipation by minimizing clock activity inside of the processor through two techniques. The first technique is clock gating—switching off the clock to circuits that are not being exercised on a cycle-by-cycle basis. All Xtensa and Diamond processor cores including the Diamond 108Mini core employ extensive functional clock gating to minimize dynamic power dissipation within the processor.

Functional clock gating creates many branches in the processor's clock tree. Sometimes, the number of clock-tree branches ranges into the hundreds. Creating such complex clock trees is the purview of automated tools because the complexity of the task easily outstrips the capacity of the human mind. Because processors are instruction-driven, it is relatively easy to determine which parts of a processor need to operate, and which do not, during each phase of each instruction's passage through the processor's pipeline.

Contrast this situation with creating similarly complex clock gating in a block of custom-designed RTL. Only by setting up exhaustive test-input conditions can all of the functional clock domains be mapped. For large, complex logic blocks, it's very nearly impossible to create such detailed maps of clock domains. For processors, such analysis is straightforward and in the case of Xtensa and Diamond processor cores, the work of inserting the clock-gating logic is done in a fully automatic fashion.

The second path the processor designer can take to low-power processor operation is to design a processor that performs tasks in fewer clock cycles. Lower clock-rate requirements earn two benefits in SOC design. First, dynamic power dissipation is directly proportional to clock rate, so processors operating at lower clock rates will inherently draw less power. Second, processors operating at lower clock rates can be operated at lower power-supply voltages, which will further reduce dynamic power dissipation and, as a side benefit, also reduces static power dissipation. Unlike personal computer processors—which have historically raced to the highest possible operating frequency—processors used in SOCs should be run as slow as the application will allow, which minimizes power dissipation.

Configurable Xtensa processor cores allow the SOC designer to create new instructions that cut the number of instructions required to execute an algorithm. All Diamond processor cores incorporate special instructions and other features to achieve the same end for certain operations. As discussed above, the Diamond 108Mini processor core incorporates local-memory interfaces that reduce the number of clock cycles needed to communicate with memory. In addition, it takes more power to drive a global bus because of the increased capacitance, so communicating with local memory inherently takes less power than communicating with memory over a global bus. In addition, the Diamond 108Mini processor core incorporates instructions for the 32-bit input and output ports. These instructions reduce the number of cycles needed to perform I/O.

The discussion of the system illustrated in Figure 7.5 has already mentioned the Run/Stall input to the Diamond 108Mini processor. This input pin can shut off nearly all of the clock trees inside of the processor. All of the Xtensa and Diamond processor cores also provide an instruction that essentially halts all processor activity. This instruction, called *WAITI* (wait for interrupt), puts the processor in a mode where most of the clocks inside of the processor are gated off. An interrupt brings the processor out of the *WAITI* mode. Execution of the *WAITI* instruction actually shuts off more of the processor's internal clocks than the Run/Stall input.

When a Diamond 108Mini slave processor enters the *WAITI* mode, it asserts its *PWaitMode* status output. In Figure 7.5, the slave processors' *PWaitMode* outputs are connected to the master processor's 32-bit input port so that the master processor can evaluate the running status of the slave processors. The Diamond 108Mini processor core's 32-bit input port makes this a glueless connection. The master processor need not consume bus bandwidth to poll the status of the other processors. Bus bandwidth can be reserved for moving instructions and data around the system.

Systems that incorporate multiple processors, such as the 4-processor system shown in Figure 7.5, can use the *WAITI* instruction to programmably shut down processors when they have completed a task or when their processing bandwidth is not needed. An interrupt quickly activates the processor when conditions warrant. Significantly, especially for the system illustrated in Figure 7.5, inbound-PIF operations can continue to occur after an Xtensa or Diamond processor core enters the *WAITI* mode, so a master processor can retrieve processed data from a waiting slave processor's data memory, fill the waiting slave processor's data memory with input data to be processed, or reprogram a waiting slave processor's local instruction memory, and then activate the slave processor with an interrupt to initiate further processing.

This sort of reserve capacity—the ability to quickly bring dormant, powered-down computing resources on line at will—opens entirely new architectural vistas to the SOC designer. Chapter 15 has more to say about these advanced system-design topics.

BIBLIOGRAPHY

Diamond Standard Processors Data Book, Tensilica, Inc., February 2006.

Leibson, S. and Kim, J., "Configurable processors: a new era in chip design," *IEEE Computer*, July 2005, pp. 51–59.

Rowen, C. and Leibson, S., *Engineering the Complex SOC*, Prentice-Hall, 2004.

CHAPTER 8

THE DIAMOND 212GP CONTROLLER CORE

The name Jeep came from the abbreviation used by the army for "General Purpose" vehicle, G.P.

The Diamond 212GP controller core is a general-purpose, 32-bit RISC processor core. Like the Diamond 108Mini core, the Diamond 212GP controller has interfaces for local instruction and data memories but it also includes a cache controller that operates 8-Kbyte, 2-way set-associative instruction and data caches for efficient execution of large programs. Even with the addition of a cache controller, the processor consumes 0.7 mm^2 of silicon and 195 μW/MHz when implemented with a 130 nm, G-type (general-purpose) process technology. The Diamond 212GP controller core brings many performance benefits of a 32-bit RISC processor to bear on the designated tasks:

- Large 4-Gbyte address space
- 32-bit computations
- Large 32-entry register file
- Separate 8-Kbyte, 2-way, set-associative instruction and data caches
- 5-stage pipelined operation resulting in a 250-MHz maximum clock rate in 130 nm technology.

The Diamond 212GP controller core has a 32-bit version of the general-purpose processor interface (PIF) bus for global SOC communications. An optional AMBA bus bridge supplied with the 232L processor core adapts the PIF to peripheral devices designed for the AMBA AHB-Lite bus. Like the Diamond 108Mini processor core, the Diamond 212GP core has interfaces for local instruction and data RAMs. However, instead of two data-RAM interfaces like the Diamond 108Mini core, the Diamond 212GP controller core has one data-RAM interface and an XLMI interface port. The XLMI port can be used as a data-RAM interface but it includes signals that permit devices with read side effects to deal with speculative read transactions. The Diamond 212GP core also incorporates direct input and output ports that can accelerate I/O.

8.1 A GENERAL-PURPOSE PROCESSOR CORE

Figure 8.1 is a block diagram of the Diamond 212GP controller core.

The Diamond 212GP processor architecture contains all of the basic elements of the Xtensa ISA. It has a 32-entry general-purpose register file, a 5-stage execution pipeline, 32-bit addressing, and a simple region-protection unit (RPU) that can be used to implement straightforward memory management. The Diamond 212GP core also incorporates the Diamond Series processor core software-debug stack, as shown on the left of Figure 8.1. This debug stack provides external access to the processor's internal, software-visible state through a 5-pin IEEE 1149.1 JTAG TAP (test access port) interface and through a trace port that provides additional program-trace data.

8.2 DIAMOND 212GP CONTROLLER CORE INTERFACES

The Diamond 212GP controller core has a number of interfaces for high-bandwidth I/O, as shown in Figure 8.2. These interfaces include a 32-bit implementation of the Xtensa PIF (main processor interface) bus and separate memory interfaces for one local instruction memory block and one local data memory block. Each of the local-memory interfaces can accommodate memories as large as 128 Kbytes. In addition, the Diamond 212GP core has an XLMI port with a 128-Kbyte address space. The XLMI port can be used to control a local memory block, which gives the Diamond 212GP controller a maximum of 384 Kbytes of local memory.

8.3 THE XLMI PORT

Diamond processor cores, like all RISC processors, perform speculative read operations that initiate read transactions on the local-memory buses. Data obtained from these read operations may or may not be used, depending on the circumstances in the processor's pipeline. For example, a branch instruction or interrupt can cause the processor to discard data that it has speculatively read.

Diamond processor load transactions that appear on the processor's interface ports (with the exception of the PIF) are speculative, so a read operation on the XLMI interface signals does not necessarily mean that the processor will consume the data it reads. A variety of internal events (such as branches) and exceptions can initiate pipeline flushes that cause the processor to discard all uncompleted instructions in its pipeline, including load instructions. The processor may later replay these loads but the data obtained from the first execution of the load instruction will

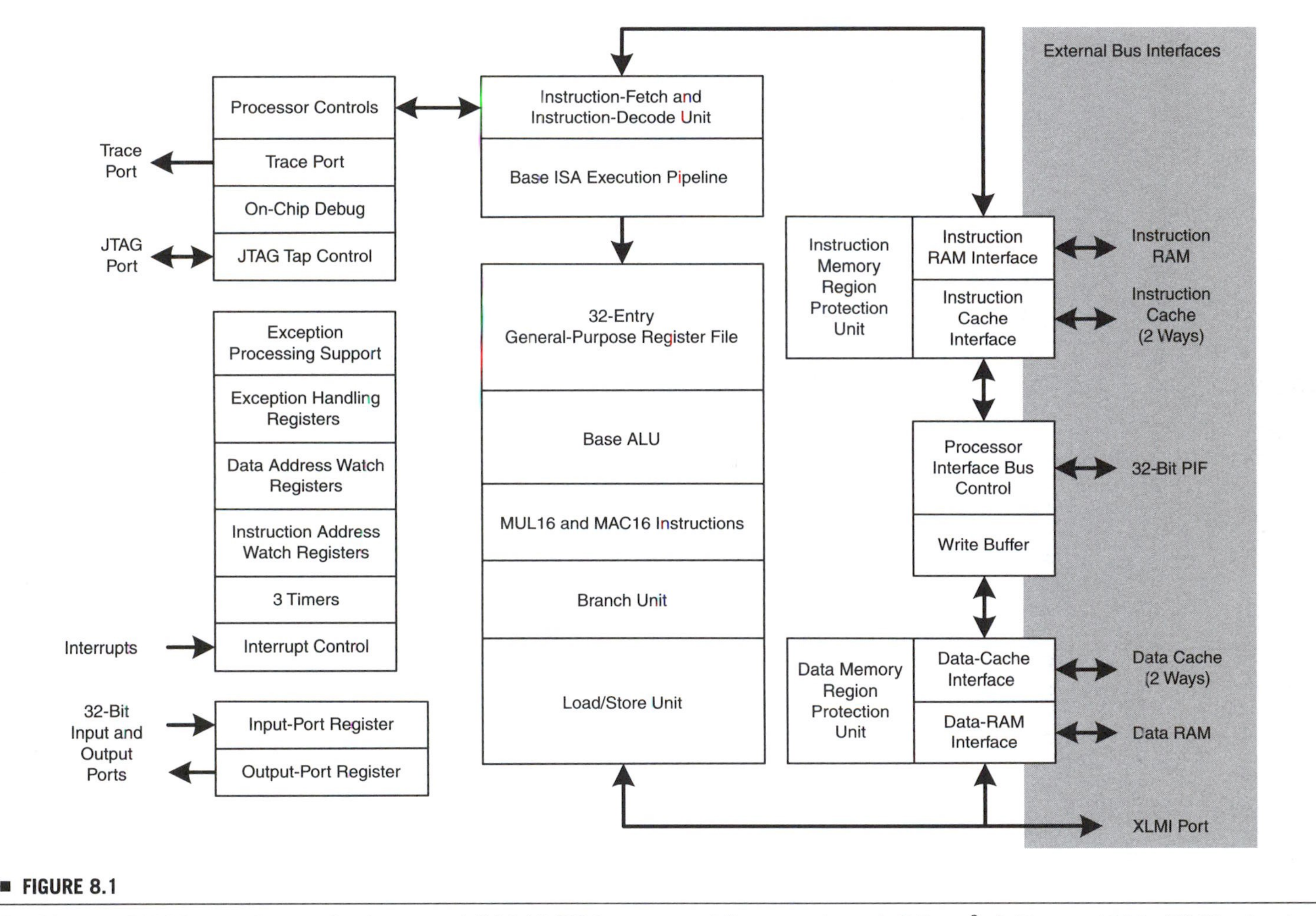

■ **FIGURE 8.1**

The Diamond 212GP controller core implements a full 32-bit RISC processor while consuming only 0.7 mm^2 of silicon and 195 μW/MHz when implemented in a 130 nm, G-type process technology.

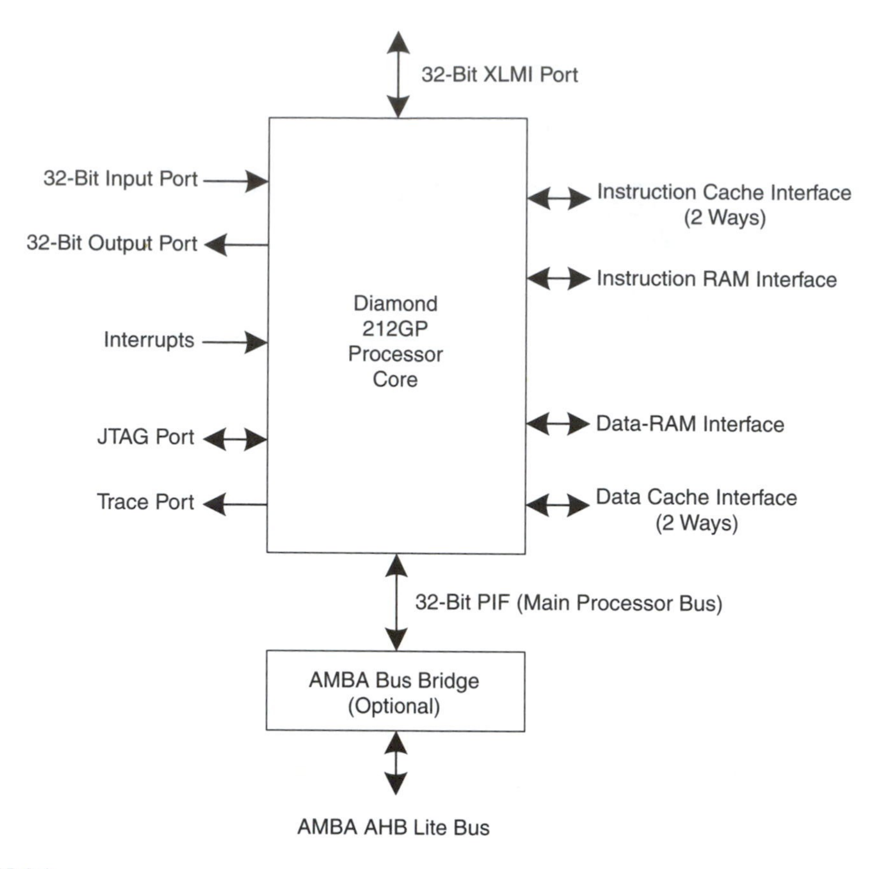

■ **FIGURE 8.2**

The Diamond 212GP controller core has two 32-bit memory-interface buses that can be used for attaching local memories (one instruction memory and one data memory), a 32-bit XLMI port, and a 32-bit implementation of the Xtensa PIF (main processor interface) bus for attaching additional memory and other system devices. The XLMI port can be used to attach a local data RAM to the Diamond 212GP controller core and it can also be used to attach other device types.

have been flushed. Consequently, all devices attached to the XLMI port must be designed to accommodate the speculative nature of processor read operations.

Simple RAMs do not have read side effects. FIFO memories and most I/O devices do have read side effects. For example, reading the status register of an I/O device often has an irreversible effect on the I/O device and reading a word from the head of a FIFO permanently removes that word from the FIFO. Attaching devices with read side effects to the local memory buses of Diamond processor cores will result in unpredictable and incorrect system behavior. Speculative transactions only occur on the Diamond cores' local-memory and XLMI buses. They do not occur on the Diamond processor cores' main PIF bus.

The Diamond 212GP processor's XLMI port is a general-purpose data port that allows for the direct attachment of:

- RAMs
- Hardware peripherals (through memory-mapped I/O registers)
- Inter-processor communication devices (such as FIFOs)

Note: The XLMI port is a data interface only. The Diamond 212GP controller does not fetch instructions from the XLMI port.

The Diamond Standard Series XLMI port enables the proper design of interfaces to devices that have read side effects, such as FIFO memories. If a device with read side effects is connected to the XLMI port, additional external logic is required to ensure that data read from the device is not lost until the XLMI port indicates that the processor has retired the load. The speculative nature of RISC processor loads makes such interface behavior mandatory for proper system operation. The XLMI port interface eases the design of this logic by signaling when the processor completes the load operation and when the pipeline has been flushed. This XLMI port feature distinguishes it from the other Diamond processor core local-memory interfaces.

The XLMI port signals the initiation of loads (normal read cycles), the completion of loads, and the flushing of speculative loads that will not be completed. Load completion is indicated by the assertion of the "*Data Port Load Retired*" signal. Assertion of this signal tells the attached device that it no longer needs to maintain a copy of the data read by the associated load operation because the processor has consumed the information.

A second XLMI port signal, "*Data Port Retire Flush,*" indicates that the processor has discarded all unretired XLMI loads. The processor asserts this signal when pipeline conditions cause instructions in the pipeline to be flushed. Devices that have read effects have difficulty in dealing with this condition, so careful design is required. These two signals, "*Data Port Load Retired*" and "*Data Port Retire Flush,*" help SOC designers take into account the speculative nature of reads directed at the Diamond 212GP controller's XLMI port.

8.4 THE DIAMOND 212GP PROCESSOR MEMORY MAP

Table 8.1 lists the address spaces assigned to the Diamond 212GP processor core's two local memory interfaces and the XLMI port. Note that the instruction RAM and data RAM memory spaces are adjacent so they form one contiguous 256-Kbyte memory block from 0x3FFE0000 to 0x4001FFFF. The address space for the Diamond 212GP's XLMI port is mapped to the same memory space as the Diamond 108Mini core's second

TABLE 8.1 ▪ Diamond 212GP controller core memory-space assignments

Local memory	Start address	End address
Local instruction RAM	0x40000000	0x4001FFFF
Local data RAM #1	0x3FFE0000	0x3FFFFFFF
XLMI port	0x3FFC0000	0x3FFDFFFF

TABLE 8.2 ▪ Diamond 212GP controller reset, interrupt, and exception vector address mapping

Vector	Address
Reset	0x50000000
Base address for register window underflow/ overflow exception vectors	0x60000000
Level 2 high-priority interrupt	0x60000180
Level 3 high-priority interrupt	0x600001C0
Level 4 high-priority interrupt	0x60000200
Level 5 high-priority interrupt	0x60000240
Debug exception	0x60000280
NMI	0x600002C0
Level 1 interrupt (Kernel mode)	0x60000300
Level 1 interrupt (User mode)	0x60000340
Double exception	0x600003C0

data memory, just below and contiguous with the address space for the data-memory block. All of the Diamond 212GP controller core's three local memory spaces thus form one contiguous 384-Kbyte address space from 0x3FFC0000 to 0x4001FFFF.

The Diamond 212GP controller core has a 32-bit, 4-Gbyte address space, so additional memory can be attached to its PIF bus if needed. However, the address spaces dedicated to local memory are hardwired to the appropriate local-memory and XLMI ports and memory transactions to those address spaces will not appear on the PIF bus. All Diamond processor cores including the Diamond 212GP automatically send memory accesses to non-local address locations out over the PIF bus.

Table 8.2 lists the Diamond 212GP controller core's assigned reset, non-maskable interrupt (NMI), and other interrupt vectors. Note that the exception vectors for the Diamond 212GP controller core are assigned to locations located in non-local memory space, so at least some memory located at addresses 0x50000000 and 0x60000000 must be attached to the Diamond 212GP controller's PIF bus to hold its reset and exception vectors. The processor's timer interrupts and external interrupt pins are pre-assigned to various high-level interrupt vectors.

Figure 8.3 shows the various important addresses mapped into the Diamond 212GP controller's address space.

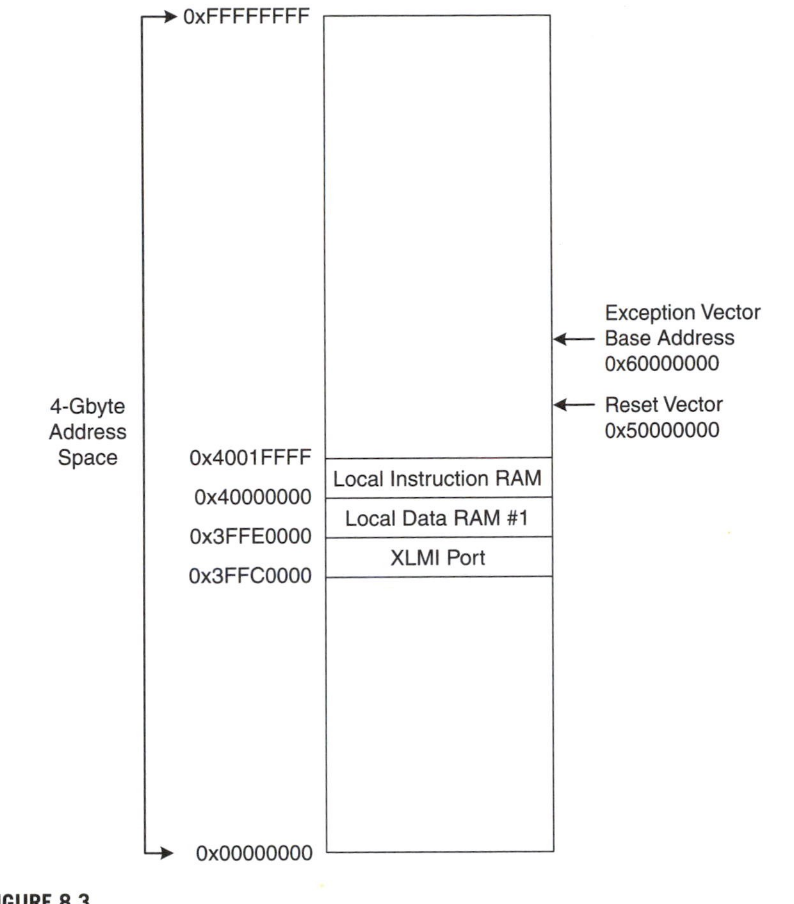

■ **FIGURE 8.3**

The Diamond 212GP controller core's pre-configured address space maps the local instruction and data memories and the XLMI port into adjacent blocks in the address space.

The PIF implementation on the Diamond 212GP controller core is 32 bits wide. The PIF bus uses a split-transaction protocol and the Diamond 212GP controller core has an 8-entry write buffer to accommodate as many as eight simultaneous outstanding write transactions. The PIF also supports inbound-PIF operations, which means that external devices connected to the Diamond 212GP controller's PIF bus can access the processor's local memories through the processor's PIF interface. This feature allows "glueless MP" systems to be built from multiple Xtensa and Diamond processor cores. The inbound-PIF portion of the Diamond 212GP controller core's PIF implementation has an 8-entry request buffer.

The Diamond 212GP controller core has three internal 32-bit timers in that can be used to generate interrupts at regular intervals. The processor also has nine external, level-triggered interrupt input pins and one edge-triggered NMI pin.

8.5 THE 212GP RPU

The Diamond 212GP controller core uses the same RPU discussed in the previous chapter. However, the Diamond RPU's cache-access modes are operational in the Diamond 212GP processor because it has instruction and data caches.

Figure 8.4 shows how the Diamond RPU divides the Diamond 212GP processor's 4-Gbyte memory space into eight equally sized, 512-Mbyte

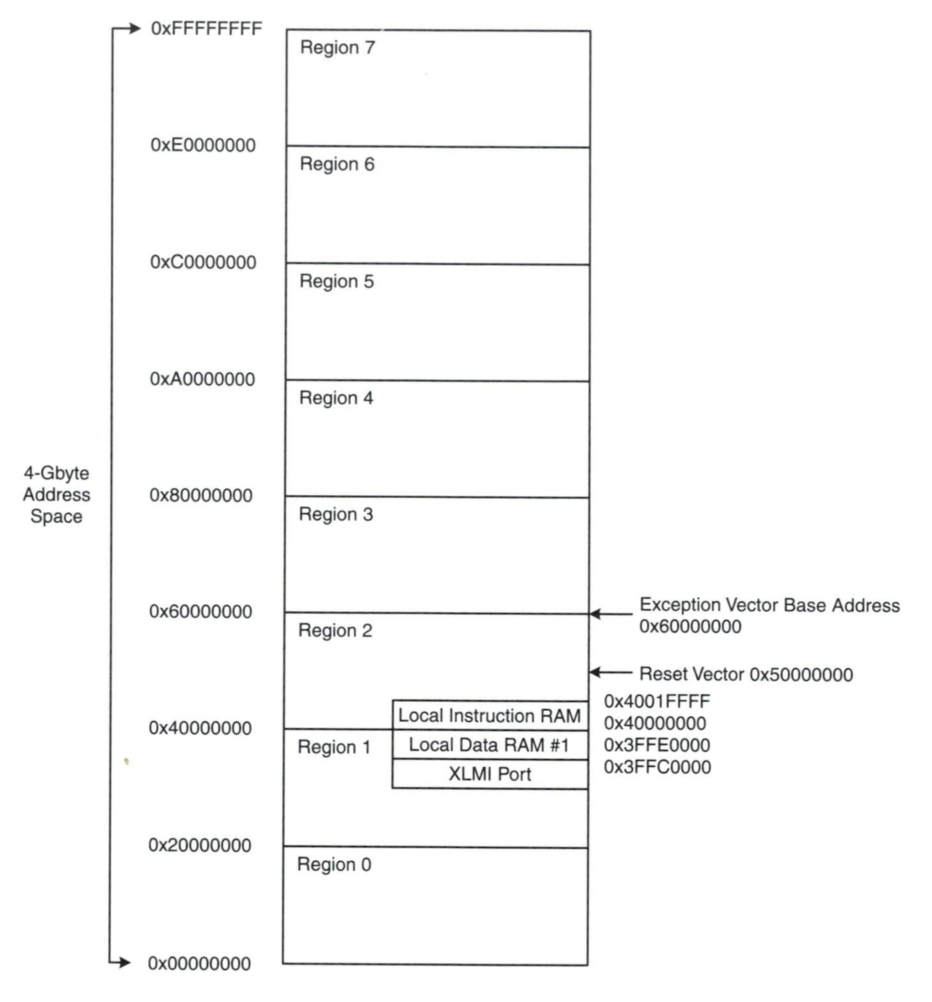

■ **FIGURE 8.4**

The Diamond 212GP core's RPU divides the processor's memory space into eight protected regions. The processor's local data memory address space and XLMI port map into memory-protection region 1 and its local instruction-memory address space maps into memory-protection region 2 so the RPU can prevent accidental writes to instruction memory through the proper use of its protection mechanisms.

regions. The Diamond 212GP controller core's local data-memory address space and the address space assigned to the XLMI port fall into memory-protection region 1. Its local instruction-memory address space falls into memory-protection region 2. Thus the RPU can prevent accidental writes to instruction memory through the proper use of its protection mechanisms. The Diamond 212GP core's non-local address space (assigned to the PIF) falls into all eight memory-protection regions, so the Diamond 212GP controller core's RPU is also useful for protecting PIF-attached memory and devices.

The Diamond 212GP controller sets the memory-protection attributes for each region independently by setting 4-bit access-mode values in separate, 8-entry instruction and data TLBs. (*Note*: the three-letter abbreviation TLB stands for *translation lookaside buffer* but, for Diamond processor core RPUs, its definition is widened to mean "translation hardware.") Each TLB has an entry for each of the eight memory-protection regions.

The access modes control both the protection level and the cache behavior for each of the eight memory-protection regions. The access modes appear in Table 8.3 and descriptions of the modes appear in Table 8.4.

Note: Tables 7.3 and 7.4 from the previous chapter are repeated in this chapter as Tables 8.3 and 8.4 because the RPUs for the Diamond 212GP and Diamond 108Mini controller cores use the same access-mode definitions but act differently due to the absence of caches in the Diamond 108Mini core.

TABLE 8.3 ■ Diamond RPU access modes

Access-mode value	Access-mode name	Instruction-fetch behavior	Load behavior	Store behavior
0000	No Allocate	Instruction-fetch exception	No allocate	Write-through/ No allocate
0001 (see Note)	Write-through/ No write allocate	Allocate	Allocate	Write-through/ No write allocate
0010	Bypass	Bypass	Bypass	Bypass
0011	Not supported	Undefined	Load exception	Store exception
0100	Write-back	Allocate	Allocate	Write-back/ Write allocate
0101–1101	Reserved	Instruction-fetch exception	Load exception	Store exception
1110	Isolate	Instruction-fetch exception	Direct processor access to memory cache	Direct processor access to memory cache
1111	Illegal	Instruction-fetch exception	Load exception	Store exception

Note: RPU access-mode 1 forces Diamond core data caches to operate in write-through mode even though all Diamond processor core data caches are pre-configured as write-back caches.

TABLE 8.4 ■ Diamond RPU access-mode descriptions

RPU access mode	Access-mode description
No allocate	Do not allocate a cache line for this address. If the address is already cached, fetch or load the cached value. If the address has an allocated cache line but the cache line is not already in the cache, fetch or load the value from main memory and place the value in the cache.
Bypass	Do not use the cache.
Write-back	Write the value to the cache and then update main memory when the cache line is evicted or when the processor forces the cache line to be written to main memory.
Isolate	Permits direct read/write access to the cache's data and tag RAM arrays.
Illegal	Any access causes an exception.

8.6 DIRECT INPUT AND OUTPUT PORTS

Two of the Diamond 212GP controller core's unique features are its 32-bit direct input and output ports, which are the same I/O ports found on the Diamond 108Mini processor core. The input and output wires on these ports connect directly to special software-visible registers in the processor core and can be used for general-purpose signaling, interrupt generation, serial communications, pulse-width modulation, and other sorts of status and control functions on the SOC. Special instructions read and manipulate the contents of these port registers. Table 8.5 lists these port instructions.

The input and output ports can be accessed in one cycle just like simple 32-bit registers attached to one of the processor's local memory buses, but without the speculation-related read side effects to deal with. (*Note*: All RISC processors must contend with speculation-related side effects; it's the nature of pipeline-based RISC processor architecture.) The special input- and output-port instructions mask the processor's speculative operations so that firmware developers can completely ignore the issue when using these ports.

The input and output ports on the Diamond 212GP controller have no handshaking lines. Devices driving the processor's input-port wires cannot determine when the processor has read the state of these wires. Similarly, there is no output signal to alert devices that the output-port wires have changed state. They simply change state. Consequently, external devices connected to the Diamond 212GP controller core's direct input and output ports are somewhat uncoupled from the timing of the firmware manipulating those ports. In this sense, the Diamond 212GP controller core's input and output ports are like the direct input and output lines on a block of custom-designed register-transfer level (RTL) logic and are intended to be used the same way.

TABLE 8.5 ■ Special input- and output-port register instructions

Instruction	Definition
CLRB_EXPSTATE	Clear any bit in the output-port register. Leave other bits unchanged.
SETB_EXPSTATE	Set any one bit of the 32-bit output-port register. Leave other bits unchanged.
READ_INPWIRE	Read the 32-bit value of the input-port register and place result in a general-purpose register-file entry.
RUR_EXPSTATE	Read the 32-bit value of the output-port register and place the result in a general-purpose register-file entry.
WRMASK_EXPSTATE	Write a masked bit pattern to the output-port register.
WUR_EXPSTATE	Write a 32-bit value to the output-port register.

8.7 THE DIAMOND 212GP CONTROLLER'S CACHE INTERFACES

Most RISC processors including the Diamond 212GP core use cache memories—fast, small RAM arrays—to buffer the processor from the slower and larger main memories generally located external to the processor core or the SOC. The Diamond 212GP core's caches store the data and instructions that a program is immediately using, while the majority of other data resides in slower main memory (RAM or ROM). In general, the Diamond 212GP core accesses instruction and data caches simultaneously, which maximizes processor bandwidth and efficiency.

The Diamond 212GP controller incorporates a pre-configured version of the Xtensa cache controller that operates separate, 2-way set-associative, 8-Kbyte instruction and data caches. The data cache employs a default write-back (as opposed to write-through) write policy although this policy can be changed under program control through the RPU access modes discussed in Section 8.5. The Diamond 212GP processor's cache memories accelerate every program the processor executes by storing local copies of executed code sequences and data in fast cache memory. The caches provide faster access to instructions and data than can memories attached to the processor's PIF.

The cache-control logic for the Diamond 212GP's instruction and data caches is woven (integrated) into the processor's core so the Diamond 212GP processor has direct interface ports for the eight RAM arrays needed to implement the two, 2-way set-associative, 8-Kbyte instruction and data caches. Figure 8.5 shows how the eight RAM arrays attach to the processor core's cache interface ports.

Four of the cache-RAM arrays store data and the other four RAM arrays store the cache tags, which are used to hold critical information about the instructions and data stored in the cache. Each cache way for the instruction and data caches requires separate RAM arrays for cache data and cache tags so eight RAM arrays are needed to complete the Diamond 212GP

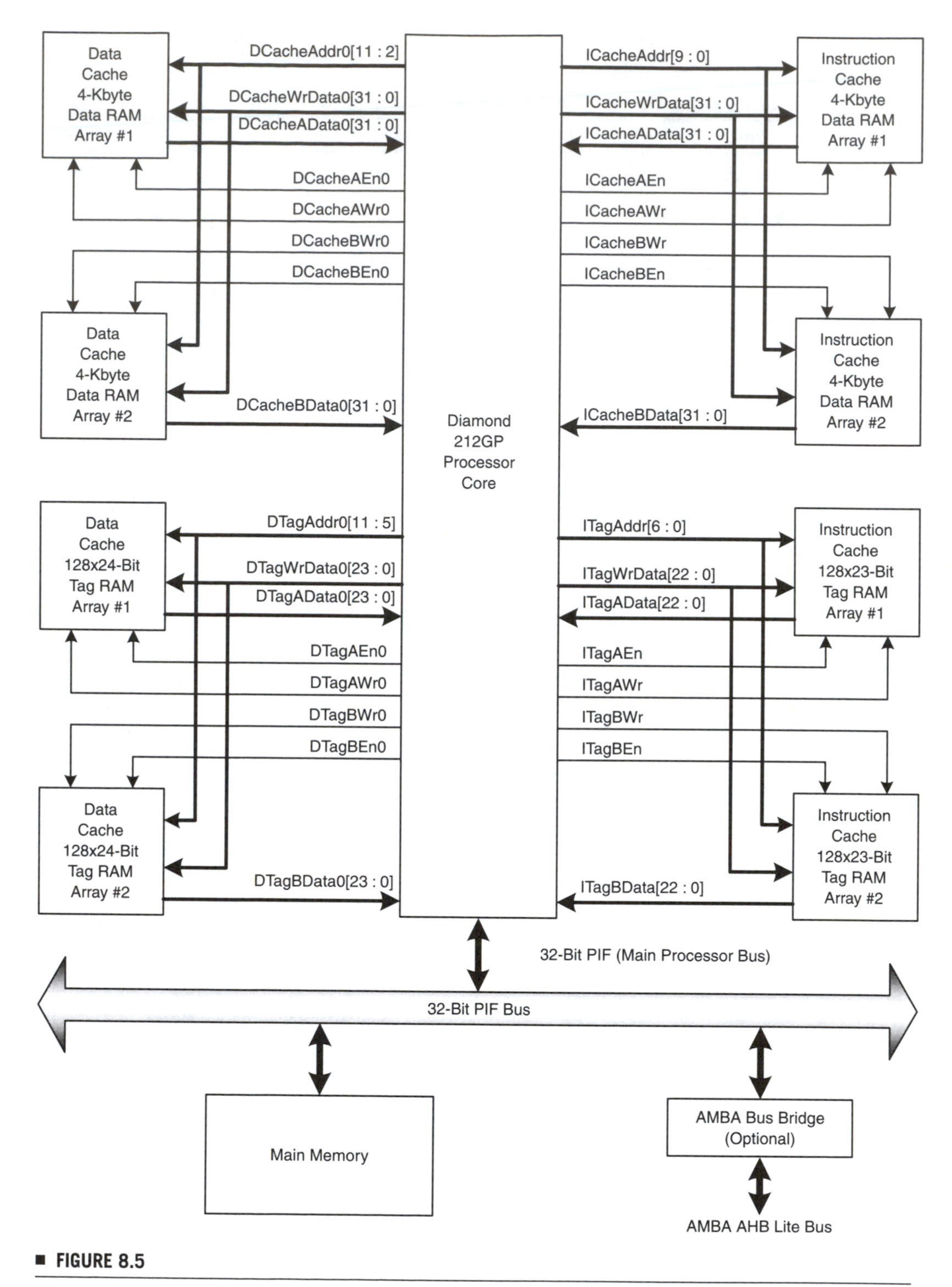

■ **FIGURE 8.5**

The Diamond 212GP controller core incorporates a cache controller that operates separate, 2-way set-associative, 8-Kbyte instruction and data caches. The processor has direct interfaces for the eight RAM arrays (used for storing cached data and cache tags) needed to implement the two caches.

controller core's cache memory. Although the cache controller is integrated into the logic of the Diamond 212GP core, the cache's RAM arrays are external to the core.

The Diamond 212GP processor accesses information directly from the cache memories when an instruction fetch or a load hits in the cache. A hit occurs when there is a match between the cache's address tags (stored in the cache's tag RAM arrays) and the target address of the transaction (instruction fetch, load, or store). When there is a cache hit, the processor fetches the required instruction or data directly from the associated instruction or data cache and the operation continues. This sort of memory transaction is as fast as fetching instructions from local instruction memory or loading data from local data memory.

A cache miss occurs when the cache's address tags do not match the address of the transaction. In the case of a cache miss, the desired instruction or data is not present in the cache and the integral cache controller retrieves a cache line worth of instructions or data from non-local memory (directly attached to the PIF or located off chip) to load the missing information into the cache. The requested information must be retrieved from the non-local memory before processor operation can continue.

The Diamond 212GP processor's cache interface is separate from but intimately related to the PIF. The first load access of a given instruction or data word will cause a cache miss (because the information is not yet stored in the cache) and will therefore generate a read cycle on the PIF directed at main memory. However, when the desired instruction or data arrives, the cache controller saves a copy of the retrieved information in the fast cache memory so that the cache will actually supply the instructions or data on subsequent accesses to the same location. This mechanism speeds system access and produces better performance than if the program had to run in main memory alone.

Because loads from and stores to the data cache compete for cache access, store transactions pass through a store buffer within the processor so that loads and stores can overlap. The processor's store buffer is a shared queue of write operations targeted at the data-cache, data-RAM, and XLMI ports. Store operations complete when the cache is not otherwise occupied with load operations.

8.8 SYSTEM DESIGN WITH THE DIAMOND 212GP PROCESSOR CORE

Figure 8.6 shows a system built with one Diamond 212GP controller core and three Diamond 108Mini processor cores. The 4-processor cores can communicate with each other and with global memory over the shared 32-bit PIF bus. A bus arbiter controls access to the PIF bus. This system closely resembles the system shown previously in Figure 7.5 but the

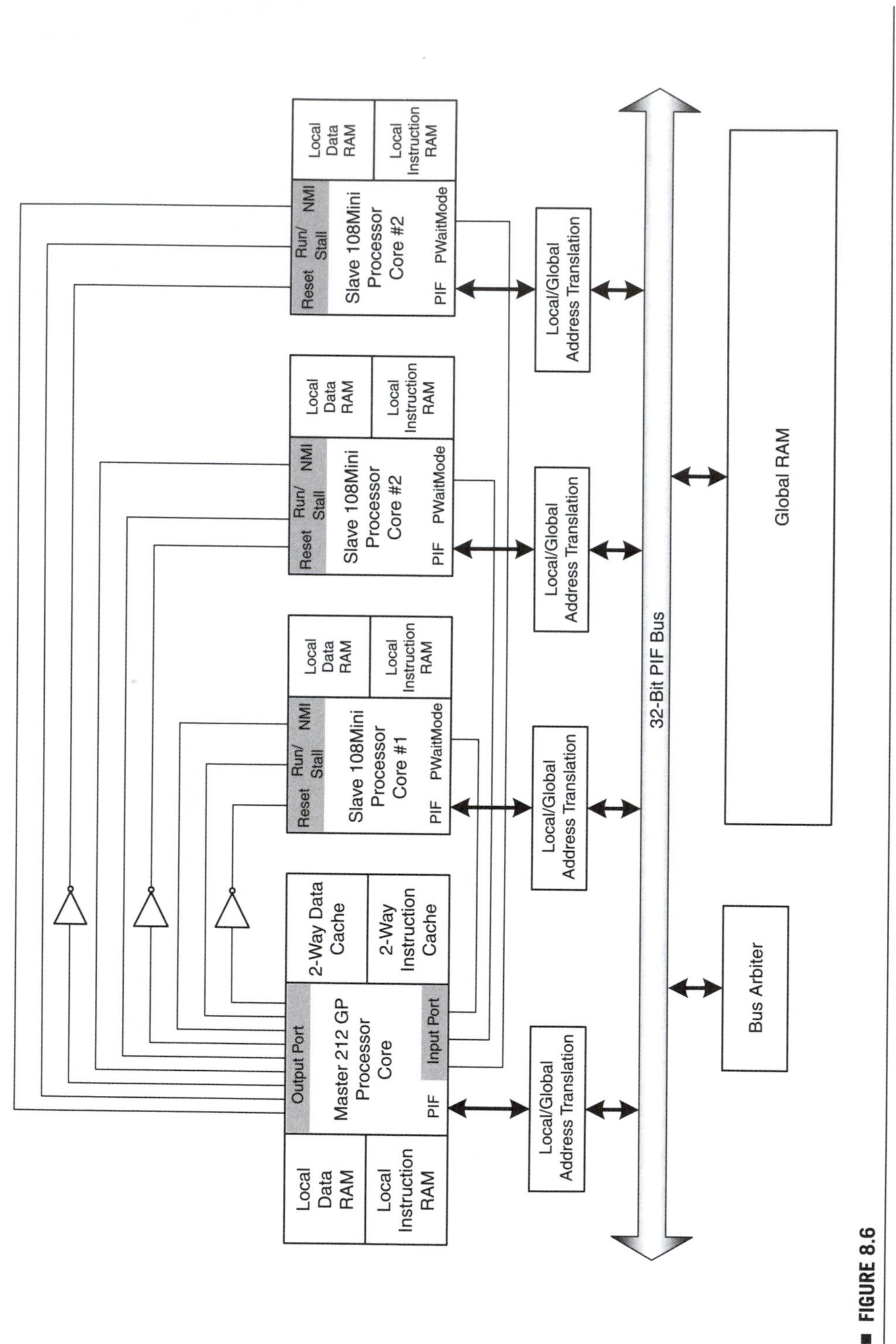

■ **FIGURE 8.6**

This multi-processor system design allows the master Diamond 212GP processor (shown on the left) to control the operation of the three Diamond 108Mini slave processors through their Reset, Run/Stall, and NMI inputs.

Diamond 108Mini master processor in that system has been replaced by a Diamond 212GP processor in the system shown in Figure 8.6. The Diamond 212GP processor on the left of the figure is configured as the system master and the three Diamond 108Mini processor cores are slaves.

As with the system shown in Figure 7.5, the four processor cores shown in Figure 8.6 can communicate with global memory over the shared 32-bit PIF bus and with each other's local memories using the Diamond processor cores' inbound-PIF feature. A bus arbiter controls access to the PIF bus. Local/Global address-translation blocks attached to each processor's PIF bus perform the critical function of mapping the attached processor's local address space into one unified global address map.

Three wires from the master processor's direct output port are connected to the Reset inputs of the three Diamond 108Mini slave processors; three more output-port wires are connected to the slave processors' Run/Stall inputs; and yet another three output-port wires are connected to the slave processors' NMI inputs. The Diamond 212GP controller core is equipped with the same 32-bit output port as the Diamond 108Mini processor core so the master/slave connections of this system are unchanged from the system discussed in the previous chapter.

This system configuration allows the master processor to independently reset and halt all three slaves. The three slave processors are automatically reset when the master processor is reset because the reset-initialized state of all port pins on the master processor's 32-bit output port is zero. Inverters attached to the master processor's output-port wires that are used for resetting the slave processors will therefore assert the slave processors' Reset inputs when the master processor is reset.

After initializing itself, the Diamond 212GP master processor can assert the Run/Stall input to each slave processor and then remove the reset signal to each Diamond 108Mini slave processor. While each slave processor is stalled, the master processor can transfer program code from the large global memory (or from its own local memory) to each slave processor's local instruction memory using inbound-PIF write operations—which can occur even while the slave processors are stalled. One advantage this system has over the 4-processor system discussed in the previous chapter is that the Diamond 212GP master processor has a cache, which gives it better performance when executing code directly from PIF-attached global memory.

When the master processor releases a slave processor's Run/Stall input, the slave processor will commence program execution starting at its reset-vector address. Because the Diamond 108Mini processor's reset vector is set to 0x50000000, each slave processor's first instruction fetch will be directed to the PIF-connected global memory. That first instruction however can set up a jump to a location in the slave processor's local instruction memory, which will then isolate the slave's operations from the PIF bus.

This scheme allows two of the Diamond 108Mini slave processors to independently execute code from their local memories concurrently while the Diamond 212GP master processor is programming or reprogramming the third slave processor's instruction memory. Together, these four processors consume approximately 2 mm^2 on the SOC die and the resulting multiprocessor system will exhibit even better processing performance than the system shown in the previous chapter because the master processor in this system has instruction and data caches.

The three Diamond 108Mini slave processor cores shown in Figure 8.6 still should not execute code directly out of the global memory attached to the PIF bus because they have no instruction or data caches and memory transactions on local-memory buses are about five times faster than the same transactions conducted over the PIF bus. However, the Diamond 212GP master processor does have instruction and data caches so its performance will be good even when executing code from PIF-connected global memory.

Although all four processors still must share the PIF bus, this system design would devote most of the PIF bandwidth to the master processor and the slave processors would use the PIF sparingly. The addition of caches to the master processor in this 4-processor system changes the use dynamics for the shared-bus architecture and allows the system to get much better performance when compared to the system consisting of four cacheless processors, at the expense of additional silicon for the slightly larger Diamond 212GP controller core plus the RAM arrays for its caches.

Similar to the discussion of the Diamond 108Mini-based, 4-processor system in Chapter 7, when a Diamond 108Mini slave processor in the system shown in Figure 8.6 enters the *WAITI* mode, it asserts its *PWaitMode* status output. The Diamond 108Mini slave processors' *PWaitMode* outputs are connected to the Diamond 212GP master processor's input port so that the master processor can evaluate the running status of the slave processors. Due to the Diamond 212GP controller core's 32-bit input port, this is a glueless connection, which means that the Diamond 212GP master processor need not consume bus bandwidth to poll the status of the other processors in the system. Bus bandwidth can be reserved for moving instructions and data around the system.

BIBLIOGRAPHY

Diamond Standard Processors Data Book, Tensilica, Inc., February 2006.

Leibson, S. and Kim, J., "Configurable Processors: A New Era in Chip Design," IEEE Computer, July 2005, pp. 51–59.

Rowen, C. and Leibson, S., *Engineering the Complex SOC*, Prentice-Hall, 2004.

CHAPTER 9

THE DIAMOND 232L CPU CORE

Some people have told me
they don't think a fat penguin
really embodies the grace of Linux.
—Linus Torvalds

The Diamond 232L CPU core is a general-purpose RISC processor core. Like the 212GP core, the Diamond 232L includes a cache controller but the Diamond 232L processor's cache controller operates larger, 16-Kbyte, 4-way set-associative instruction and data caches. The Diamond 232L CPU also includes a demand-paged memory-management unit (MMU) with translation lookaside buffer (TLB) and memory protection to support large operating systems such as Linux. Even with these features, the Diamond 232L CPU runs at 233 MHz, consumes only 0.8 mm^2 of silicon, and dissipates approximately 212 µW/MHz when implemented in a 130 nm, G-type process technology.

The Diamond 232L core brings the performance benefits of a full-featured 32-bit RISC CPU to bear on the designated tasks:

- Large 4-Gbyte address space
- 32-bit computations
- Large 32-entry register file
- Separate 16-Kbyte, 4-way, set-associative instruction and data caches
- Demand-paged MMU with TLB
- 5-stage pipelined operation resulting in a 266-MHz maximum clock rate in 130 nm LV technology

The Diamond 232L processor core has a 32-bit version of the general-purpose processor interface (PIF) bus for global system-on-chip (SOC) communications. An optional AMBA bus bridge supplied with the Diamond 232L processor core adapts the PIF to peripheral devices designed for the AMBA AHB-Lite bus. Unlike the Diamond 108Mini and 212GP controller cores, the Diamond 232L CPU core has no local instruction- or data-RAM interfaces. Instead, the Diamond 232L CPU's large instruction and data caches act like local memories.

9.1 THE DIAMOND 232L: A FULL-FEATURED CPU CORE

The Diamond 232L CPU architecture contains all of the basic elements of the Xtensa ISA. It has a 32-entry general-purpose register file, a 5-stage execution pipeline, 32-bit addressing. In addition, it has a demand-paged MMU with TLB that provides advanced memory management for operating systems such as Linux. The Diamond 232L CPU core also incorporates the Diamond Series processor core software-debug stack, as shown on the left of Figure 9.1. This debug stack provides external access to the processor's internal, software-visible state through a 5-pin IEEE 1149.1 JTAG TAP (test access port) interface and through a trace port that provides additional program-trace data. Figure 9.1 is a block diagram of the Diamond 232L processor core.

9.2 DIAMOND 232L CPU CORE INTERFACES

As shown in Figure 9.2, the Diamond 232L CPU core has a 32-bit implementation of the Xtensa PIF (main processor interface) bus and separate interfaces for the 4-way set-associative instruction and data caches. Unlike the 108Mini and 212GP controller cores, the Diamond 232L CPU core does not have interfaces for local memories or an XLMI port. All memories (except for the instruction and data caches) and any other system devices attach to the Diamond 232L CPU through its PIF bus.

9.3 THE DIAMOND 232L CPU MEMORY MAP

The Diamond 232L CPU core's entire address space is mapped to the PIF bus. The processor directs all instruction fetches, loads, stores, and cache spills and fills through the PIF. Table 9.1 lists the Diamond 232L CPU core's assigned reset, non-maskable interrupt (NMI), and other interrupt vectors. The Diamond 232L's reset and exception vectors are assigned to locations located in high memory, so some memory must be located at addresses 0xD0000000 to 0xD00003FF and 0xFE000000. In reality, the Diamond 232L CPU will be used for running large operating systems, so a large part of this upper address space will be populated with PIF-attached RAM.

Figure 9.3 shows the various important addresses mapped into the high memory addresses of the Diamond 232L CPU's uniform address space.

The PIF implementation on the Diamond 232L CPU core is 32 bits wide. The PIF bus uses a split-transaction protocol and the Diamond 232L processor core has an 8-entry write buffer to accommodate as many as eight simultaneous outstanding write transactions.

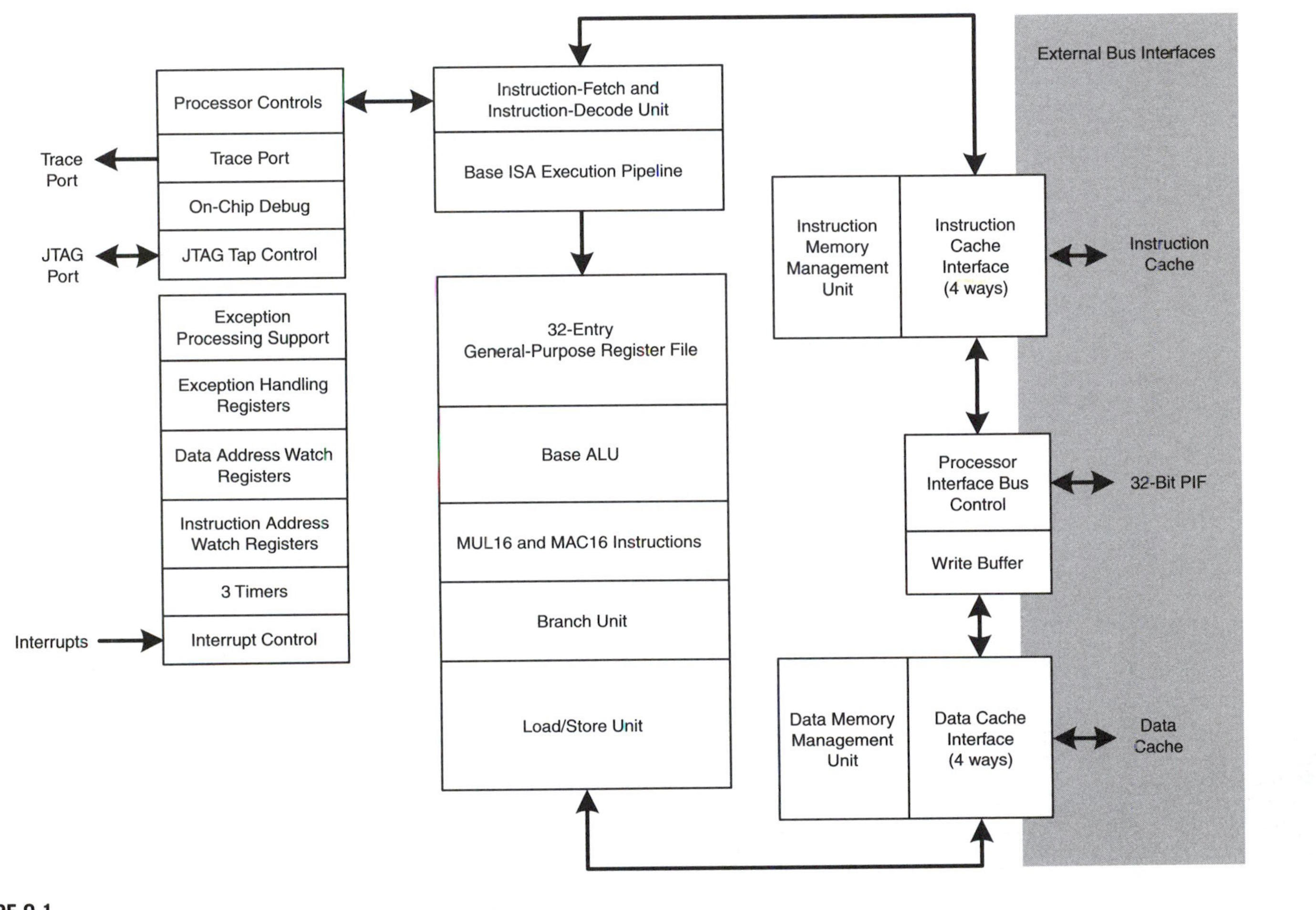

■ **FIGURE 9.1**

The Diamond 232L processor core implements a full-featured 32-bit RISC CPU while consuming only 0.8 mm^2 of silicon and approximately 212 μW/MHz when implemented in a 130 nm, G-type process technology.

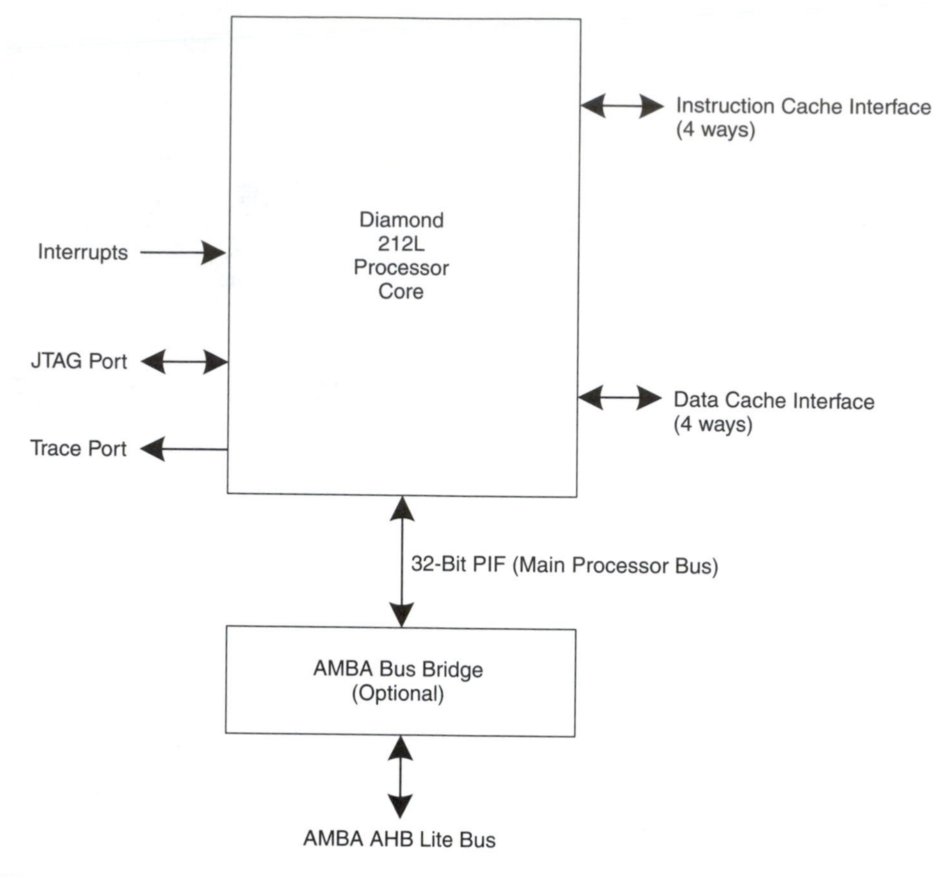

■ **FIGURE 9.2**

The Diamond 232L CPU core has a 32-bit implementation of the Xtensa PIF (main processor interface) bus and separate interfaces for the 4-way set-associative instruction and data caches.

TABLE 9.1 ■ Diamond 232L processor reset, interrupt, and exception vector address mapping

Vector	Address
Reset	0xFE000000
Base address for register window underflow/ overflow exception vectors	0xD0000000
Level 2 high-priority interrupt	0xD0000180
Level 3 high-priority interrupt	0xD00001C0
Level 4 high-priority interrupt	0xD0000200
Level 5 high-priority interrupt	0xD0000240
Debug exception	0xD0000280
NMI	0xD00002C0
Level 1 interrupt (Kernel mode)	0xD0000300
Level 1 interrupt (User mode)	0xD0000340
Double exception	0xD00003C0

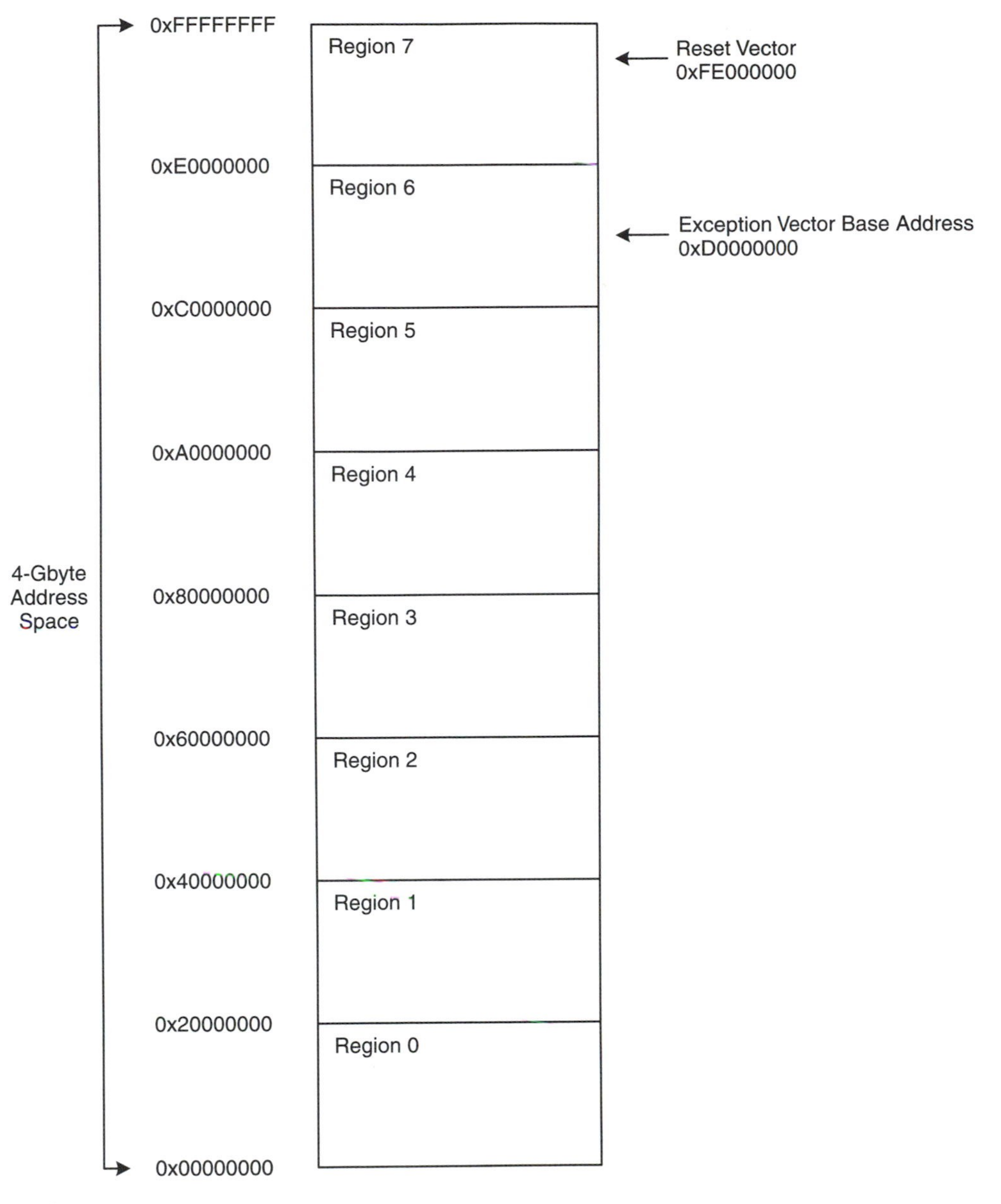

■ **FIGURE 9.3**

The Diamond 232L CPU core's pre-configured address space maps the reset and interrupt vectors to the high addresses in the processor's memory map.

The 212GP processor core has three internal 32-bit timers in that can be used to generate interrupts at regular intervals. The processor also has nine external, level-triggered interrupt input pins and one edge-triggered, NMI pin.

9.4 THE DIAMOND 232L CACHE INTERFACES

Processor cache memories use fast, small RAM arrays to buffer the processor from the slower and larger memories generally located external to the processor core or the SOC. The Diamond 232L CPU core's data and instruction caches store data and instructions that a program is immediately using, while the rest of the data and program reside in slower main memory—RAM, or ROM. In general, the Diamond 232L CPU core can access instruction and data caches simultaneously, which maximizes processor bandwidth and efficiency.

The Diamond 232L processor incorporates a pre-configured version of the Xtensa cache controller that operates separate, 16-Kbyte, 4-way, set-associative, instruction and data caches. The data cache employs a write-back (as opposed to write-through) write policy. The Diamond 232L processor's cache memories accelerate any program the processor executes by storing copies of executed code sequences and data in fast cache memory. The caches provide faster access to instructions and data than can memories attached to the processor's PIF.

The cache controller is woven (integrated) into the Diamond 232L processor core so the processor has direct interface ports to the sixteen RAM arrays used for storing cached data and cache tags. Each cache way requires separate RAM arrays for cache data and cache tags so sixteen RAM arrays are needed to complete the Diamond 232L CPU core's cache memory. These RAM arrays are external to the processor's core. Figure 9.4 shows how the eight RAM arrays attach to the processor core's cache interface ports. In all respects except for the number of ways, the Diamond 232L CPU's caches work identically to those of the 212GP controller core, as discussed in Chapter 8.

9.5 THE DIAMOND 232L MMU

The Diamond 232L CPU incorporates a pre-configured version of the Xtensa MMU, which has independent instruction and data TLBs. By providing sophisticated address-translation and memory-protection mechanisms, the MMU allows an operating system to manage tasks as completely independent programs running in their own memory spaces.

Tasks written for processors with MMUs can run anywhere in memory and are prevented from interfering with the memory spaces of other running tasks. This feature is especially helpful when tasks are independently written by individual programmers or software-development teams that have limited or no knowledge of the other tasks being written for the same hardware. The most common environment where such independently written tasks run is under the control of a sophisticated operating system such as Linux.

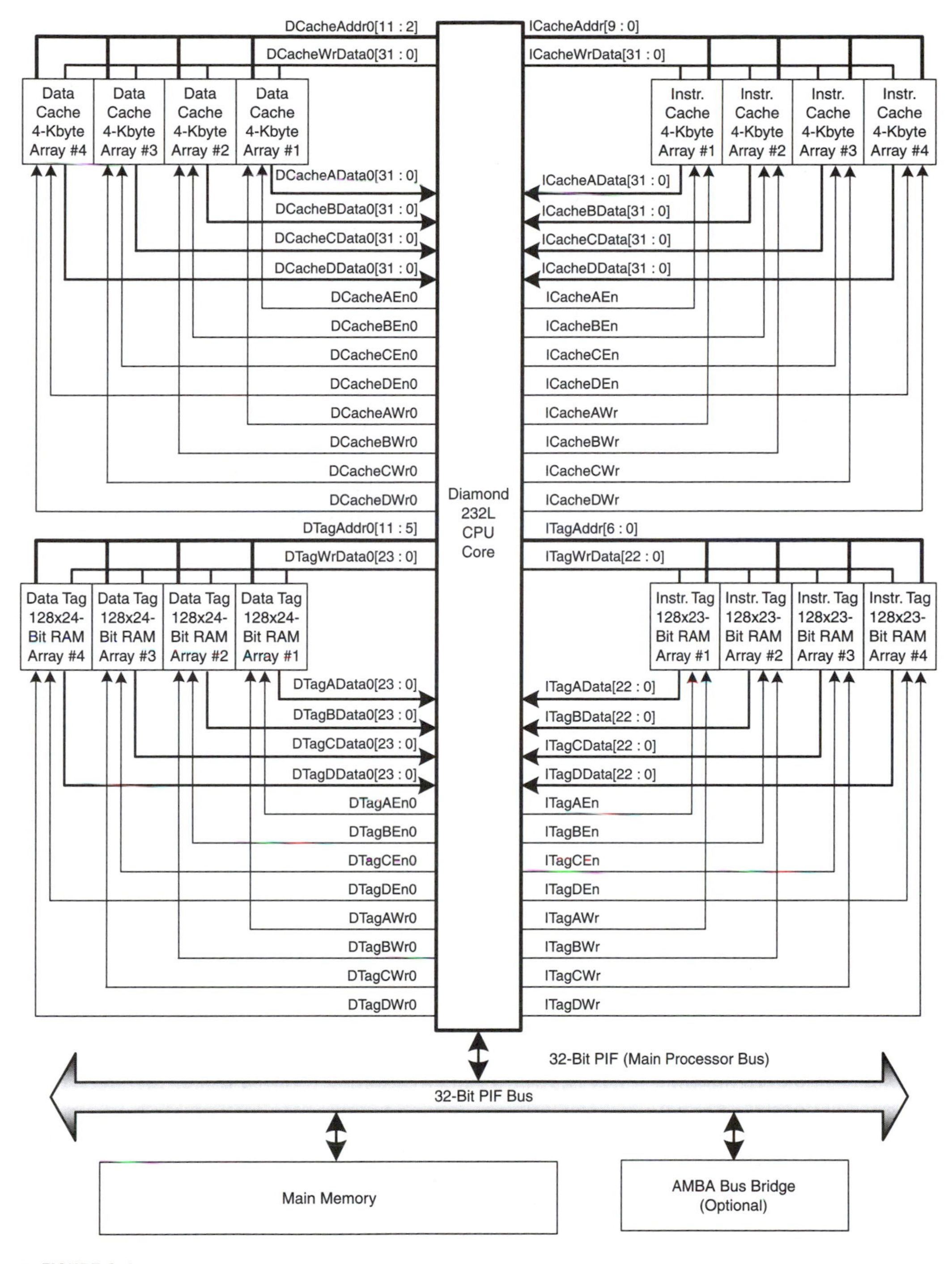

■ **FIGURE 9.4**

The 232L processor core incorporates a cache controller that runs separate, 4-way set-associative, 16-Kbyte instruction and data caches. The processor has direct interfaces for the sixteen RAM arrays—used for storing cached data and cache tags.

The Diamond 232L's MMU creates a *virtual-memory* space that is mapped onto the processor's *physical memory* space. All program accesses undergo address translation in the MMU before being issued over the CPU's PIF bus. The MMU translates virtual-space addresses into physical-space addresses using its TLBs, which use *page-table entries* (PTEs) to describe the virtual-to-physical address translation. The MMU's page table employs a 2-level structure that resides in virtual-memory address space. A hardware mechanism built into the CPU core performs the first-level autoloading of PTEs into the TLBs. If a second level of PTE access is needed, an exception occurs and software completes the PTE loading.

The Diamond 232L MMU has separate instruction and data TLBs (called the ITLB and DTLB). These TLBs are larger versions of the TLB used in the region-protection units (RPUs) of other Diamond processor cores. The ITLB translates virtual addresses associated with instruction fetches and the DTLB translates virtual addresses associated with loads and stores.

Although a complete discussion of MMU operation is far beyond the scope of this book (the topic of memory management requires a book unto itself), it's important to take a quick look at the MMU's internals to give some sense of its ability to meet the needs of complex operating systems. Unlike the TLB associated with the RPUs in the other Diamond cores, the Diamond 232L core's ITLB has seven ways and the DTLB has 10 ways and most of the ways contain multiple PTEs. The Diamond 232L CPU automatically refills the PTEs held in ways 0–3 of each TLB when needed (when the required PTE is not present in the TLB). The other ways in each TLB are static and are loaded under program control. Table 9.2 lists the various TLB ways in the Diamond 232L's MMU.

Figure 9.5 shows how the translation process works within the MMU and the first four ways of the ITLB and DTLB. The processor generates a virtual target address for an instruction fetch, a load operation, or a store operation. This virtual address serves as the input to the TLB. The lower 12 bits of the virtual address, called the *offset*, pass through the TLB unchanged and are not translated. These address bits point to a location within a 4-Kbyte memory page. The next two bits of the virtual address are used as *index* bits and specify one of the four entries in a TLB way (limited to ways 0–3).

The upper 18 bits of the virtual address identify a *virtual page number* (VPN), which is used to query the specified entries in the TLB. The 18-bit VPN and an 8-bit ASID (address-space identifier, described below) are used to locate a specific PTE in the appropriate TLB. If a corresponding PTE matches the combined values of the VPN and ASID, there is a match and the TLB provides a 20-bit *physical page number* (PPN) and four access-mode bits that are used to access physical memory. Note that the address-translation process works similarly for the other TLB ways but the bit assignments can be somewhat different.

TABLE 9.2 ■ The Diamond 232L CPU's TLB ways

Way(s)	TLB	Purpose	Number of entries in TLB way
0–3	ITLB, DTLB	Holds 4-Kbyte PTE entries, autorefillable	4 entries per way
4	ITLB, DTLB	Large page support (1-, 4-, 16-, or 64-Mbytes/page)	4
5	ITLB, DTLB	Static-mapped, 128-Mbyte pages for Ring 0 (KSEG_Cached and KSEG_Bypass)	2
6	ITLB, DTLB	Static-mapped, 256-Mbyte pages for Ring 0 (KSEG_Cached and KSEG_Bypass)	2
7–9	DTLB	4-Kbyte page wired entries	1 entry per way

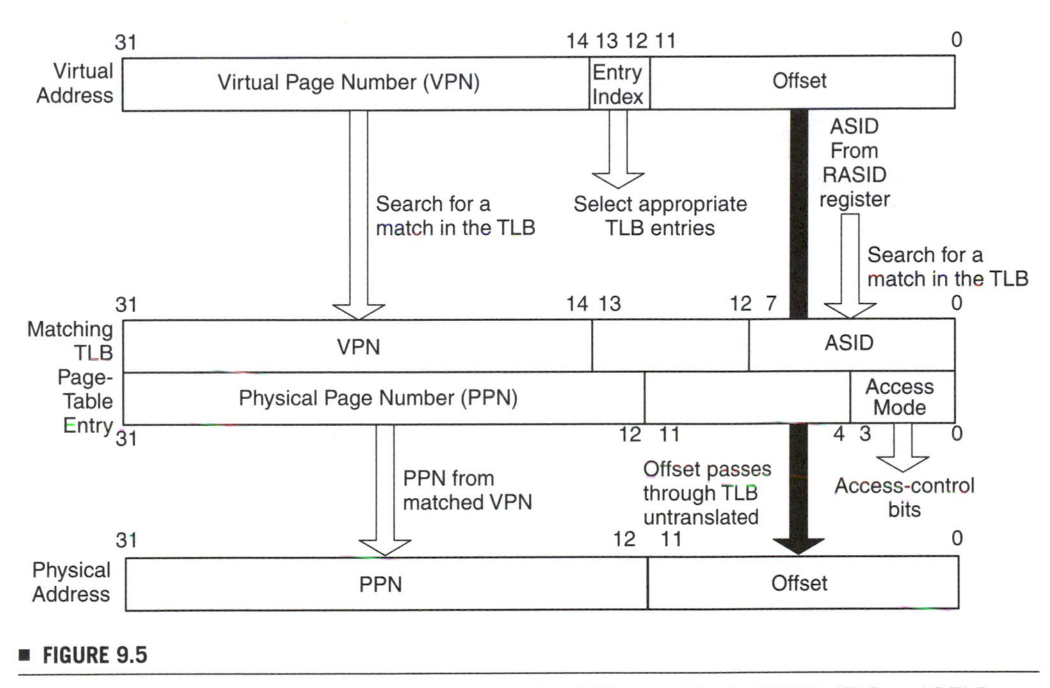

■ **FIGURE 9.5**

The MMU translates virtual to physical addresses using PTEs stored in the MMU's ITLB and DTLB.

The four access-mode bits control processor access to memory. Table 9.3 lists the MMU's memory-access modes and Table 9.4 describes the MMU's access-mode behaviors. Assertion of the X (executable) access-mode bit allows instruction fetches and assertion of the W (writable) bit allows write accesses. The two CM (cache-mode) bits determine the CPU's cache behavior, as described in the table.

TABLE 9.3 ■ Diamond 232L MMU access modes

Access mode CM	W	X	Cache behavior mode	Fetch behavior	Load behavior	Store behavior
00	0	0	Bypass cache	Exception	Bypass cache	Exception
00	0	1		Bypass cache	Bypass cache	Exception
00	1	0		Exception	Bypass cache	Bypass cache
00	1	1		Bypass cache	Bypass cache	Bypass cache
01	0	0	Write-back cache	Exception	Cached	Exception
01	0	1		Cached	Cached	Exception
01	1	0		Exception	Cached	Write-back
01	1	1		Cached	Cached	Write-back
10	0	0	Write-through cache	Exception	Cached	Exception
10	0	1		Cached	Cached	Exception
10	1	0		Exception	Cached	Write-through
10	1	1		Cached	Cached	Write-through
11	0	0	Invalid	Exception	Exception	Exception
11	0	1	Direct cache access	Exception	Isolate	Isolate
11	1	0	Invalid	Exception	Exception	Exception
11	1	1	Reserved (Do not use)	Exception	Exception	Exception

TABLE 9.4 ■ Diamond 232L MMU access mode behaviors

Cache behavior mode	Description
No allocate	Do not allocate a cache line for this address. If the address is already cached, fetch or load the cached value. If the address has an allocated cache line but the cache line is not already in the cache, fetch or load the value from main memory and place the value in the cache.
Bypass	Do not use the cache.
Write-back	Write the value to the cache and then update main memory when the cache line is evicted or when the processor forces the cache line to be written to main memory.
Isolate	Permits direct read/write access to the cache's data and tag RAM arrays.
Reserved	Any access causes an exception.

If a matching PTE is not present in the appropriate TLB, the resulting TLB miss causes the MMU hardware to automatically attempt to load a PTE from the page table's first level in the processor's virtual-memory space. If the MMU hardware cannot locate the appropriate PTE by walking the PTEs stored in the page table's first level, a second miss occurs, which causes an exception that activates a software routine (called a second-level miss handler). This software routine must then either locate the appropriate PTE in the page table's second level or create one. Second-level TLB miss handling is defined by the handler software, not by the processor. The second-level miss handler must set up a page in the page table so that the first-level, hardware-driven TLB autorefill can then succeed.

Using this translation mechanism, the Diamond 232L CPU core maps multiple virtual address spaces into the one physical address space. An example of such a mapping for a Linux operating system in a single-processor environment appears in Figure 9.6.

9.6 PRIVILEGE LEVELS AND RINGS

The simplest sort of privilege mechanism, implemented by many CPUs, employs two privilege levels. These two levels are often called the *user* and *supervisor* or *user* and *kernel* levels. The MMU makes higher-level instruction and data spaces inaccessible to user-level code to protect the most privileged (supervisor or kernel) code and data.

The Diamond 232L CPU's MMU operates with four privilege levels called *rings*. In the Diamond 232L CPU, kernel code resides at Ring 0. Code running at Ring 0 can access the address spaces in all of the other privilege rings. The privilege levels are called rings because conceptually, lower-numbered levels (with higher privilege) envelop or "ring" higher-numbered levels. Figure 9.7 illustrates this concept of nested privilege levels.

Figure 9.7 shows the one kernel task at Ring 0 running two other processes at Ring 1. The process running at Ring 1 on the left is running one task at Ring 2, which in turn is running one task at Ring 3. The task running at Ring 1 on the right is running two tasks at Ring 2. One of those tasks at Ring 2 is running three tasks at Ring 3 and the other task running at Ring 2 is running one task at Ring 3. At any given moment, the Diamond 232L CPU can be executing one task at each of the ring levels.

A simple way (but not the only way) to think about the four privilege rings is:

1. Ring 0: Shared kernel address space
2. Ring 1: Per-process kernel address space
3. Ring 2: Shared application address space
4. Ring 3: Thread-level address space

Each of the running tasks is differentiated by a value held in the Diamond 232L CPU's ASID registers. There is a separate ASID register for each of the CPU's four privilege rings. Thus at any time, four separate tasks running can be running on the processor—one at each of the four privilege or ring levels.

Each task has a different 8-bit ASID assigned to it by the operating system kernel. The task running at Ring 0 always has an ASID value of 1, so there can only be one such task. An assigned ASID value of 0 is invalid. (At reset, the processor initializes some entries in the TLB as invalid but

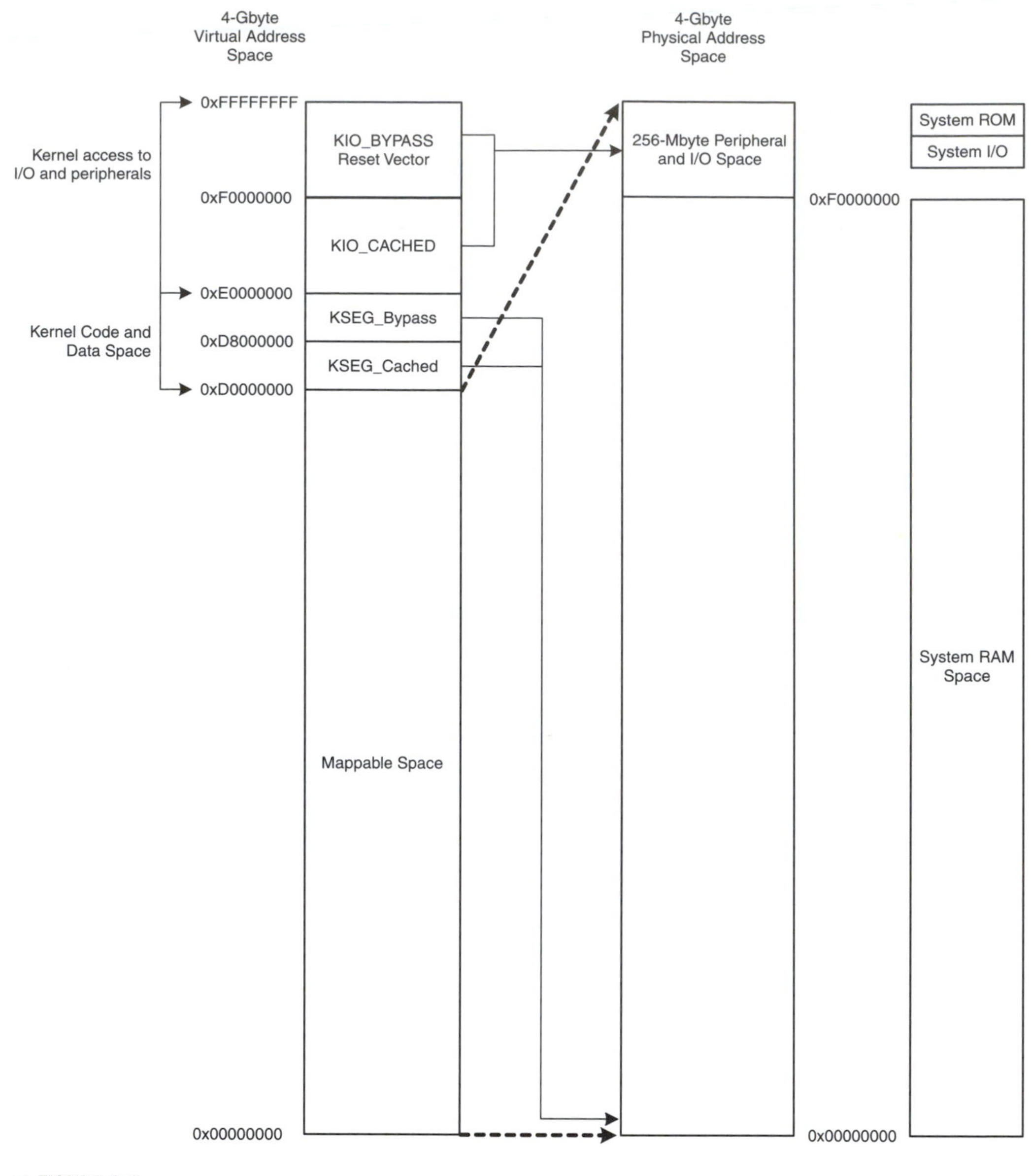

■ **FIGURE 9.6**

The Diamond 232L CPU core's MMU allows a virtual-to-physical address mapping that's consistent with the needs of a sophisticated operating system such as Linux.

the ASID-management software should not use this value.) The other 254 ASID values are assigned to tasks running in Rings 1, 2, and 3.

The ASID values for the four active tasks (one active task running in each of the four privilege rings) are stored in the Diamond 232L CPU's 32-bit *RASID* special register, shown in Figure 9.8. When the OS kernel

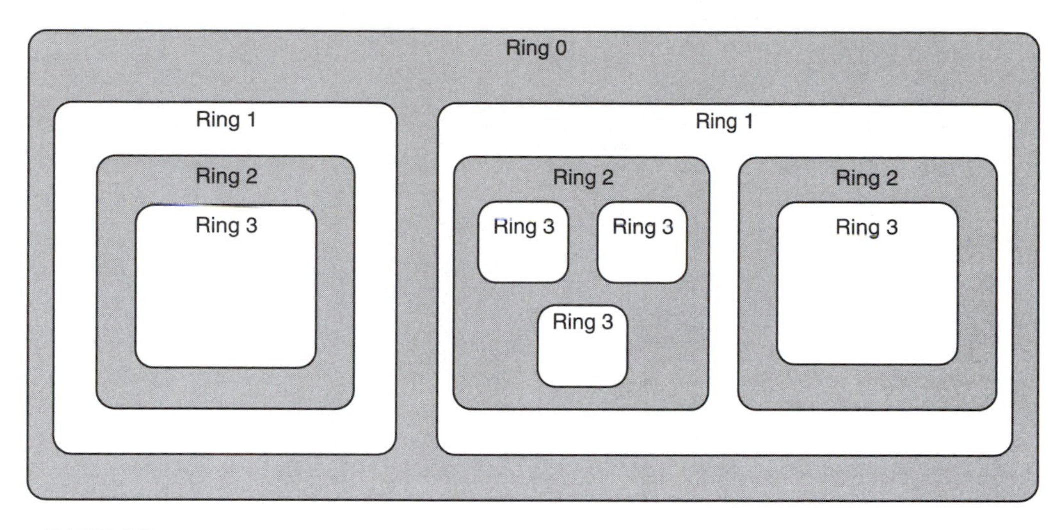

■ **FIGURE 9.7**

The Diamond 232L CPU core's MMU can manage as many as four privilege rings.

31 24	23 16	15 8	7 0	
Ring 3 ASID	Ring 2 ASID	Ring 1 ASID	Ring 0 ASID (Always 1)	232L Processor RASID Register Format
8 Bits	8 Bits	8 Bits	8Bits	

■ **FIGURE 9.8**

The Diamond 232L CPU core's RASID register holds the four ASID values for the four actively running tasks. The ASID for the ring-0 task is always set to 1.

swaps out a task running at Ring 1, 2, or 3, it must also change the corresponding ASID stored in the *RASID* register. The MMU uses the ASID values as an input to the address-translation process.

9.7 SYSTEM DESIGN WITH THE DIAMOND 232L CPU CORE

Figure 9.9 shows a software-coherent, symmetric-multiprocessing (SMP) system built with four Diamond 232L CPU cores (each with their own local memory) and a large global memory. The four processor cores communicate with each other through the global memory over the shared 32-bit PIF bus. A bus arbiter and locking device control access to the PIF bus. This hardware design can be used to implement either a message-passing or shared-memory SMP system.

This system allows a group of cooperating Diamond 232L CPU cores to execute multiple independent tasks under the control of an SMP operating system. Access to shared global memory must be performed under

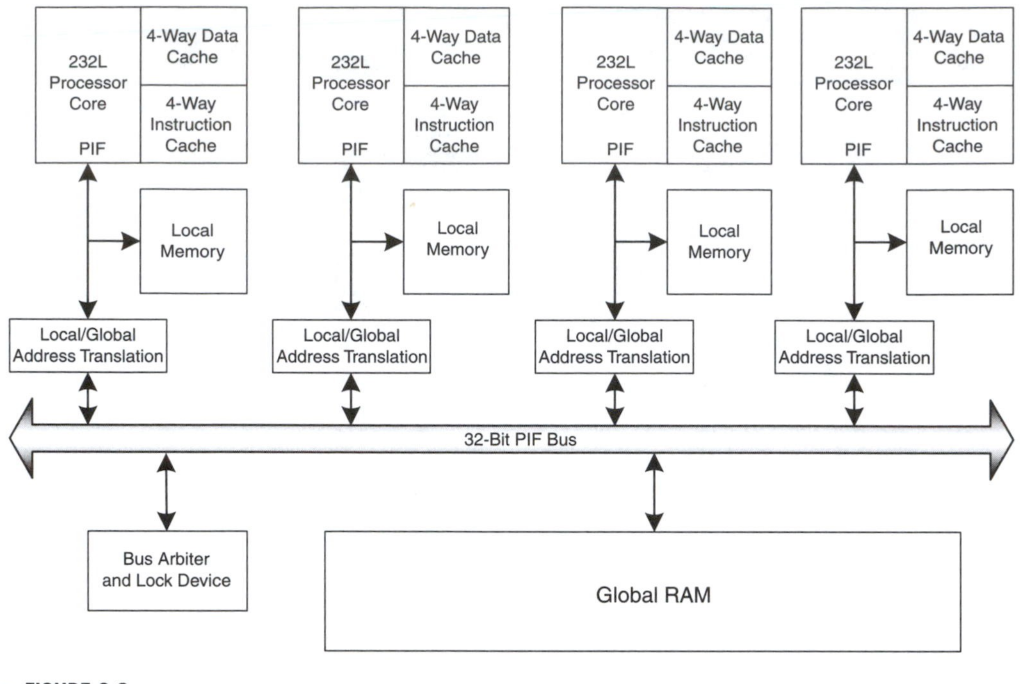

▪ **FIGURE 9.9**

This multi-processor system design allows four Diamond 232L CPU cores to form an SMP system.

some sort of mutual exclusion (mutex) mechanism. Figure 9.9 shows a combination hardware bus arbiter and locking device that provides mutex-governed access to global memory.

Many SMP systems are designed with hardware cache coherency. In such systems, each processor snoops the transactions occurring on the main bus and tracks the memory traffic of the other processors, invalidating its own cache when needed. Hardware cache coherency requires substantial extra silicon to implement the cache-snooping logic for each processor. It is possible to realize most of the benefits of hardware-cache-coherent systems without implementing the cache-snooping hardware by using software cache coherency. Software-cache-coherent systems require that the processors in the system have explicit control of their caches, which the Diamond 232L core does.

Together, the four Diamond 232L processor cores in this system require only 3.2 mm^2 on the SOC, implemented in a 130 nm, G-type process technology. The various cache, local, and global memory arrays naturally consume additional silicon depending on size.

The hardware system shown in Figure 9.9 can easily be scaled using a step-and-repeat design process. For example, Figure 9.10 shows an 8-processor version of the same system. All of the same SMP code and mutex

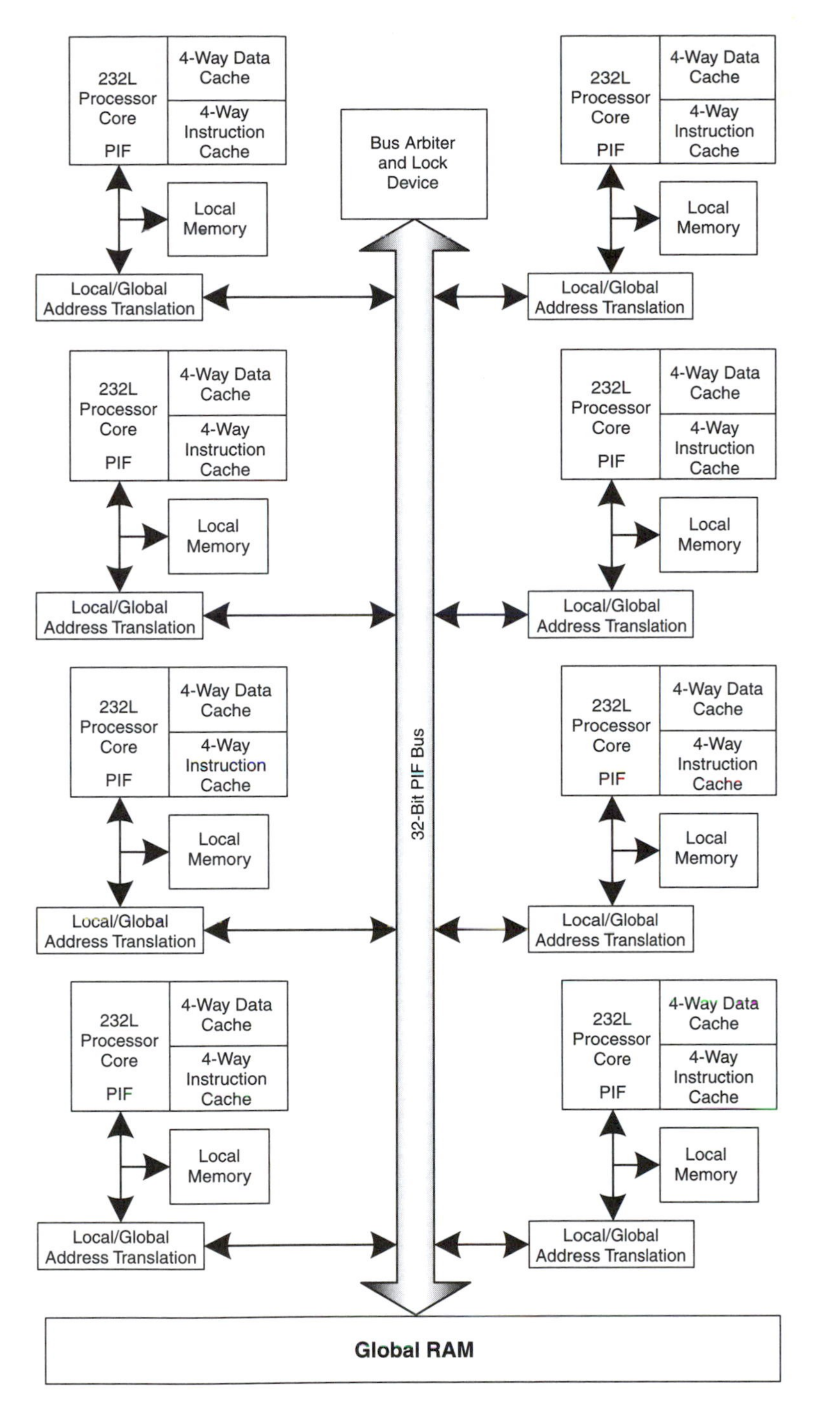

■ **FIGURE 9.10**

This multi-processor system design allows eight Diamond 232L CPU cores to form an SMP system.

mechanisms developed for the 4-processor system can be adapted to the 8-processor system so the operating-system code should need only minimal changes to accommodate the four extra CPUs. The eight Diamond 232L processor cores shown in this system require only 6.4 mm^2 on the SOC (when implemented in a 130 nm, G-type process technology), so this 8-processor system is still rather small.

BIBLIOGRAPHY

Diamond Standard Processors Data Book, Tensilica, Inc., February 2006.

Kontothanassis, L.I. and Scott, M.L., "High performance software coherence for current and future architectures," *Journal of Parallel and Distributed Computing*, 29(2), September 1995, pp. 179–195.

Leibson, S. and Kim, J., "Configurable Processors: A New Era in Chip Design," *IEEE Computer*, July 2005, pp. 51–59.

Rowen, C. and Leibson, S., *Engineering the Complex SOC*, Prentice-Hall, 2004.

Xtensa Microprocessor Programmer's Guide, Tensilica, Inc., July 2005.

CHAPTER 10

THE DIAMOND 570T SUPERSCALAR CPU CORE

Superscalar processors are the natural next step in the evolution of general-purpose processors.
—Mike Johnson

Superscalar Microprocessor Design

The Diamond 570T CPU core is a high-performance, 3-way static-superscalar RISC CPU core. The term superscalar describes a processor that achieves superlative performance by executing multiple scalar instructions per clock cycle. Just as warp engines allow the Starship Enterprise to break the speed-of-light barrier, superscalar execution pipelines allow the Diamond 570T CPU to break the 1-instruction-per-clock barrier that limits conventional scalar RISC architectures. Scalar instructions are the instructions found in the ISAs of most general-purpose processors including the Xtensa ISA. Superscalar processor implementations are a viable alternative to increasing clock rates (which unacceptably increase power dissipation). Rather than execute more instructions per second through clock-rate escalation, superscalar processors execute multiple instructions per clock cycle through parallel execution pipelines.

Most superscalar microprocessor designs are of the dynamic superscalar type. Dynamic superscalar processors contain logic that examines the incoming scalar instruction stream and seeks to group independent scalar instructions into bundles of instructions that can be executed concurrently. Processor designers take the dynamic superscalar approach to accelerate any scalar code that might have been written for scalar versions of the target ISA.

The Diamond 570T CPU core is a static-superscalar design. It adds two 64-bit instruction formats to the existing 16- and 24-bit Xtensa ISA instruction formats, as shown in Figure 10.1. The Diamond 570T processor's first 64-bit instruction format contains two operation slots. The instructions' first slot accommodates base Xtensa instructions. The second slot accommodates a special set of wide-branch instructions, listed in Table 10.1.

The Diamond 570T is the only Diamond processor core that implements "wide branching." Wide-branch instructions improve code performance

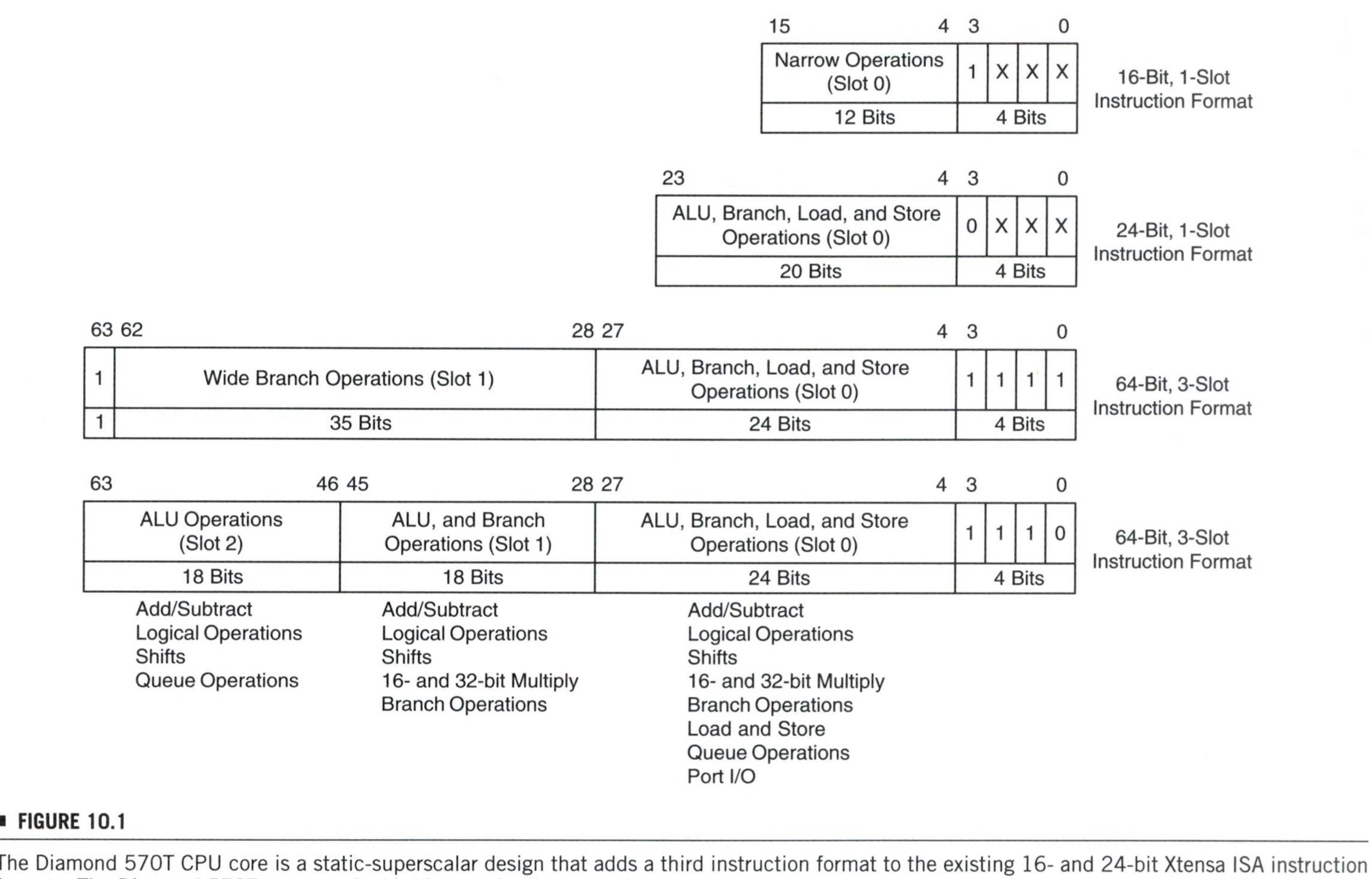

■ FIGURE 10.1

The Diamond 570T CPU core is a static-superscalar design that adds a third instruction format to the existing 16- and 24-bit Xtensa ISA instruction formats. The Diamond 570T processor's third instruction format is 64 bits wide, and this wide format bundles three independent operations.

TABLE 10.1 ■ Diamond 570T CPU wide-branch instructions

Instruction	Definition
BALL.W18	Branch if All Bits Set
BANY.W18	Branch if Any Bit Set
BBC.W18	Branch if Bit Clear
BBCI.W18	Branch if Bit Clear Immediate
BBS.W18	Branch if Bit Set
BBSI.W18	Branch if Bit Set Immediate
BEQ.W18	Branch if Equal
BEQI.W18	Branch if Equal Immediate
BEQZ.W18	Narrow Branch if Equal Zero
BGE.W18	Branch if Greater Than or Equal
BGEI.W18	Branch if Greater Than or Equal Immediate
BGEU.W18	Branch if Greater Than or Equal Unsigned
BGEUI.W18	Branch if Greater Than or Equal Unsigned Immediate
BGEZ.W18	Branch if Greater Than or Equal to Zero
BLT.W18	Branch if Less Than
BLTI.W18	Branch if Less Than Immediate
BLTU.W18	Branch if Less Than Unsigned
BLTUI.W18	Branch if Less Than Unsigned Immediate
BLTZ.W18	Branch if Less Than Zero
BNALL.W18	Branch if Not-All Bits Set
BNE.W18	Branch if Not Equal
BNEI.W18	Branch if Not Equal Immediate
BNEZ.W18	Narrow Branch if Not Equal Zero
BNONE.W18	Branch if No Bit Set

for large programs that can exploit conditional branching over larger address spaces than can be handled by the reach of the standard Xtensa ISA branch instructions. The Diamond 570T CPU's wide-branch instructions reduce the number of required jump instructions in large programs because the wide-branch instructions' address range is 18 bits instead of the 8- or 12-bit range of the branch instructions in the base Xtensa ISA. A 35-bit wide-branch operation encoding in slot 1 of the Diamond 570T CPU's first 64-bit instruction format permits the wide-branch instructions' enlarged branch-target address range.

The Diamond 570T CPU's second 64-bit instruction format bundles three independent operations. The XCC compiler paired with the Diamond 570T CPU core performs the instruction bundling that is performed by hardware in dynamic superscalar processor designs.

All three instruction formats can be freely intermixed and tightly packed in memory. The Xtensa architecture's inherent ability to handle differently sized instructions was a key enabler in the development of the Diamond 570T CPU core's design.

The static-superscalar design approach taken in developing the Diamond 570T CPU offers three major advantages to SOC designers. First, the XCC compiler has a broader reach and a much more global perspective on the code than does the processor. Consequently, the compiler can be far more effective in optimizing the compiled code for superscalar execution and in creating instruction bundles. Second, instruction optimization and bundling takes place on host computers running at multiple GHz, far faster than the target embedded processor will run. This speeds the creation of the superscalar code. Third, there is no target hardware overhead in the SOC required for the superscalar instruction bundling, which results in cost savings in every fabricated SOC.

Even with the addition of two more execution pipelines, the Diamond 570T CPU runs at 233 MHZ, consumes only 1.46 mm^2 of silicon, and dissipates approximately 275 μW/MHz when implemented in a 130 nm, G-type (general-purpose) process technology. In a 130 nm LV process technology, the Diamond 570T CPU can run at 233 MHz and consumes less than 1 mm^2 of silicon.

The Diamond 570T CPU core brings the performance benefits of a full-featured 32-bit RISC CPU to bear on the designated tasks:

- Large 4-Gbyte address space
- 32-bit computations
- Large 32-entry register file
- Separate 16-Kbyte, 2-way, set-associative instruction and data caches
- 5-stage pipelined operation resulting in a 233-MHz maximum clock rate in 130 nm LV technology.

The Diamond 570T CPU core has a 64-bit version of the general-purpose PIF bus for global SOC communications. An optional AMBA bus bridge supplied with the Diamond 570T CPU core adapts the PIF to peripheral devices designed for the AMBA AHB-Lite bus. The Diamond 570T CPU core has local instruction- and data-RAM interfaces and separate 2-way, set-associative, 16-Kbyte instruction and data caches.

10.1 THE DIAMOND 570T: A HIGH-PERFORMANCE CPU CORE

The Diamond 570T CPU architecture contains all of the basic elements of the Xtensa ISA. It has a 32-entry general-purpose register file, a 5-stage execution pipeline and 32-bit addressing. It has two additional execution pipelines giving the processor core the ability to execute three independent instructions per clock. The Diamond 570T CPU's three execution pipelines are not symmetric, as shown in Figure 10.2. The first execution pipeline, associated with operation slot 0, contains the

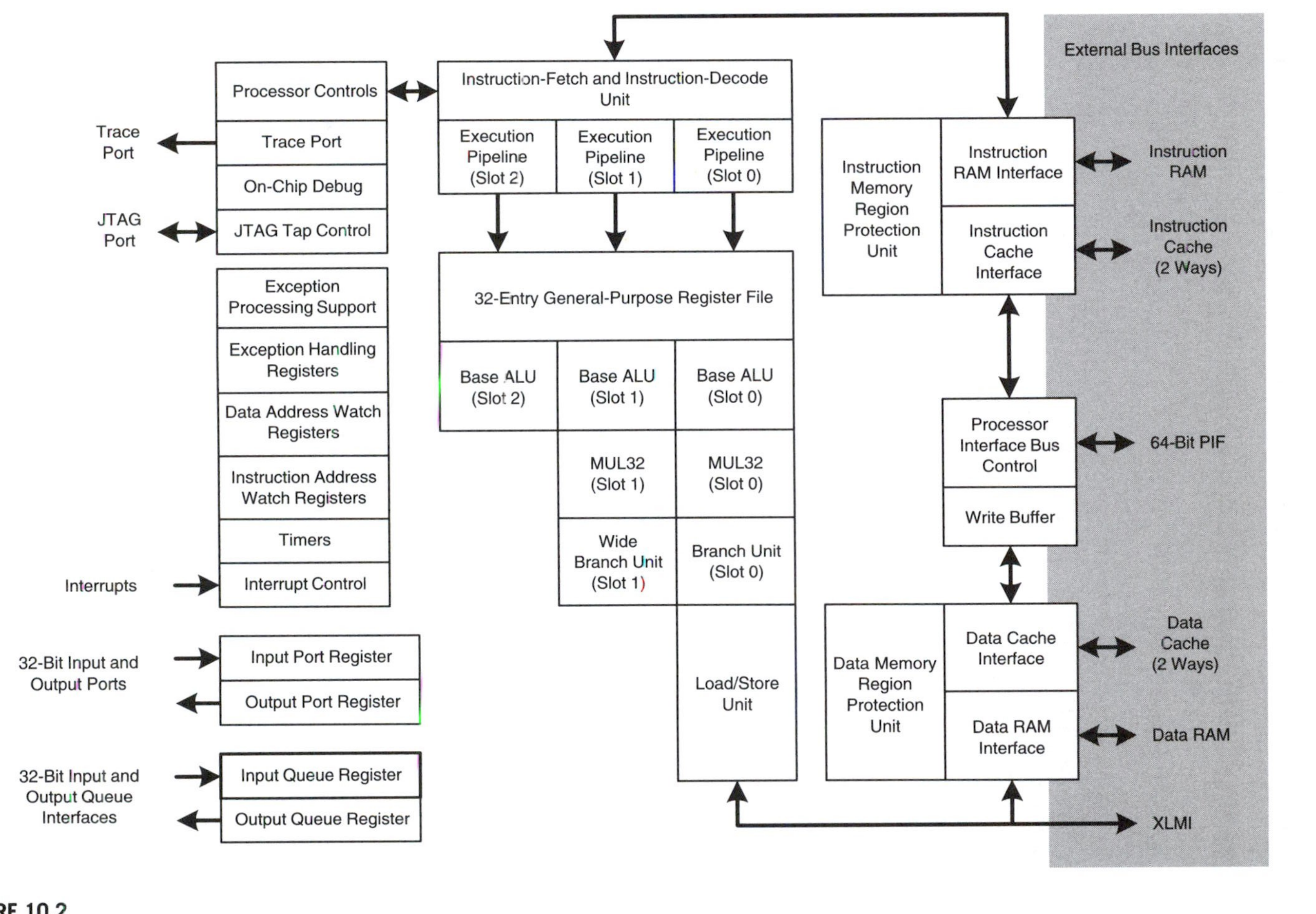

■ **FIGURE 10.2**

The Diamond 570T processor core implements a high-performance, 3-way superscalar, 32-bit RISC CPU while consuming only 1.46 mm^2 of silicon and approximately 275 μW/MHz when implemented in a 130 nm, G-type process technology. In a 130 nm LV process technology, the Diamond 570T CPU can run at 233 MHz and consumes less than 1 mm^2 of silicon.

base Xtensa ALU, a 32-bit multiplier, a branch unit, and the processor's load/store unit.

The Diamond 570T's second execution pipeline contains a second base Xtensa ALU, a second multiplier, and a second branch unit. Thus both of the first two Diamond 570T execution pipelines can execute the full Xtensa arithmetic and logical instruction set including branches and can also perform 32-bit multiplication. The first execution pipeline alone is responsible for executing load and store instructions. The Diamond 570T's third execution pipeline contains a third Xtensa ALU and can execute the full Xtensa arithmetic and logical instruction set. The XCC compiler for the Diamond 570T CPU core understands the unique nature of each of the three Diamond 570T execution pipelines and automatically bundles instructions accordingly.

The Diamond 570T CPU core also incorporates the Diamond Series processor core software-debug stack, as shown on the left of Figure 10.2. This debug stack provides external access to the processor's internal, software-visible state through a 5-pin IEEE 1149.1 JTAG TAP (test access port) interface and through a trace port that provides additional program-trace data.

10.2 DIAMOND 570T CPU CORE INTERFACES

As shown in Figure 10.3, the Diamond 570T CPU core has a 64-bit implementation of the Xtensa PIF (main processor interface) bus and separate interfaces for the 2-way set-associative instruction and data caches. The Diamond 570T CPU core also has 64-bit interfaces for local instruction and data memories, and a 64-bit XLMI port.

10.3 THE DIAMOND 570T CPU MEMORY MAP

The Diamond 570T CPU core's address space is mapped to the local memories and the PIF bus as shown in Table 10.2. Note that the Diamond 570T's instruction- and data-RAM memory spaces are adjacent so they form one contiguous 256-Kbyte memory block from 0x3FFE0000 to 0x4001FFFF. The address space for the Diamond 570T's XLMI port is mapped just below and contiguous with the address space for the data-memory block. All of the Diamond 570T CPU core's three local-memory spaces thus form one contiguous 384-Kbyte address space from 0x3FFC0000 to 0x4001FFFF.

The Diamond 570T CPU core has a 32-bit, 4-Gbyte address space, so additional memory can be attached to its PIF bus if needed. However, the address spaces dedicated to local memory are hardwired to the appropriate local-memory and XLMI ports, and memory transactions to those address spaces will not appear on the PIF bus. All Diamond processor

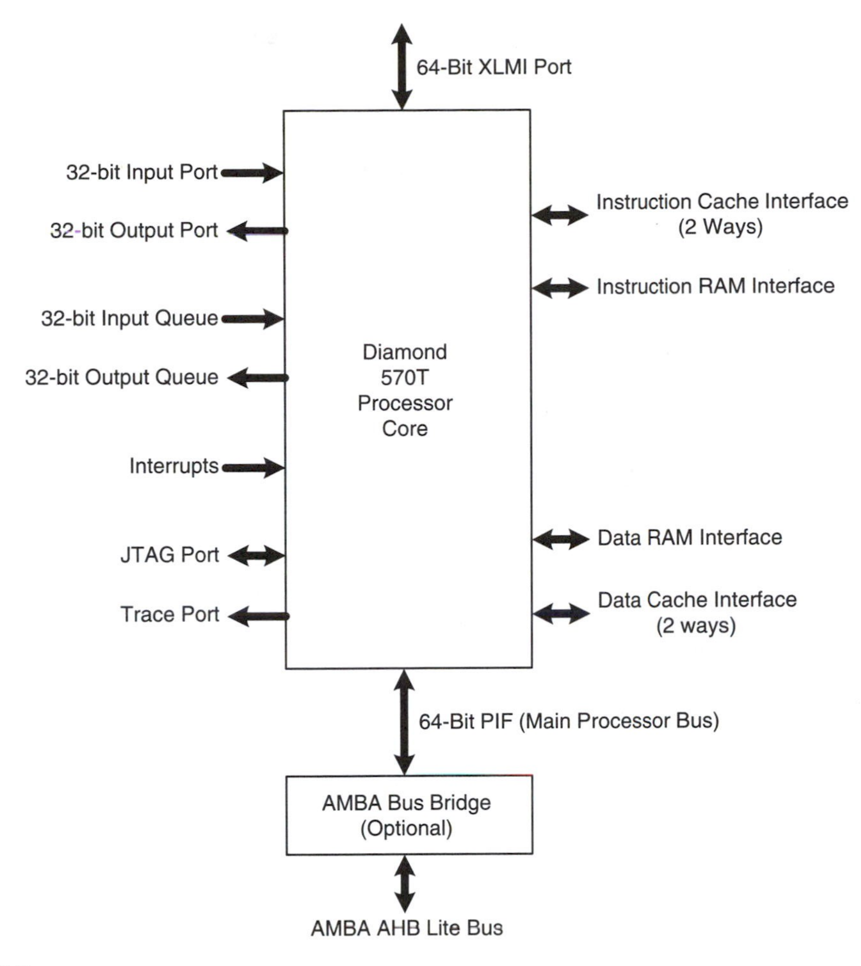

■ **FIGURE 10.3**

The Diamond 570T CPU core has a 64-bit implementation of the Xtensa PIF (main processor interface) bus and separate 64-bit interfaces for the 2-way, set-associative instruction and data caches; 64-bit interfaces for local instruction and data memories; and a 64-bit XLMI port.

TABLE 10.2 ■ Diamond 570T CPU core memory-space assignments

Local memory	Start address	End address
Local instruction RAM	0x40000000	0x4001FFFF
Local data RAM #1	0x3FFE0000	0x3FFFFFFF
XLMI port	0x3FFC0000	0x3FFDFFFF

cores including the Diamond 570T automatically send memory accesses to non-local address locations out over the PIF bus.

Table 10.3 lists the Diamond 570T CPU core's assigned reset, NMI, and other interrupt vectors. The Diamond 570T's reset and exception vectors

TABLE 10.3 ■ Diamond 570T CPU reset, interrupt, and exception vector address mapping

Vector	Address
Reset	0x50000000
Base address for register window underflow/ overflow exception vectors	0x60000000
Level 2 high-priority interrupt	0x60000180
Level 3 high-priority interrupt	0x600001C0
Level 4 high-priority interrupt	0x60000200
Level 5 high-priority interrupt	0x60000240
Debug exception	0x60000280
NMI	0x600002C0
Level 1 interrupt (Kernel mode)	0x60000300
Level 1 interrupt (User mode)	0x60000340
Double exception	0x600003C0

are assigned to locations in the PIF's address space, so some memory must be located on the main bus.

Figure 10.4 shows the various important addresses mapped into the Diamond 570T's uniform address space.

The PIF implementation on the Diamond 570T CPU core is 64 bits wide. The PIF bus uses a split-transaction protocol and the Diamond 570T processor core has an 8-entry write buffer to accommodate as many as eight simultaneous outstanding write transactions. The Diamond 570T's inbound-PIF transaction buffer is 16 entries deep.

The Diamond 570T CPU core has three internal 32-bit timers in that can be used to generate interrupts at regular intervals. The processor also has nine external, level-triggered interrupt input pins and one edge-triggered, NMI pin.

10.4 THE DIAMOND 570T CPU's CACHE INTERFACES

The Diamond 570T CPU core's data and instruction caches store data and instructions that a program is immediately using, while the rest of the data and program reside in slower main memory—RAM or ROM. In general, the Diamond 570T CPU core can access instruction and data caches simultaneously, which maximizes processor bandwidth and efficiency.

The Diamond 570T CPU incorporates a pre-configured version of the Xtensa cache controller that operates separate, 16-Kbyte, 2-way, set-associative instruction and data caches. Note that the width of the data buses to the Diamond 570T CPU's instruction and data caches is 64 bits rather than the 32-bit widths for the caches discussed in earlier chapters. The data cache employs a write-back (as opposed to write-through) write

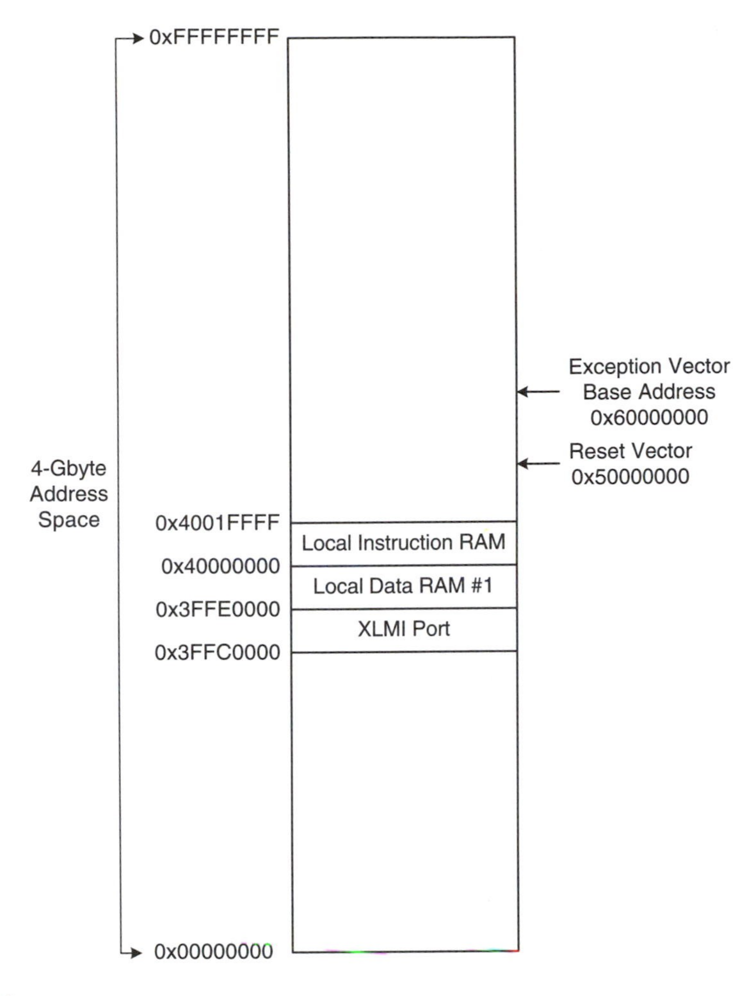

■ **FIGURE 10.4**

The Diamond 570T CPU core's pre-configured address space maps the reset and interrupt vectors to the high addresses in the processor's memory map.

policy. The Diamond 570T CPU's cache memories accelerate any program the processor executes by storing copies of executed code sequences and data in fast cache memory. The caches provide faster access to instructions and data than can memories attached to the processor's PIF.

The cache controller is woven (integrated) into the Diamond 570T CPU core so the processor has direct interface ports to the eight RAM arrays used for storing cached data and cache tags. Each cache way requires separate RAM arrays for cache data and cache tags so eight RAM arrays are needed to complete the Diamond 570T CPU core's cache memory. These RAM arrays are external to the processor's core. Figure 10.5 shows how the eight RAM arrays attach to the processor core's cache interface ports.

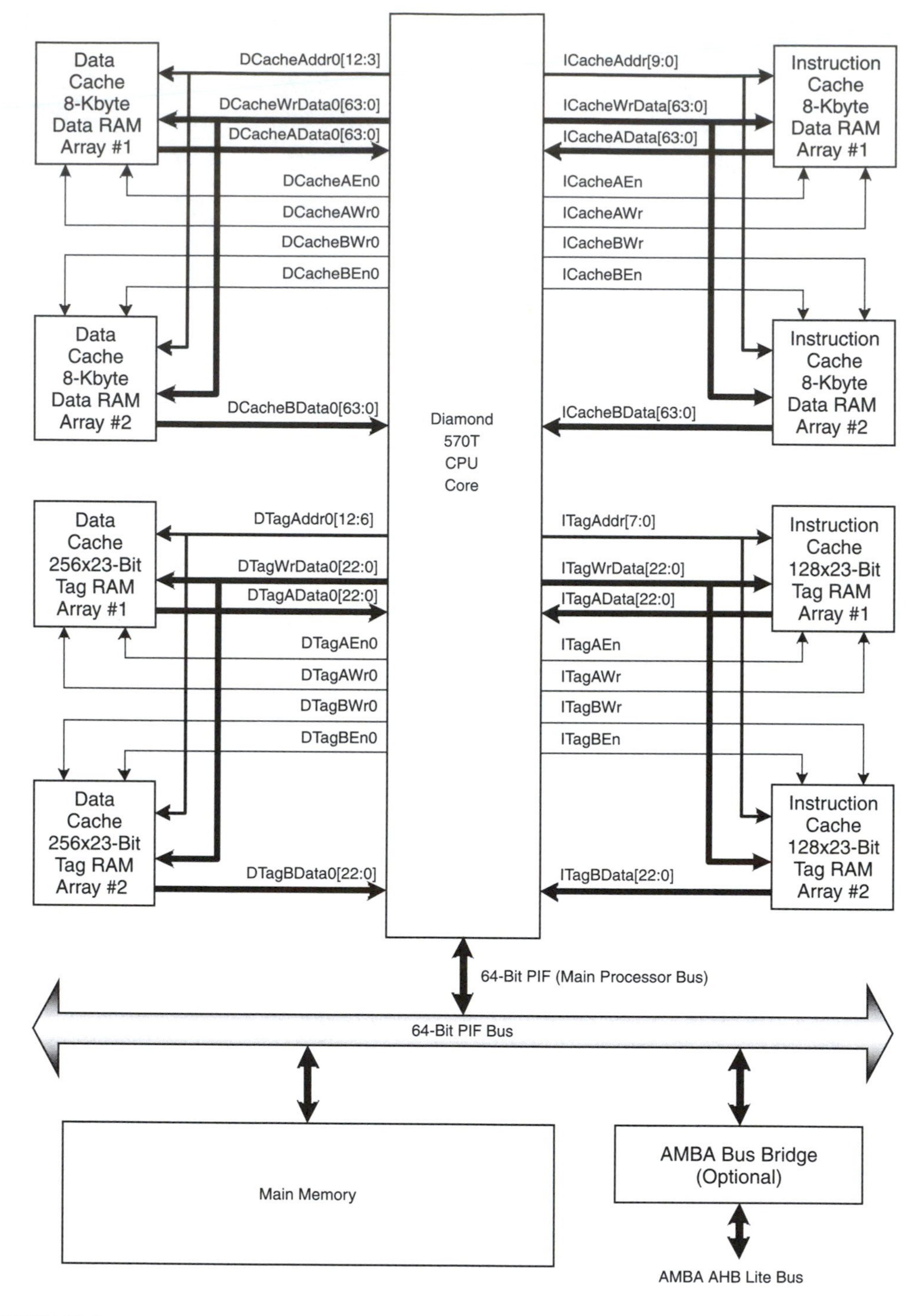

FIGURE 10.5

The Diamond 570T processor core incorporates a cache controller that runs separate, 16-Kbyte, 2-way, set-associative instruction and data caches. The processor has direct interfaces for the eight RAM arrays—used for storing cached data and cache tags.

In all respects except for the size and number of ways, the Diamond 570T CPU's caches work identically to those of the Diamond 212GP controller core, as discussed in Chapter 8.

10.5 THE DIAMOND 570T CPU's RPU

The Diamond 570T CPU core uses the same region-protection unit (RPU) discussed in Chapter 8 on the 212GP processor.

Figure 10.6 shows how the Diamond RPU divides the Diamond 570T CPU's 4-Gbyte memory space into eight equally sized, 512-Mbyte regions. The Diamond 570T CPU core's local data-memory address space and the address space assigned to the XLMI port fall into memory-protection region 1. Its local instruction-memory address space falls into memory-protection region 2. Thus the RPU can prevent accidental writes to instruction memory through the proper use of its protection mechanisms. The Diamond 570T CPU core's non-local address space (assigned to the PIF) falls into all eight memory-protection regions, so the Diamond 570T CPU core's RPU is also useful for protecting PIF-attached memory and devices.

The Diamond 570T CPU sets the memory-protection attributes for each region independently by setting 4-bit access-mode values in separate, 8-entry instruction and data TLBs. (*Note*: The three-letter abbreviation TLB stands for translation lookaside buffer but, for Diamond processor core RPUs, its definition is widened to mean "translation hardware.") Each TLB has an entry for each of the eight memory-protection regions.

The access modes control both the protection level and the cache behavior for each of the eight memory-protection regions. The access modes appear in Table 10.4 and descriptions of the modes appear in Table 10.5. (*Note*: Tables 7.3 and 7.4 from Chapter 7 are repeated in this chapter as Tables 10.4 and 10.5.)

10.6 DIRECT INPUT AND OUTPUT PORTS

Two of the Diamond 570T CPU core's unique features are its 32-bit direct input and output ports, which are the same I/O ports found on the Diamond 108Mini processor core. The input and output wires on these ports connect directly to special software-visible registers in the processor core and can be used for general-purpose signaling, interrupt generation, serial communications, pulse-width modulation, and other sorts of status and control functions on the SOC. Special instructions read and manipulate the contents of these port registers. Table 10.6 lists these port instructions.

The input and output ports can be accessed in one cycle just like simple 32-bit registers attached to one of the processor's local-memory buses, but

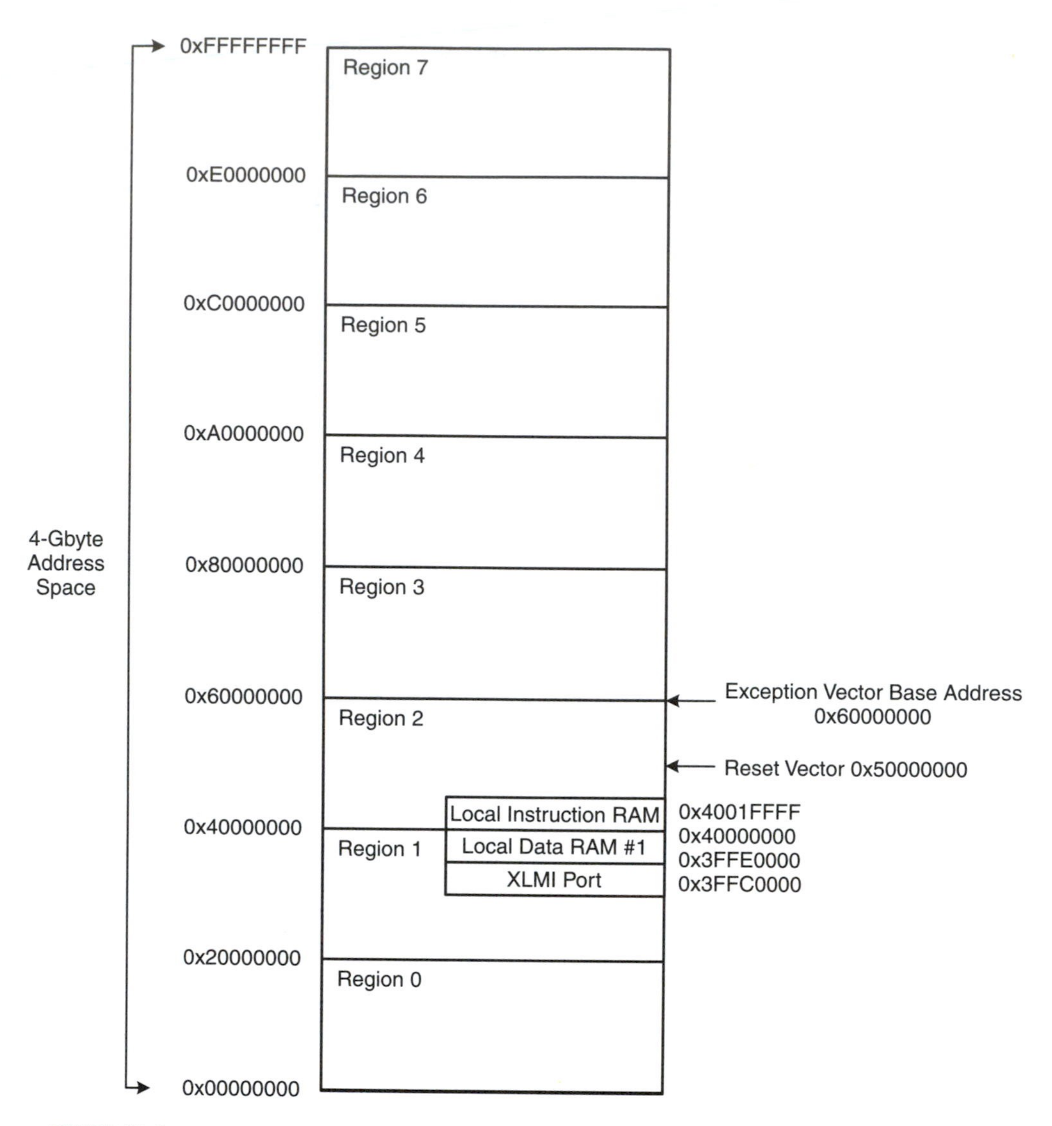

▪ **FIGURE 10.6**

The Diamond 570T CPU core's RPU divides the processor's memory space into eight protected regions. The processor's local data memory address space and XLMI port map into memory-protection region 1 and its local instruction-memory address space maps into memory-protection region 2 so the RPU can prevent accidental writes to instruction memory through the proper use of its protection mechanisms.

without the speculation-related read side effects to deal with. (*Note*: All RISC processors must contend with speculation-related side effects; it's the nature of pipeline-based RISC processor architecture.) The special input- and output-port instructions mask the processor's speculative operations so that firmware developers can completely ignore the issue when using these ports.

TABLE 10.4 ■ Diamond RPU access modes

Access-mode value	Access-mode name	Instruction-fetch behavior	Load behavior	Store behavior
0000	No allocate	Instruction-fetch exception	No allocate	Write-through/ No allocate
0001 (see Note)	Write-through/ No write allocate	Allocate	Allocate	Write-through/ No write allocate
0010	Bypass	Bypass	Bypass	Bypass
0011	Not supported	Undefined	Load exception	Store exception
0100	Write-back	Allocate	Allocate	Write-back/Write allocate
0101–1101	Reserved	Instruction-fetch exception	Load exception	Store exception
1110	Isolate	Instruction-fetch exception	Direct processor access to memory cache	Direct processor access to memory cache
1111	Illegal	Instruction-fetch exception	Load exception	Store exception

Note: RPU access-mode 1 forces Diamond core data caches to operate in write-through mode even though all Diamond processor core data caches are pre-configured as write-back caches.

TABLE 10.5 ■ Diamond RPU access-mode descriptions

RPU access mode	Access mode description
No allocate	Do not allocate a cache line for this address. If the address is already cached, fetch or load the cached value. If the address has an allocated cache line but the cache line is not already in the cache, fetch or load the value from main memory and place the value in the cache.
Bypass	Do not use the cache.
Write-back	Write the value to the cache and then update main memory when the cache line is evicted or when the processor forces the cache line to be written to main memory.
Isolate	Permits direct read/write access to the cache's data and tag RAM arrays.
Illegal	Any access causes an exception.

The 32-bit input and output ports on the Diamond 570T CPU have no handshaking lines. Devices driving the processor's input-port wires cannot determine when the processor has read the state of these wires. Similarly, there is no output signal to alert devices that the output-port wires have changed state. They simply change state. Consequently, external devices connected to the Diamond 570T processor core's direct input and output ports are somewhat uncoupled from the timing of the firmware manipulating those ports. In this sense, the Diamond 570T processor core's input

TABLE 10.6 ■ Input- and output-port register instructions

Instruction	Definition
CLRB_EXPSTATE	Clear any bit in the output-port register. Leave other bits unchanged.
SETB_EXPSTATE	Set any one bit of the 32-bit output-port register. Leave other bits unchanged.
READ_INPWIRE	Read the 32-bit value of the input-port register and place result in a general-purpose register-file entry.
RUR_EXPSTATE	Read the 32-bit value of the output-port register and place the result in a general-purpose register-file entry.
WRMASK_EXPSTATE	Write a masked bit pattern to the output-port register.
WUR_EXPSTATE	Write a 32-bit value to the output-port register.

TABLE 10.7 ■ Input- and output-queue-interface instructions

Instruction	Definition
CHECK _IPQ	Check ready status of input queue
CHECK_OPQ	Check ready status of output queue
READ_IPQ	Read value on input queue
SETB_EXPSTATE	Set any bit in the output port, parameter is immediate bit index
WRITE_OPQ	Write value to output queue

and output ports are like the direct input and output lines on a block of custom-designed RTL logic and are intended to be used the same way.

10.7 INPUT AND OUTPUT QUEUE INTERFACES

In addition to the 32-bit input and output ports discussed in the previous section, the Diamond 570T CPU core has a 32-bit input-queue interface and a 32-bit output queue interface. These queue interfaces, respectively, connect directly to the head or tail of a FIFO memory. Each queue interface has a pair of hardware handshaking wires that automatically sequence the flow of data between the attached FIFO memory and the processor. The input-queue handshake signals are called *PopReq* (Pop Request) and *Empty*. The output-queue handshake signals are called *PushReq* (Push Request) and *Full*.

Table 10.7 lists the instructions added to the Diamond 570T CPU to manage the input and output queue interfaces.

The Diamond 570T CPU's input- and output-queue interfaces have been designed to mate directly to synchronous FIFO devices. The queue interfaces are synchronous in that all data transfers between the Diamond 570T CPU core and the external FIFOs occur on the rising edge of the clock signal. There should be no combinational paths from *PopReq* to *Empty* in the input-queue interface logic or from *PushReq* to *Full* in the output-queue interface logic.

If an input queue becomes empty as a result of a pop request from the Diamond 570T CPU, the queue should assert its *Empty* signal in the next clock cycle, following the clock edge marking the data transfer. Similarly, if an output queue becomes full as a result of a push request, it should assert its *Full* signal in the next clock cycle after the clock edge, when it accepts the data from the processor.

The Diamond 570T CPU can assert its *PopReq* signal even when an input queue is empty and it can assert its *PushReq* signal even when an output queue is full. External queue designs must protect the queue FIFOs against input-queue underflows and output-queue overflows by ignoring the processor's requests when an input queue is empty or when an output queue is full.

The Diamond 570T CPU core should be the only device connected to the read port of an input FIFO or the write port of an output FIFO because the Diamond core's queue interfaces are not designed to connect to external queues that are shared between multiple clients. During normal queue operation, the status of an input queue must change from "not empty" to "empty" only in response to a pop request from the Diamond 570T CPU core. Similarly, the status of an output queue must change from "not full" to "full" only in response to a push request from the Diamond 570T CPU core. The only situation where it may be appropriate to violate this particular restriction (for both input and output queues) is during a system flush. Such a flush might occur, for example, before data processing begins on a new stream of data.

System designers must ensure that the queue interfaces are used in a manner consistent with the recommendations outlined in this section. Not doing so may result in unexpected or incorrect queue behavior.

External queues can also be connected to the Diamond 570T CPU core through its 64-bit XLMI port. To choose between the XLMI port and the queue interfaces to connect to external queues, an SOC architect should consider the following points:

- FIFO queues connected to the Diamond 570T CPU's XLMI port are accessed via memory-mapped load and store instructions. Access to the FIFO queues attached to the processor through the output- and input-queue interfaces is through the push and pop instructions, because the queue interfaces do not exist within the processor's address space. The designer should consider whether memory-mapped or queue push and pop operations are more appropriate for a specific application. The Diamond 570T CPU can transfer data to or from only one XLMI-attached queue during any given cycle (because there's only one XLMI port), but it can transfer data to and from both the input and output queues during each clock cycle using slots 0 and 2 of the processor's second 64-bit instruction-word format.

- If a FIFO queue is attached to the XLMI port, it must share that port with any other XLMI-attached devices, which may result in bandwidth issues. The Diamond 570T CPU's queue interfaces are not shared, and therefore an external FIFO attached to a queue interface receives the full bandwidth of that interface.
- The Diamond 570T's XLMI port is 64 bits wide and can therefore conduct 64-bit I/O transactions. The Diamond 570T CPU's input- and output-queue interfaces are 32 bits wide, which might limit bandwidth in some applications.
- If the Diamond 570T CPU tries to read from an input queue when it's empty, an XLMI-attached FIFO queue will stall the I/O operation. The stalled I/O read transaction will stall the processor's pipeline as well and this stalled state is not interruptible. The queue interfaces can also stall the processor's pipeline but interrupts can still occur and will be serviced when a queue transaction is stalled.
- If the processor executes a store to the XLMI port immediately followed by a load from an XLMI-attached FIFO queue, the store is buffered and the load occurs first. If this load causes an I/O stall because the addressed FIFO queue is empty, a resource deadlock results that will freeze the processor's pipeline. The XLMI load cannot complete because the FIFO queue is empty and the store to the FIFO queue cannot take place because the pipeline is stalled. No such resource conflicts exist with the queue interfaces.
- All XLMI-attached devices including FIFO queues must handle speculative reads over the XLMI port. To do this, XLMI-attached devices must observe and use the processor control signals that indicate whether a read has committed or flushed. The queue interfaces do not conduct speculative operations.

10.8 SYSTEM DESIGN WITH THE DIAMOND 570T CPU CORE

Figure 10.7 shows a simple one-dimensional systolic-processing system built with three Diamond 570T CPU cores (each with its own local memory), a large global memory, and two FIFO queues linking the three processor cores. The three Diamond 570T CPU cores communicate with each other through the global memory over the shared 64-bit PIF bus and they can pass information from CPU core #1 to CPU core #2 and then to CPU core #3 through the attached FIFOs. A bus arbiter and locking device control access to the PIF bus.

This system allows the three Diamond 570T CPU cores to split a tough processing problem. CPU #1 executes the first third of a complex algorithm and passes intermediate data to CPU #2 over the attached FIFO. CPU #2

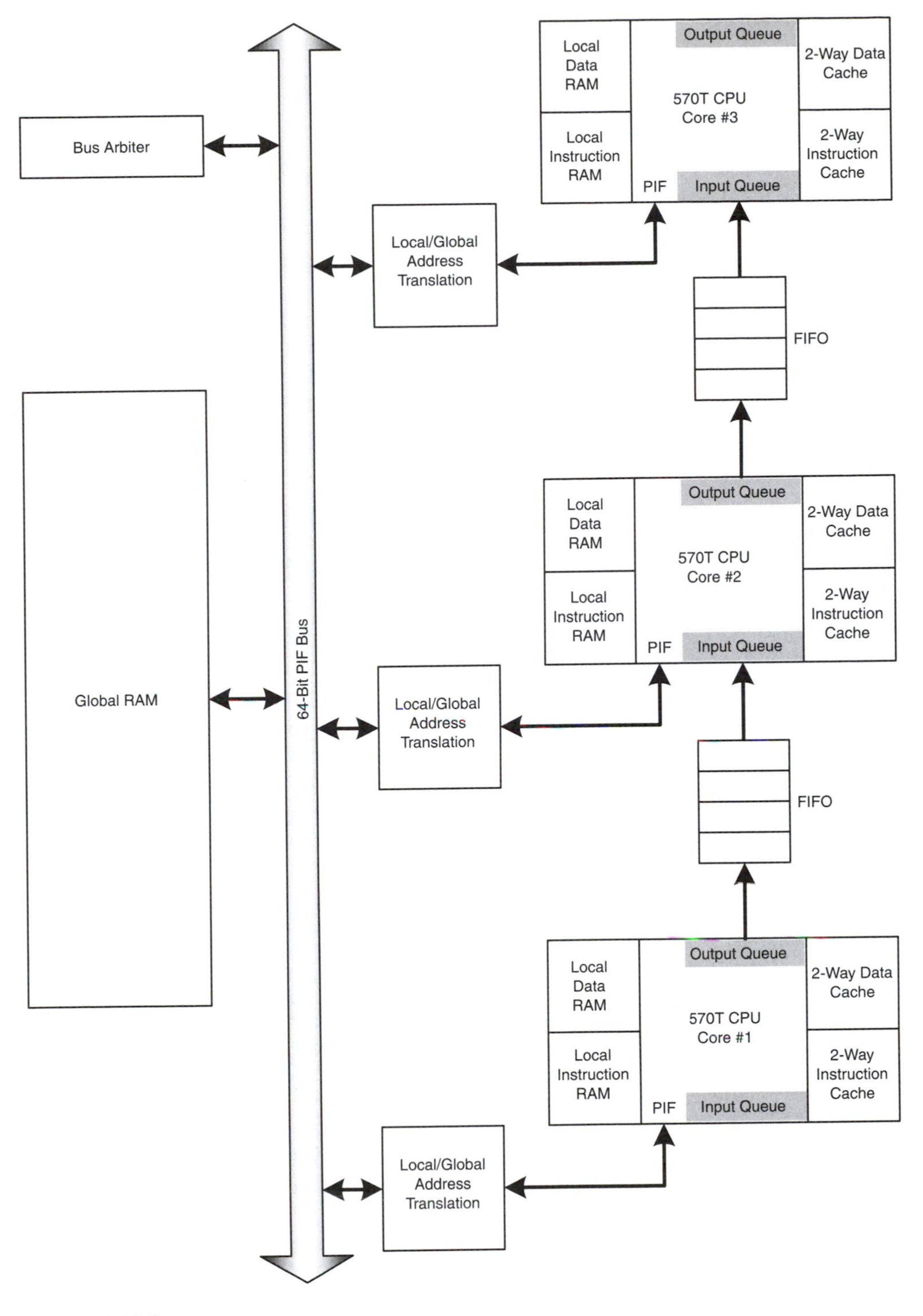

FIGURE 10.7

This systolic-processing system uses three Diamond 570T CPU cores linked with FIFOs to create a high-speed, flow-through processing system.

continues the processing and hands intermediate results to CPU #3, which finishes the algorithm. Due to the 3-slot, 3-operations/clock design of the Diamond 570T CPU's 64-bit instruction format, the processor can accept a 32-bit data word on its input-queue interface during the same clock cycle that it's outputting a processed 32-bit data word on its output-queue interface. Such transactions can occur on each clock cycle. This architecture produces high-performance flow-through systems with very high I/O bandwidth.

Complex, multi-step algorithms including various video-coding, image-processing, and network packet-processing applications with high data-rate requirements can employ systolic processing very effectively. Such algorithms exploit a problem's inherent regularity and parallelism to achieve high performance and low I/O requirements by distributing the work over several independent processors.

Systolic-processing arrays generally consist of one type—or at most a few types—of simple processors (scalar, not array processors, for example) connected in a regular pattern. The concept was first proposed in 1978 by H.T. Kung and C.E. Leiserson who were at Carnegie Mellon University at the time. Since then, systolic algorithms have been proposed as solutions to many computationally demanding problems including digital-signal and image processing, linear algebra, pattern recognition, linear and dynamic programming, and graph problems. Nanometer silicon and 21st-century semiconductor lithography is finally able to produce systolic-processing systems economically.

The advantages of a systolic-processing array diminish if the processors used to implement the array can only be used to solve a narrow set of problems. For example, the design cost of such an array cannot be amortized over a large number of units. However, a systolic-processing array on an SOC built from multiple identical copies of a high-performance, general-purpose processor such as the superscalar Diamond 570T can be programmed to implement many different systolic algorithms, thus widening the application reach of the systolic-array design.

The benefit of this approach to system design is that clock rates remain low while data-throughput capabilities are high because the application task is distributed across multiple independent processors. Implemented in a 130 nm LV process technology, the three Diamond 570T CPUs shown in Figure 10.7 consume less than 3 mm^2 of silicon. The multiple-processor hardware system shown in Figure 10.7 can easily be scaled to bigger processing loads using a relatively simple step-and-repeat design process.

BIBLIOGRAPHY

Diamond Standard Processors Data Book, Tensilica, Inc., February 2006.

Fisher, A.L., Kung, H.T., Monier, L. and Dohi, Y., "Architecture of the PSC: A Programmable Systolic Chip," *Proceedings of the 10th International Symposium on Computer Architecture* (ISCA 1983), pp. 48–53.

Johnson, M., *Superscalar Microprocessor Design*, Prentice-Hall, 1991.

Koren, I., Mendelson, B., Peled, I. and Silberman, G.M., "A data-driven VLSI array for arbitrary algorithms," *IEEE Computer*, 21(10), October 1988, pp. 30–43.

Leibson, S. and Kim, J., "Configurable Processors: A New Era in Chip Design," *IEEE Computer*, July 2005, pp. 51–59.

Rowen, C. and Leibson, S., *Engineering the Complex SOC*, Prentice-Hall, 2004.

Xtensa Microprocessor Programmer's Guide, Tensilica, Inc., July 2005.

CHAPTER 11

THE DIAMOND 330HIFI AUDIO DSP CORE

Elwood : What kind of music do you usually have here?
Claire : Oh, we got both kinds. We got country and western.
—The Blues Brothers, 1980

The Diamond 330HiFi audio DSP core is a high-performance, 2-way superscalar RISC CPU core that has been optimized for recording, encoding, decoding, and playing music and other audio files. Like the Diamond 570T CPU core described in Chapter 10, the term superscalar as applied to the Diamond 330HiFi core describes a processor that achieves superlative performance by executing multiple scalar instructions per clock cycle. The Diamond 330HiFi DSP core's two superscalar execution pipelines and 64-bit FLIX-format instructions allow the processor to break the 1-instruction-per-clock barrier. The processor's architecture has been sculpted specifically to optimally run audio codecs (coders/decoders) in the fewest possible clock cycles to achieve the lowest possible operating power. Several codecs for widely used digital-audio formats are available from Tensilica.

11.1 300 INSTRUCTIONS BOOST AUDIO PERFORMANCE

Tensilica added more than 300 new instructions and two new register files to the base Xtensa ISA to create the Diamond 330HiFi audio DSP. The new instructions work directly with 24-bit audio data types and allow the processor to more efficiently process audio data streams, which in turn allows the Diamond 330HiFi DSP to perform the same work at a lower clock rate and to therefore dissipate less power while executing the audio codecs. The two new register files are respectively an 8-entry file named "P" with 48 bits/entry (each entry can hold two 24-bit values) and a 4-entry file named "Q" with 56-bit entries. The 56-bit values are generated from a set of instructions that control a dual multiplier/accumulator (MAC). Each of the two pipelined multipliers can perform a 24×24-bit or a 32×16-bit multiplication with a throughput of one multiplication per multiplier per cycle. The results obtained from the two multipliers are accumulated in the Q registers. The 300 new audio-specific

TABLE 11.1 ■ Specialized operation groups for the Diamond 330HiFi audio DSP

Diamond 330HiFi audio operation groups	
1	Dual multiply with 56-bit accumulate
2	Add/subtract and variable/immediate shifts
3	Huffman encode/decode and bit-stream support
4	Convert/round/truncate
5	2-way SIMD Arithmetic and Boolean operations

operations added to the basic Xtensa ISA to create the Diamond 330HiFi audio DSP core are grouped as shown in Table 11.1.

- Group 1 includes the operations that drive the DSP core's dual MAC. Codecs use these instructions to perform audio-stream transforms between the time and frequency domains; for windowing and frequency-band splitting; for sample-rate conversion and windowing; and for special effects such as reverb and three-dimensional sound simulation.
- Group 2 includes shift operations—used for normalization that maximizes the dynamic range of the Diamond 330HiFi audio DSP's fixed-point calculations—and arithmetic functions that operate on the P and Q register files and the dual MAC's 56-bit accumulator.
- Group 3 includes Huffman encoding and decoding operations—used for noiseless coding and decoding—and bit-stream support operations that pick unencoded bits from the audio bit stream for use in various parts of the audio-codec algorithms.
- Group 4 includes conversion, rounding, and truncation functions for the P and Q register files. These operations also permit data exchange between the P and Q register files and the processor's base 32-bit AR register file.
- Group 5 operations perform 2-way single-instruction, multiple-data (SIMD) arithmetic and Boolean functions on the paired 24-bit data words stored in the 48-bit P register file.

Adding 300 instructions to a processor for any application (such as audio in this case) makes the compiled target application code very efficient, but it is an un-RISC-like concept and it would be prohibitively expensive—essentially impractical—to create such an application-specific processor without an automated tool like the Xtensa Processor Generator.

A detailed examination of all 300 operations in the five operation groups requires a book unto itself but a closer look at just the MAC operation group illustrates the flexibility of the Diamond 330HiFi audio DSP's

FLIX-format instructions. An instruction in this group can include the following functions:

- Single or dual multiplication.
- Fractional or integer arithmetic.
- 24×24-bit, 16×16-bit, or 32×16-bit operands.
- Overwrite, add, or subtract accumulation with or without saturation.
- Signed or unsigned arithmetic.

Complex instructions can be built from these MAC primitives by bundling two operations into a wide instruction word, as shown in Figure 11.1. The Diamond 330HiFi audio DSP core's 64-bit instruction format bundles two independent operations. The XCC compiler paired with the Diamond 330HiFi audio DSP core automatically performs the needed instruction bundling to use this 2-operation instruction word.

All three instruction formats can be freely intermixed and tightly packed in memory. The Xtensa architecture's inherent ability to handle differently sized instructions was a key enabler in the development of the Diamond 330HiFi audio DSP core's design.

A block diagram of the Diamond 330HiFi DSP core's two execution pipelines appears in Figure 11.2. Even with the addition of an audio execution pipeline, the Diamond 330HiFi audio DSP core runs at 233 MHZ. Even with three ALUs and two audio multipliers, the Diamond 330HiFi audio DSP consumes only 1.86 mm^2 of silicon, and dissipates approximately 255 μW/MHz (running the MP3 decoder) when implemented in a 130 nm, G-type (general-purpose) process technology. Although the Diamond 330HiFi audio DSP core can run at 233 MHz, the intent of adding the audio instructions is to allow the processor to run at a few tens of MHz and draw very little power while playing digital audio.

The Diamond 330HiFi audio DSP core brings the performance benefits of a full-featured 32-bit RISC CPU to bear on the designated tasks:

- Large 4-Gbyte address space
- 32-bit computations
- Large 32-entry register file
- 2-way, set-associative, 4-Kbyte instruction cache
- 2-way, set-associative, 8-Kbyte data cache
- 64-bit interface ports for one local instruction RAM and two local data RAMs
- 5-stage pipelined operation resulting in a 233-MHz maximum clock rate in 130 nm G technology.

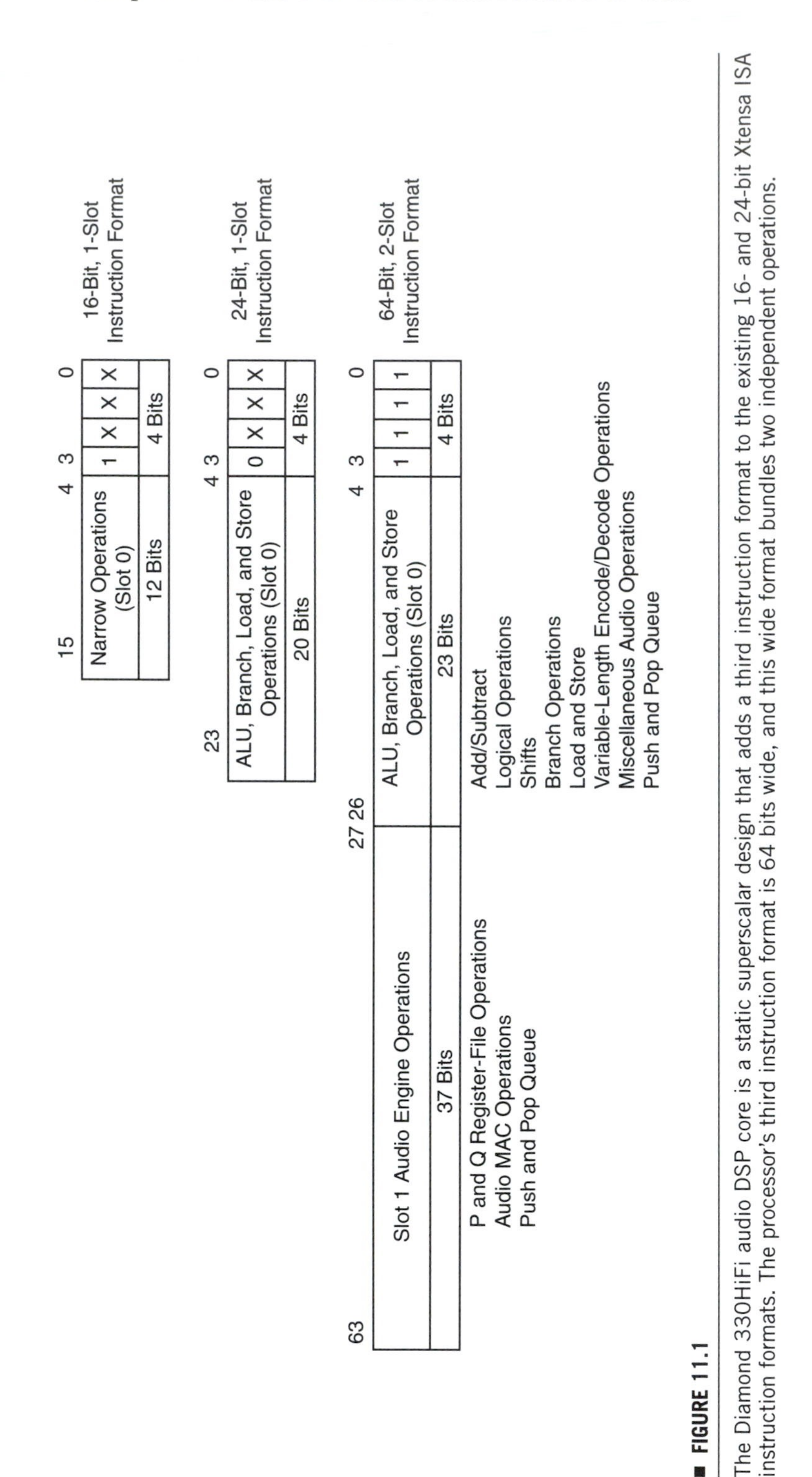

■ **FIGURE 11.1**

The Diamond 330HiFi audio DSP core is a static superscalar design that adds a third instruction format to the existing 16- and 24-bit Xtensa ISA instruction formats. The processor's third instruction format is 64 bits wide, and this wide format bundles two independent operations.

The Diamond 330HiFi audio DSP core has a 64-bit version of the general-purpose PIF bus for global SOC communications. An optional AMBA bus bridge supplied with the Diamond 330HiFi audio DSP core adapts the PIF to peripheral devices designed for the AMBA AHB-Lite bus.

11.2 THE DIAMOND 330HiFi: A HIGH-PERFORMANCE AUDIO DSP CORE

The Diamond 330HiFi audio DSP architecture contains all of the basic elements of the Xtensa ISA. It has a 32-entry general-purpose register file, a 5-stage execution pipeline, and 32-bit addressing. This execution pipeline has been augmented with several audio instructions to reduce the number of cycles needed to execute audio-codec firmware. In addition, the Diamond 330HiFi audio DSP has a second set of register files and execution pipeline giving the processor core the ability to execute two independent instructions per clock. The Diamond 330HiFi audio DSP's two execution pipelines are not symmetric, as shown in Figure 11.2. The first execution pipeline, associated with operation slot 0, contains the base Xtensa ALU and some additional function units for audio instructions. The second pipeline contains two audio-specific ALUs and two audio-specific multipliers so that it can execute additional audio-oriented instructions. A full block diagram of the Diamond 330HiFi audio DSP appears in Figure 11.3.

The Diamond 330HiFi audio DSP core also incorporates the Diamond Series processor core software-debug stack, as shown on the left of Figure 11.3. This debug stack provides external access to the processor's internal, software-visible state through a 5-pin IEEE 1149.1 JTAG TAP (test access port) interface and through a trace port that provides additional program-trace data.

11.3 DIAMOND 330HiFi AUDIO DSP CORE INTERFACES

As shown in Figure 11.4, the Diamond 330HiFi audio DSP core has a 64-bit implementation of the Xtensa PIF (main processor interface) bus and separate interfaces for the one local instruction RAM, two local data RAMs, and the 2-way set-associative instruction and data caches.

11.4 THE DIAMOND 330HiFi AUDIO DSP CORE'S MEMORY MAP

The Diamond 330HiFi audio DSP core's address space is mapped to the local memories and the PIF bus as shown in Table 11.2. Note that the

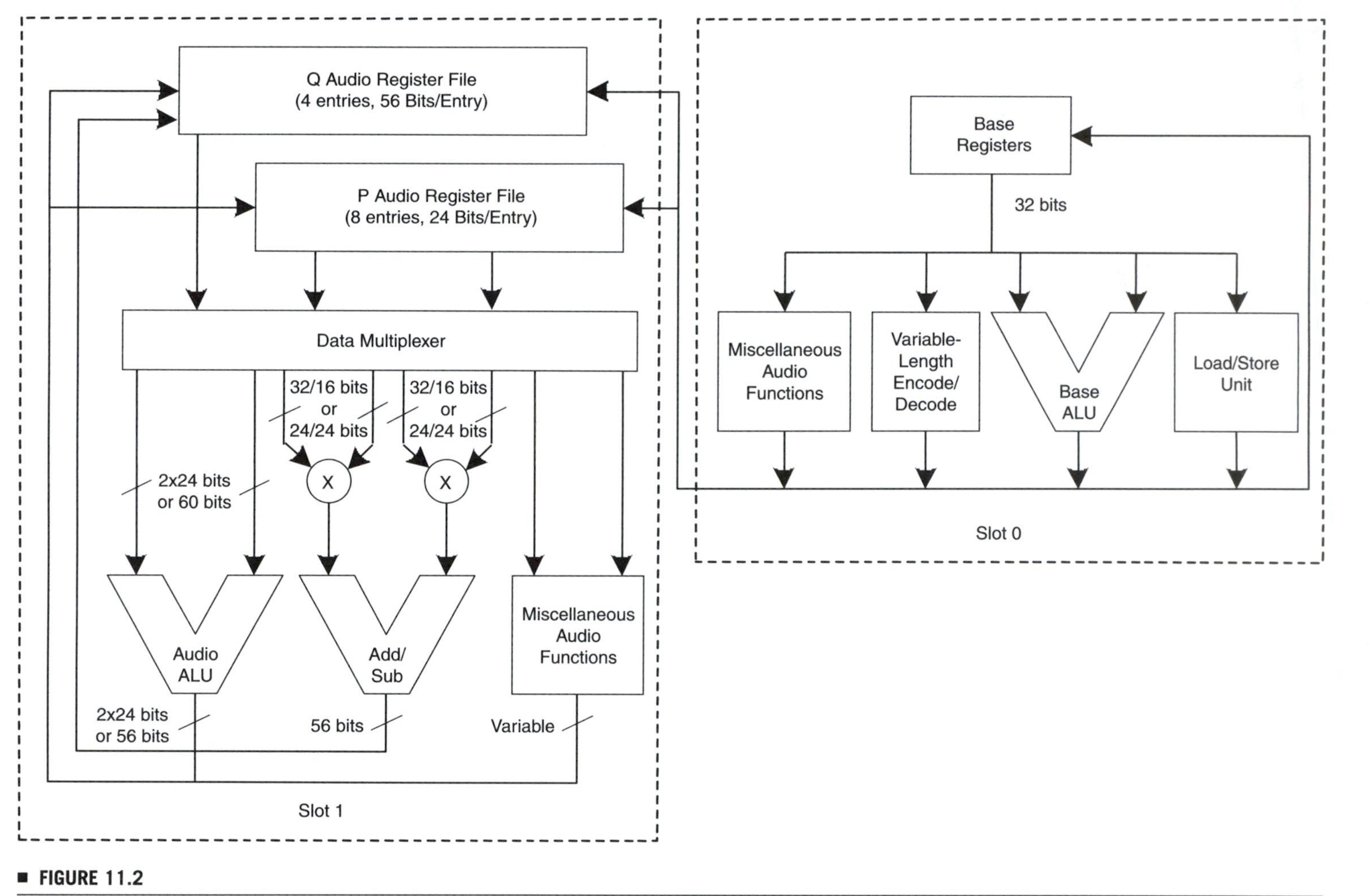

■ **FIGURE 11.2**

The Diamond 330HiFi audio DSP core has two execution pipelines for efficient audio-codec execution.

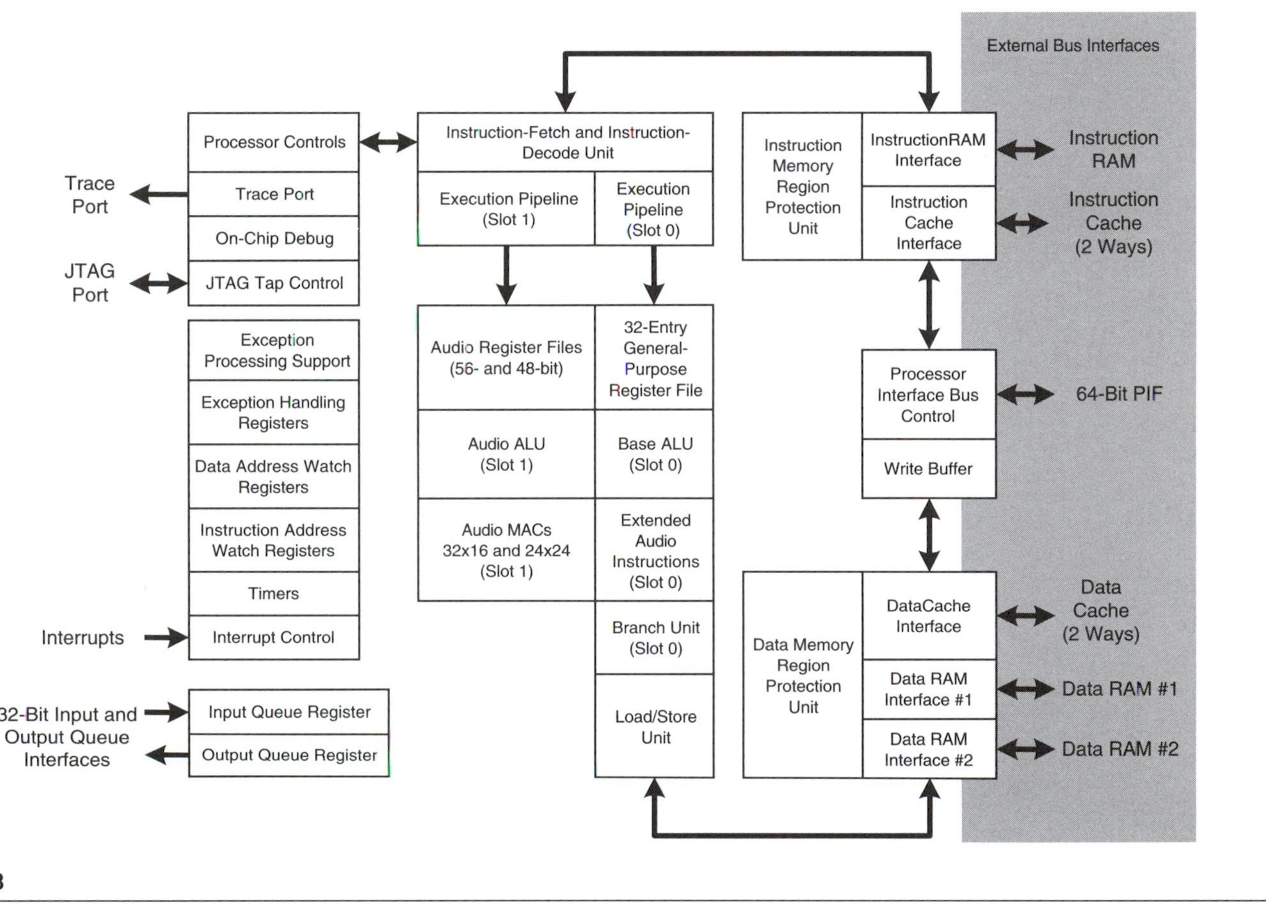

■ **FIGURE 11.3**

The Diamond 330HiFi audio DSP core implements a high-performance, 2-way superscalar, 32-bit RISC CPU with two audio MACs while consuming only 1.86 mm^2 of silicon and approximately 255 μW/MHz when implemented in a 130 nm, G-type process technology.

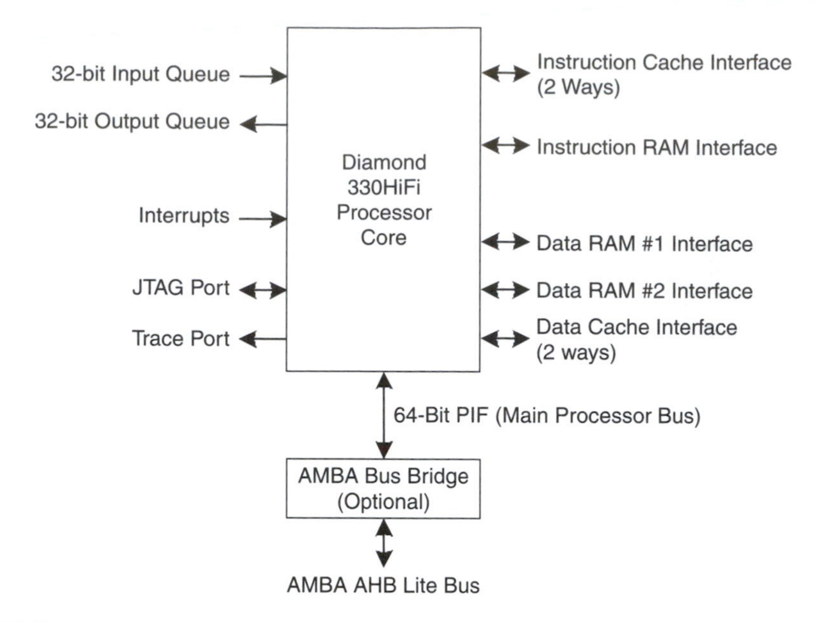

■ **FIGURE 11.4**

The Diamond 330HiFi audio DSP core has a 64-bit implementation of the Xtensa PIF (main processor interface) bus and separate interfaces for the one local instruction RAM, two local data RAMs, and the 2-way set-associative instruction and data caches.

TABLE 11.2 ■ Diamond 330HiFi audio DSP core memory-space assignments

Local memory	Start address	End address
Local instruction RAM	0x40000000	0x4001FFFF
Local data RAM #1	0x3FFE0000	0x3FFFFFFF
Local data RAM #2	0x3FFC0000	0x3FFDFFFF

Diamond 330HiFi's instruction RAM and data RAM memory spaces are adjacent so they form one contiguous 384-Kbyte memory block from 0x3FFC0000 to 0x4001FFFF.

The Diamond 330HiFi audio DSP core has a 32-bit, 4-Gbyte address space, so additional memory can be attached to its PIF bus if needed. However, the address spaces dedicated to local memory are hardwired to the appropriate local memory and XLMI ports and memory transactions to those address spaces will not appear on the PIF bus. All Diamond processor cores including the Diamond 330HiFi automatically send memory accesses to non-local address locations out over the PIF bus.

Table 11.3 lists the Diamond 330HiFi audio DSP core's assigned reset, NMI, and other interrupt vectors. The Diamond 330HiFi's reset and exception vectors are assigned to locations in the PIF's address space, so some memory must be located on the PIF bus.

TABLE 11.3 ■ Diamond 330HiFi DSP core reset, interrupt, and exception vector address mapping

Vector	Address
Reset	0x50000000
Base address for register window underflow/overflow exception vectors	0x60000000
Level 2 high-priority interrupt	0x60000180
Level 3 high-priority interrupt	0x600001C0
Level 4 high-priority interrupt	0x60000200
Level 5 high-priority interrupt	0x60000240
Debug exception	0x60000280
NMI	0x600002C0
Level 1 interrupt (Kernel mode)	0x60000300
Level 1 interrupt (User mode)	0x60000340
Double exception	0x600003C0

Figure 11.5 shows the various important addresses mapped into the Diamond 330HiFi audio DSP core's uniform address space.

The PIF implementation on the Diamond 330HiFi audio DSP core is 64 bits wide. The PIF bus uses a split-transaction protocol and the Diamond 330HiFi audio DSP core has an 8-entry write buffer to accommodate as many as eight simultaneous outstanding write transactions. The Diamond 330HiFi audio DSP core's inbound-PIF transaction buffer is 16 entries deep.

The Diamond 330HiFi audio DSP core has three internal 32-bit timers in that can be used to generate interrupts at regular intervals. The processor also has nine external, level-triggered interrupt input pins and one edge-triggered, NMI pin.

11.5 THE DIAMOND 330HiFi AUDIO DSP CORE'S CACHE INTERFACES

The Diamond 330HiFi audio DSP core's data and instruction caches store data and instructions that a program is immediately using, while the rest of the data and program reside in slower main memory—RAM, or ROM. In general, the Diamond 330HiFi audio DSP core can access instruction and data caches simultaneously, which maximizes processor bandwidth and efficiency.

The Diamond 330HiFi audio DSP incorporates a pre-configured version of the Xtensa cache controller that operates separate but asymmetric, 2-way, set-associative instruction and data caches. The instruction cache is a 2-way, set-associative, 4-Kbyte cache and the data cache is a 2-way, set-associative, 8-Kbyte cache. Note that the width of the data buses

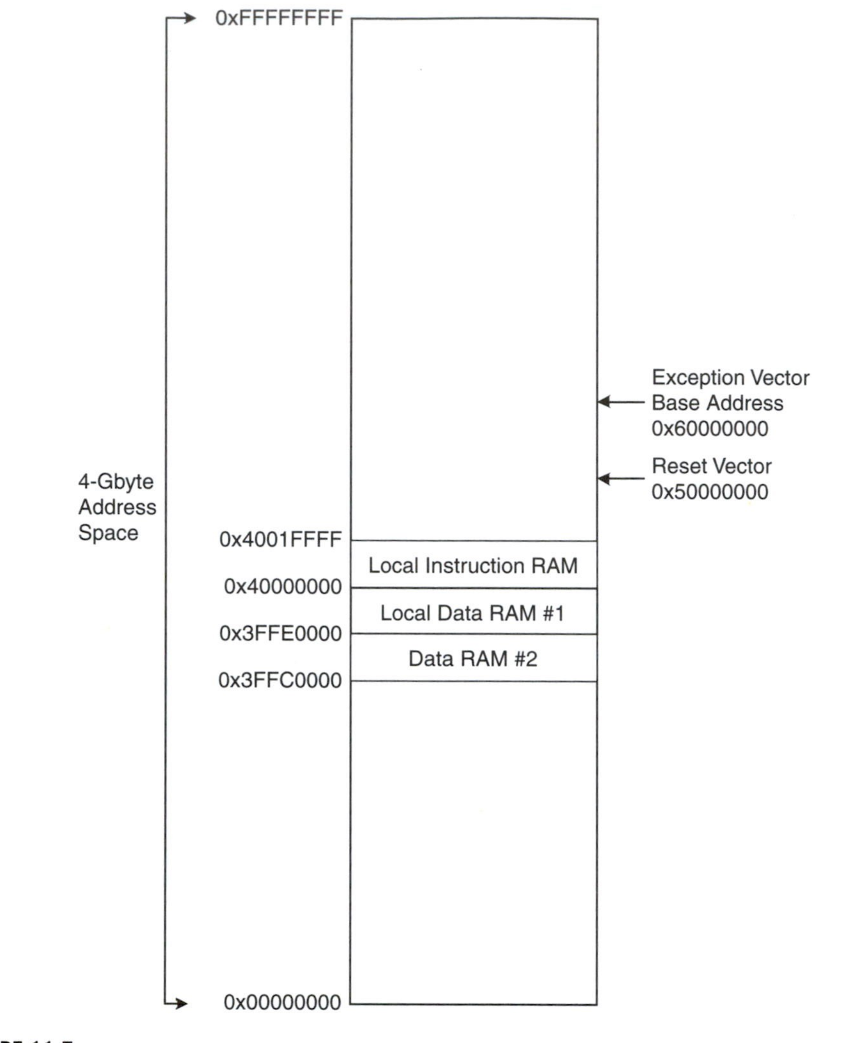

■ **FIGURE 11.5**

The Diamond 330HiFi audio DSP core's pre-configured address space maps the reset and interrupt vectors to memory located in the PIF (non-local) memory-address space.

to the Diamond 330HiFi audio DSP core's instruction and data caches is 64 bits. The data cache employs a write-back (as opposed to write-through) write policy. The Diamond 330HiFi audio DSP core's cache memories accelerate any program the processor executes by storing copies of executed code sequences and data in fast cache memory. The caches provide faster access to instructions and data than can memories attached to the processor's PIF.

The cache controller is woven (integrated) into the Diamond 330HiFi audio DSP core so the processor has direct interface ports to the eight

RAM arrays used for storing cached data and cache tags. Each cache way requires separate RAM arrays for cache data and cache tags so eight RAM arrays are needed to complete the Diamond 330HiFi audio DSP core's cache memory. These RAM arrays are external to the processor's core. Figure 11.6 shows how the eight RAM arrays attach to the processor core's cache interface ports. In all respects except for the size and number of ways, the Diamond 330HiFi audio DSP core's caches work identically to those of the Diamond 212GP controller core, as discussed in Chapter 8.

11.6 THE DIAMOND 330HiFi AUDIO DSP CORE'S (REGION-PROTECTION UNIT)

The Diamond 330HiFi audio DSP core uses the same region-protection unit (RPU) discussed in Chapter 8 on the Diamond 212GP processor.

Figure 11.7 shows how the Diamond RPU divides the Diamond 330HiFi audio DSP core's 4-Gbyte memory space into eight equally sized, 512-Mbyte regions. The Diamond 330HiFi audio DSP core's local data-memory address space and the address space assigned to the XLMI port fall into memory-protection region 1. Its local instruction-memory address space falls into memory-protection region 2. Thus the RPU can prevent accidental writes to instruction memory through the proper use of its protection mechanisms. The Diamond 330HiFi audio DSP core's non-local address space (assigned to the PIF) falls into all eight memory-protection regions, so the RPU is also useful for protecting PIF-attached memory and devices.

The Diamond 330HiFi audio DSP core sets the memory-protection attributes for each region independently by setting 4-bit access-mode values in separate, 8-entry instruction and data TLBs. (*Note*: the three-letter abbreviation TLB stands for translation lookaside buffer but, for Diamond processor core RPUs, its definition is widened to mean "translation hardware.") Each TLB has an entry for each of the eight memory-protection regions.

The access modes control both the protection level and the cache behavior for each of the eight memory-protection regions. The access modes appear in Table 11.4 and descriptions of the modes appear in Table 11.5. (*Note*: Tables 7.3 and 7.4 from Chapter 7 are repeated in this chapter as Tables 11.4 and 11.5.)

11.7 INPUT- AND OUTPUT-QUEUE INTERFACES

The Diamond 330HiFi audio DSP core has a 32-bit input-queue interface and a 32-bit output-queue interface. These queue interfaces respectively

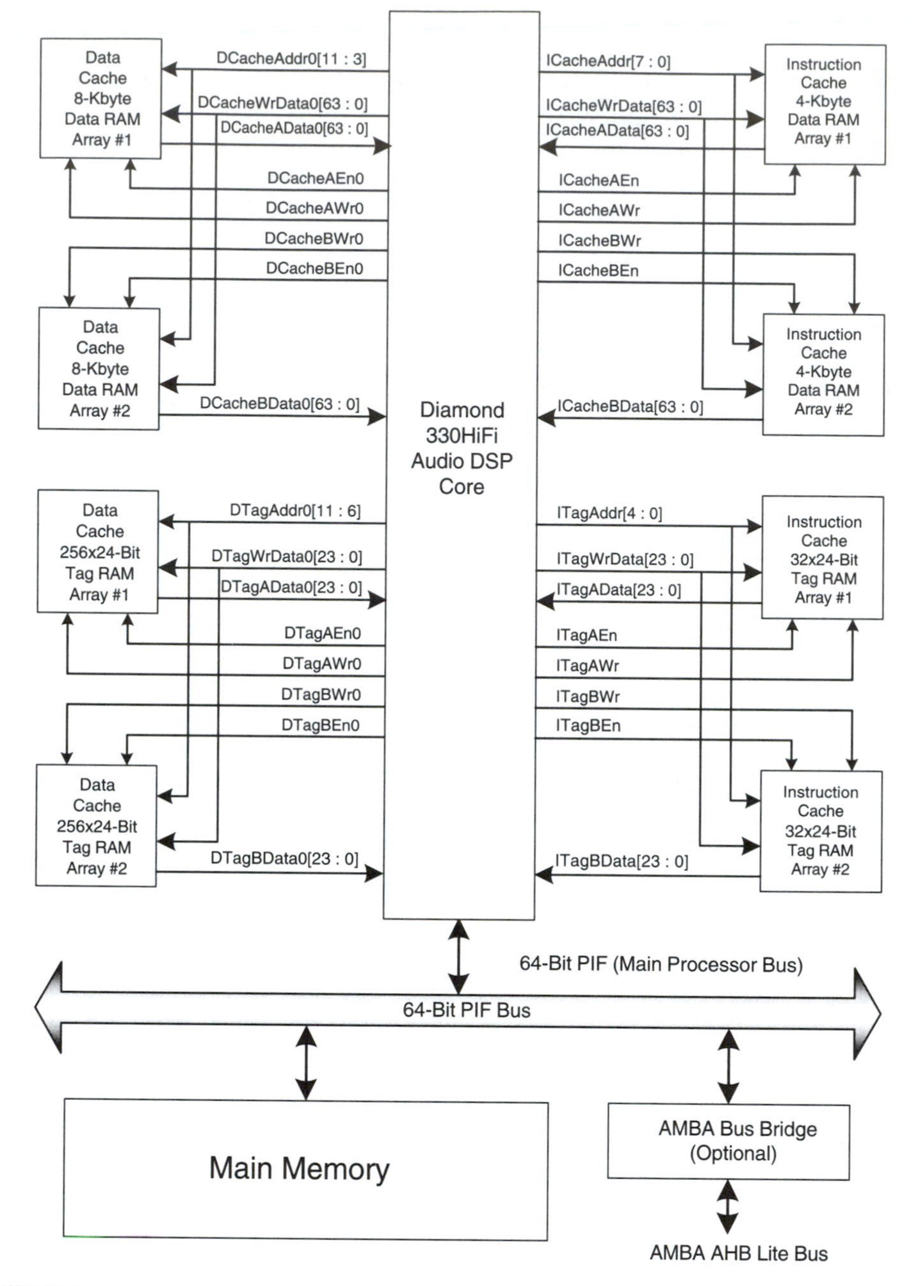

■ **FIGURE 11.6**

The Diamond 330HiFi audio DSP core incorporates a cache controller that runs separate, 2-way, set-associative instruction and data caches. The instruction cache is a 2-way, set-associative, 4-Kbyte cache and the data cache is a 2-way, set-associative, 8-Kbyte cache.

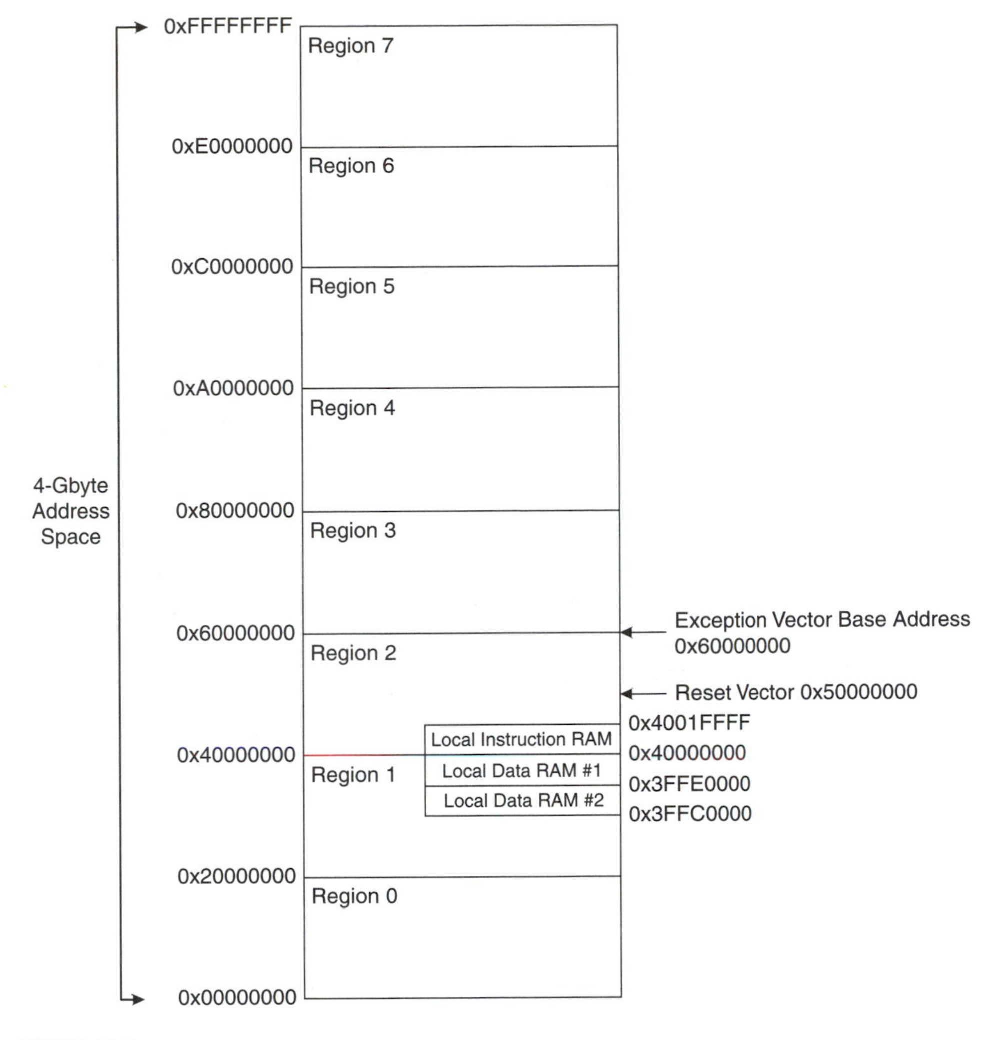

■ **FIGURE 11.7**

The Diamond 330HiFi audio DSP core's RPU divides the processor's memory space into eight protected regions. The processor's two local data-memory address blocks map into memory-protection region 1 and its local instruction-memory address space maps into memory-protection region 2 so the RPU can prevent accidental writes to instruction memory through the proper use of its protection mechanisms.

connect directly to the head or tail of a FIFO memory. Each queue interface has a pair of hardware handshaking wires that automatically sequence the flow of data between the attached FIFO memory and the processor. The input-queue handshake signals are called *PopReq* (Pop Request) and *Empty*. The output-queue handshake signals are called *PushReq* (Push Request) and *Full*.

TABLE 11.4 ■ Diamond RPU access modes

Access-mode value	Access-mode name	Instruction-fetch behavior	Load behavior	Store behavior
0000	No allocate	Instruction-fetch exception	No allocate	Write-through/No allocate
0001 (see Note)	Write-through/No write allocate	Allocate	Allocate	Write-through/ No write allocate
0010	Bypass	Bypass	Bypass	Bypass
0011	Not supported	Undefined	Load exception	Store exception
0100	Write-back	Allocate	Allocate	Write-back/Write allocate
0101–1101	Reserved	Instruction-fetch exception	Load exception	Store exception
1110	Isolate	Instruction-fetch exception	Direct processor access to memorycache	Direct processor access to memory cache
1111	Illegal	Instruction-fetch exception	Load exception	Store exception

Note: RPU access-mode 1 forces Diamond core data caches to operate in write-through mode even though all Diamond processor core data caches are pre-configured as write-back caches.

TABLE 11.5 ■ Diamond RPU access-mode descriptions

RPU access mode	Access-mode description
No allocate	Do not allocate a cache line for this address. If the address is already cached, fetch or load the cached value. If the address has an allocated cache line but the cache line is not already in the cache, fetch or load the value from main memory and place the value in the cache.
Bypass	Do not use the cache.
Write-back	Write the value to the cache and then update main memory when the cache line is evicted or when the processor forces the cache line to be written to main memory.
Isolate	Permits direct read/write access to the cache's data and tag RAM arrays.
Illegal	Any access causes an exception.

Table 11.6 lists the instructions added to the Diamond 330HiFi audio DSP core to manage the input- and output-queue interfaces.

The Diamond 330HiFi audio DSP core's input- and output-queue interfaces have been designed to mate directly to synchronous FIFO devices. The queue interfaces are synchronous in that all data transfers between the Diamond 330HiFi audio DSP core and the external FIFOs occur on the rising edge of the clock signal. There should be no combinational paths from *PopReq* to *Empty* in the input-queue interface logic or from *PushReq* to *Full* in the output-queue interface logic.

TABLE 11.6 ■ Input- and output-queue interface instructions

Instruction	Definition
CHECK _IPQ	Check ready status of input queue
CHECK_OPQ	Check ready status of output queue
READ_IPQ	Read value on input queue
SETB_EXPSTATE	Set any bit in the output port, parameter is immediate bit index
WRITE_OPQ	Write value to output queue

If an input queue becomes empty as a result of a pop request from the Diamond 330HiFi audio DSP core, the queue should assert its *Empty* signal in the next clock cycle, following the clock edge marking the data transfer. Similarly, if an output queue becomes full as a result of a push request, it should assert its *Full* signal in the next clock cycle after the clock edge, when it accepts the data from the processor.

The Diamond 330HiFi audio DSP core can assert its *PopReq* signal even when an input queue is empty and it can assert its *PushReq* signal even when an output queue is full. External queue designs must protect the queue FIFOs against input-queue underflows and output-queue overflows by ignoring the processor's requests when an input queue is empty or when an output queue is full.

The Diamond 330HiFi audio DSP core should be the only device connected to the read port of an input FIFO or the write port of an output FIFO because the Diamond core's queue interfaces are not designed to connect to external queues that are shared between multiple clients. During normal queue operation, the status of an input queue must change from "not empty" to "empty" only in response to a pop request from the Diamond 330HiFi audio DSP core. Similarly, the status of an output queue must change from "not full" to "full" only in response to a push request from the Diamond 330HiFi audio DSP core. The only situation where it may be appropriate to violate this particular restriction (for both input and output queues) is during a system flush. Such a flush might occur, for example, before data processing begins on a new stream of data.

System designers must ensure that the queue interfaces are used in a manner consistent with the recommendations outlined in this section. Not doing so may result in unexpected or incorrect queue behavior.

11.8 SYSTEM DESIGN WITH THE DIAMOND 330HiFi AUDIO DSP CORE

Figure 11.8 shows a simple stereo audio system that uses the Diamond 330HiFi audio engine's queue output to create a low-overhead, dual 16-bit interface to the system's audio DACs. The DSP can output a value to the

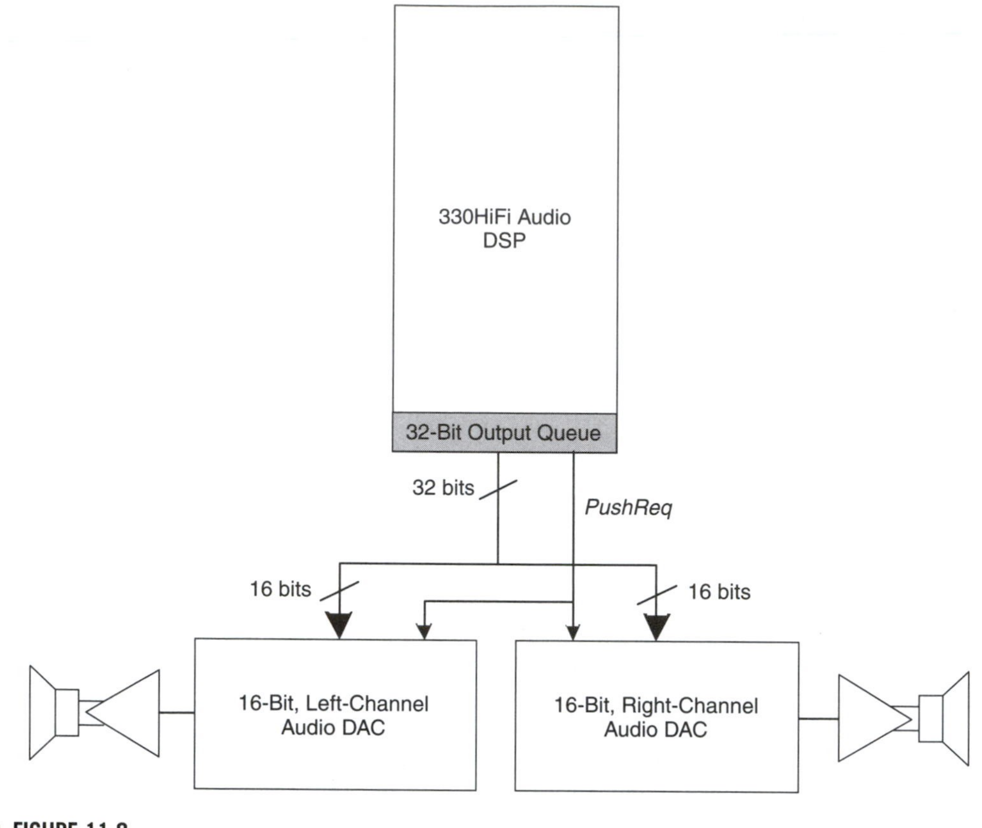

■ **FIGURE 11.8**

This stereo audio system uses Diamond 330HiFi audio DSP core's output queue to create a low-overhead, dual 16-bit interface to the system's two stereo output DACs.

two audio DACs in one cycle, which greatly reduces the I/O overhead associated with the operation. Because the audio DACs are essentially "ready" all the time, the Diamond 330HiFi can use this single-wire handshake mechanism to the audio DACs.

BIBLIOGRAPHY

Diamond Standard Processors Data Book, Tensilica, Inc., February 2006.

Kennedy, R. and Leibson, S., "Audio Gets Configurable Processor," www.audiodesignline.com.

Leibson, S and Kim, J., "Configurable Processors: A New Era in Chip Design," *IEEE Computer*, July 2005, pp. 51–59.

Rowen, C. and Leibson, S., *Engineering the Complex SOC*, Prentice-Hall, 2004.

Xtensa Microprocessor Programmer's Guide, Tensilica, Inc., July 2005.

CHAPTER 12

The Diamond 545CK DSP Core

...the 545CK has powerful DSP extensions
that push it past every licensable processor core
ever benchmarked by Berkeley Design Technology (BDTI).
—Tom Halfhill, *Microprocessor Report*

The Diamond 545CK vector DSP core is a high-performance, 3-way superscalar RISC CPU core that has been tailored to deliver very high performance on a wide range of DSP tasks. This DSP core is based on the Xtensa LX processor with a modified version of the Vectra LX vector DSP extensions and additional instructions that further enhance the core's DSP performance. Other instructions add control over the core's input- and output-queue interfaces. The Diamond 545CK DSP core's three superscalar execution pipelines, dual 128-bit load/store units, and 64-bit FLIX-format instructions allow the processor to break the 1-instruction-per-clock barrier.

12.1 THE DIAMOND 545CK DSP CORE'S INSTRUCTION FORMAT

The Diamond 545CK DSP has several ISA extensions to the base Xtensa architecture that enhance DSP performance. These additions include a 16-entry, 160-bit vector register file, a vector DSP ALU with 20-bit precision per vector, eight 18×18-bit multiplier/accumulator (MAC) units, special-purpose function units for bit packing and unpacking and Viterbi decoding, a second load/store unit required for XY memory operations, and 32-bit input- and output-queue interfaces for high-speed I/O. These extensions were developed in conjunction with Berkeley Design Technology, Inc (BDTI), a firm that specializes in analyzing and benchmarking DSPs. The extensions improve the performance of a wide range of DSP tasks as reflected in the BDTImark2000 benchmark functions, listed in Table 12.1.

Complex instructions can be built from the Diamond 545CK DSP's operation primitives by bundling three operations into a wide instruction word, as shown in Figure 12.1. The Diamond 545CK DSP core's 64-bit instruction format bundles two independent operations. The XCC (the

TABLE 12.1 ■ BDTImark2000 DSP functions

Function	Description	Example application
Real Block FIR	Finite impulse response filter that operates on a block of real (not complex) data	Speech processing, for example, G.728 speech compression
Single-Sample FIR	FIR filter that operates on a single sample of real data	Speech processing, general filtering
Complex Block FIR	FIR filter that operates on a block of complex data	Modem channel equalization
LMS Adaptive FIR	Least-mean-square adaptive filter; operates on a single sample of real data	Channel equalization, servo control, linear predictive coding
Two-Biquad IIR	Infinite impulse response filter that operates on a single sample of real data	Audio processing, general filtering
Vector Dot Product	Sum of the point-wise multiplication of two vectors	Convolution, correlation, matrix multiplication, multi-dimensional signal processing
Vector Add	Point-wise addition of two vectors, producing a third vector	Graphics, combining audio signals or images, vector search
Vector Maximum	Find the value and location of the maximum value in a vector	Error control coding, algorithms using block-floating point
Viterbi Decoder	Decodes a convolutionally encoded bit stream	Wired and wireless communications, for example, cellular phones
Control	A contrived series of control (test, branch, push, pop) and bit manipulation instructions	Virtually all signal processing applications include some "control" code
256-Point FFT	The Fast Fourier Transform converts a normal time-domain signal into the frequency domain	Radar, sonar, MPEG audio compression, spectral analysis
Bit Unpack	Unpacks words of varying length from a continuous bit stream	Audio and speech decompression

Xtensa C/C++) compiler paired with the Diamond 545CK DSP core automatically performs the needed instruction bundling to use the core's three-operation instruction word. XCC also automatically vectorizes code to take advantage of the processor's 8-way MAC unit.

All three instruction formats can be freely intermixed and tightly packed in memory. The Xtensa architecture's inherent ability to handle differently sized instructions was a key enabler in the development of the Diamond 545CK DSP core's design.

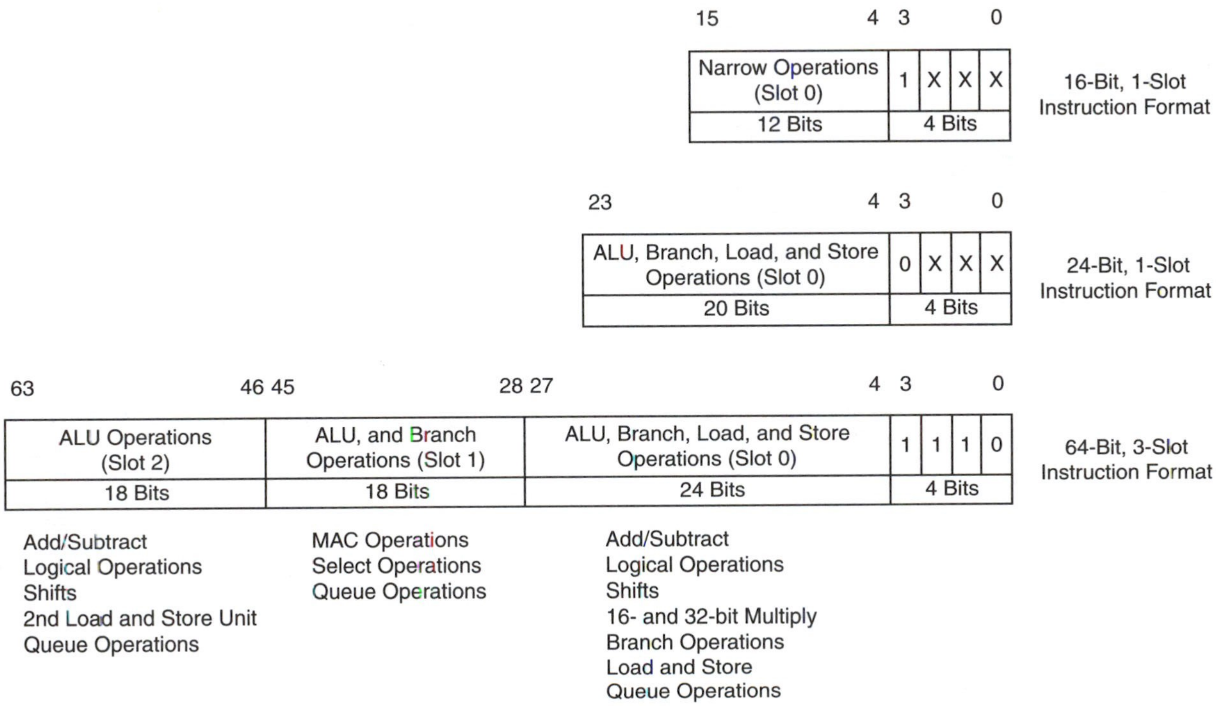

16-Bit, 1-Slot Instruction Format

15–4	3			0
Narrow Operations (Slot 0)	1	X	X	X
12 Bits	4 Bits			

24-Bit, 1-Slot Instruction Format

23–4	3			0
ALU, Branch, Load, and Store Operations (Slot 0)	0	X	X	X
20 Bits	4 Bits			

64-Bit, 3-Slot Instruction Format

63–46	45–28	27–4	3			0
ALU Operations (Slot 2)	ALU, and Branch Operations (Slot 1)	ALU, Branch, Load, and Store Operations (Slot 0)	1	1	1	0
18 Bits	18 Bits	24 Bits	4 Bits			
Add/Subtract Logical Operations Shifts 2nd Load and Store Unit Queue Operations	MAC Operations Select Operations Queue Operations	Add/Subtract Logical Operations Shifts 16- and 32-bit Multiply Branch Operations Load and Store Queue Operations				

■ **FIGURE 12.1**

The Diamond 545CK DSP core is a static superscalar design that adds a third instruction format to the existing 16- and 24-bit Xtensa ISA instruction formats. The processor's third instruction format is 64 bits wide, and this wide format bundles three independent operations.

12.2 A HIGH-PERFORMANCE DSP CORE

The Diamond 545CK DSP core brings the performance benefits of a full-featured 32-bit RISC CPU to bear on the designated tasks:

- Large 4-Gbyte address space
- 32-bit computations
- Large 32-entry register file
- 128-bit interface ports for one local instruction RAM and two local data RAMs
- 5-stage pipelined operation resulting in a 233-MHz maximum clock rate in 130 nm G-type (general-purpose) technology
- Power dissipation is 255 μW/MHz and area is 5.14 mm^2 in 130 nm LV (high-performance) technology.

The Diamond 545CK DSP core has a 128-bit version of the general-purpose PIF bus for global SOC communications. An optional AMBA bus bridge supplied with the Diamond 545CK DSP core adapts the PIF to peripheral devices designed for the AMBA AHB-Lite bus.

The Diamond 545CK DSP architecture contains all of the basic elements of the Xtensa ISA. It has a 64-entry general-purpose register file, 5-stage pipeline, and 32-bit addressing. This execution pipeline has been augmented with several vector DSP instructions to reduce the number of cycles needed to execute DSP code. In addition, the Diamond 545CK DSP has a second register file designed to support vector DSP algorithms and a second load/store unit so that the DSP can implement algorithms that use XY memory addressing. The vector register file is 160 bits wide and has 16 entries.

Three execution pipelines give the Diamond 545CK DSP core the ability to execute three independent instructions per clock. A full block diagram of the Diamond 545CK DSP appears in Figure 12.2. The three execution pipelines are not symmetric. The first execution pipeline, associated with operation slot 0, contains the base Xtensa ALU and a 16 × 16-bit multiplier. The second pipeline contains the 8-way MAC unit and the select units, which manipulate data stored in the vector register file. The third execution pipeline contains the vector DSP ALU functions and the second load/store unit. All three pipelines can issue instructions to the input- and output-queue interfaces.

The Diamond 545CK DSP core also incorporates the Diamond Series processor core software-debug stack, as shown on the left of Figure 12.2. This debug stack provides external access to the processor's internal, software-visible state through a 5-pin IEEE 1149.1 JTAG TAP (test access

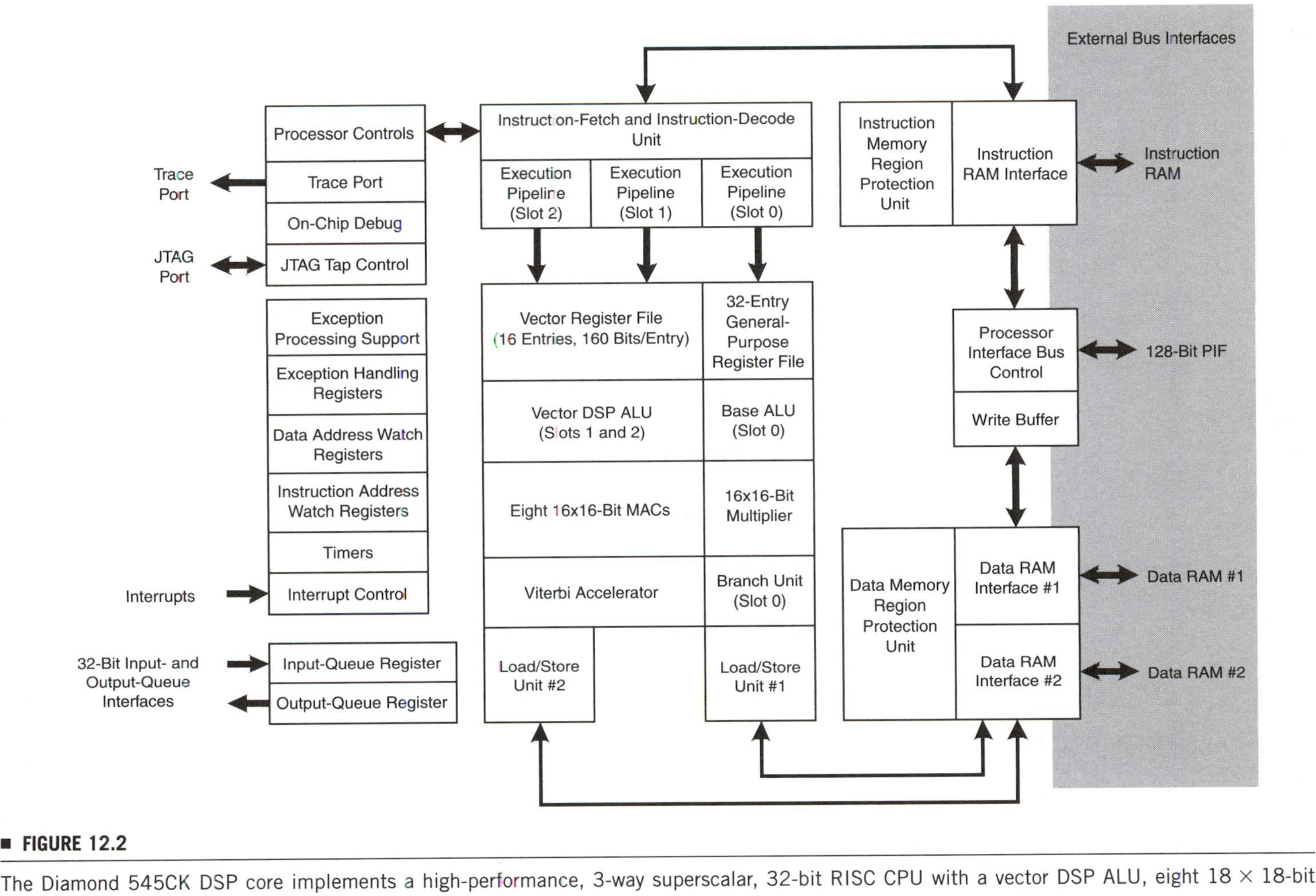

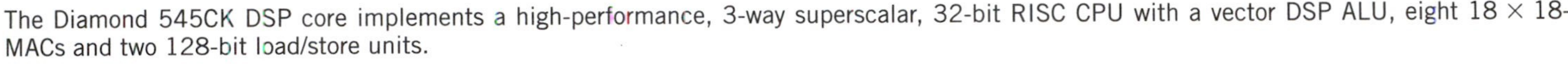

■ **FIGURE 12.2**

The Diamond 545CK DSP core implements a high-performance, 3-way superscalar, 32-bit RISC CPU with a vector DSP ALU, eight 18 × 18-bit MACs and two 128-bit load/store units.

port) interface and through a trace port that provides additional program-trace data.

12.3 DIAMOND 545CK DSP CORE INTERFACES

As shown in Figure 12.3, the Diamond 545CK DSP core has a 128-bit implementation of the Xtensa PIF (main processor interface) bus and separate interfaces for the one local instruction RAM and two local data RAMs. Note that each of the two load/store units in the Diamond 545CK DSP core has its own interface port to each local data memory. External circuitry must connect both local ports to the associated local data memory and must arbitrate between the two load/store ports when simultaneous access occurs. There is only one instruction-fetch unit in the core, so there is only one port to local instruction memory.

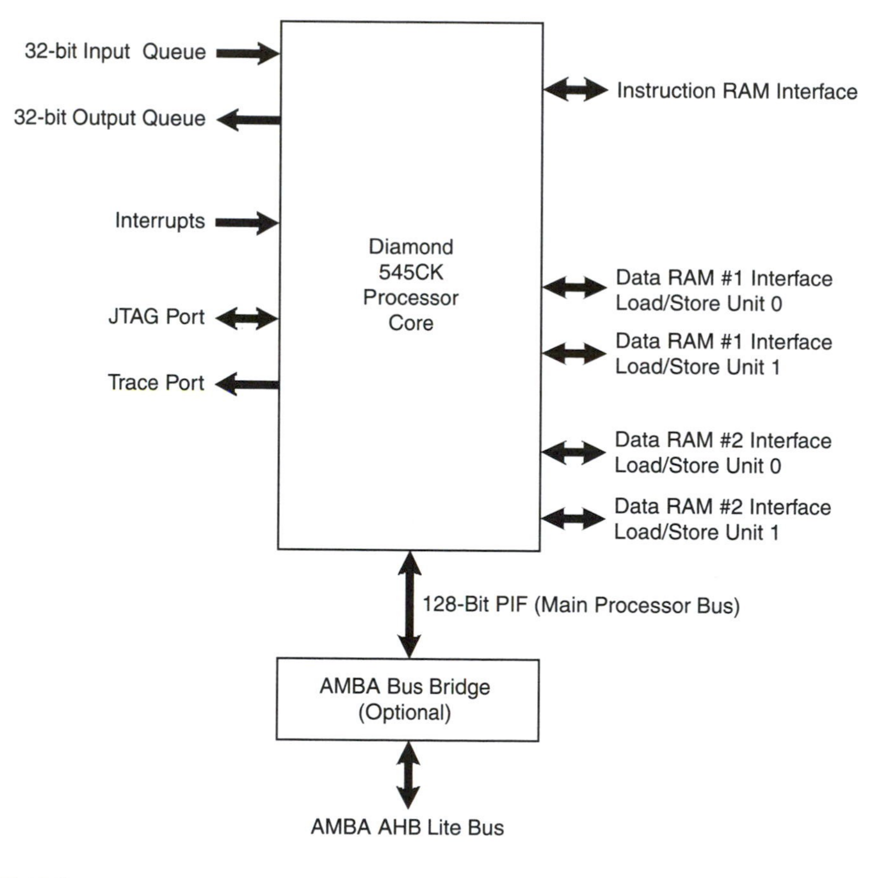

■ **FIGURE 12.3**

The Diamond 545CK DSP core has a 128-bit implementation of the Xtensa PIF (main processor interface) bus and separate interfaces for the one local instruction RAM and two local data RAMs.

12.4 THE DIAMOND 545CK DSP CORE'S MEMORY MAP

The Diamond 545CK DSP core's address space is mapped to the local memories and the PIF bus as shown in Table 12.2. Note that the Diamond 545CK's instruction-RAM and two data-RAM memory spaces are adjacent so they form one contiguous 384-Kbyte memory block from 0x3FFC0000 to 0x4001FFFF.

The Diamond 545CK DSP core has a 32-bit, 4-Gbyte address space, so additional memory can be attached to its PIF bus if needed. However, the address spaces dedicated to local memory are hardwired to the appropriate local-memory ports and memory transactions to those address spaces will not appear on the PIF bus. All Diamond processor cores including the Diamond 545CK automatically send memory accesses to non-local address locations out over the PIF bus.

Table 12.3 lists the Diamond 545CK DSP core's assigned reset, NMI, and other interrupt vectors. The Diamond 545CK's reset and exception vectors are assigned to locations in the PIF's address space, so some memory must be located on the main bus.

Figure 12.4 shows the various important addresses mapped into the Diamond 545CK DSP core's uniform address space.

The PIF implementation on the 545CK DSP core is 128 bits wide. The PIF bus uses a split-transaction protocol and the 545CK DSP core has an

TABLE 12.2 ■ Diamond 545CK DSP core memory-space assignments

Local memory	Start address	End address
Local instruction RAM	0x40000000	0x4001FFFF
Local data RAM #1	0x3FFE0000	0x3FFFFFFF
Local data RAM #2	0x3FFC0000	0x3FFDFFFF

TABLE 12.3 ■ Diamond 545CK DSP reset, interrupt, and exception vector address mapping

Vector	Address
Reset	0x50000000
Base address for register window underflow/overflow exception vectors	0x40000000
Level 2 high-priority interrupt	0x40000180
Level 3 high-priority interrupt	0x400001C0
Level 4 high-priority interrupt	0x40000200
Level 5 high-priority interrupt	0x40000240
Debug exception	0x40000280
NMI	0x400002C0
Level 1 interrupt (Kernel mode)	0x40000300
Level 1 interrupt (User mode)	0x40000340
Double exception	0x400003C0

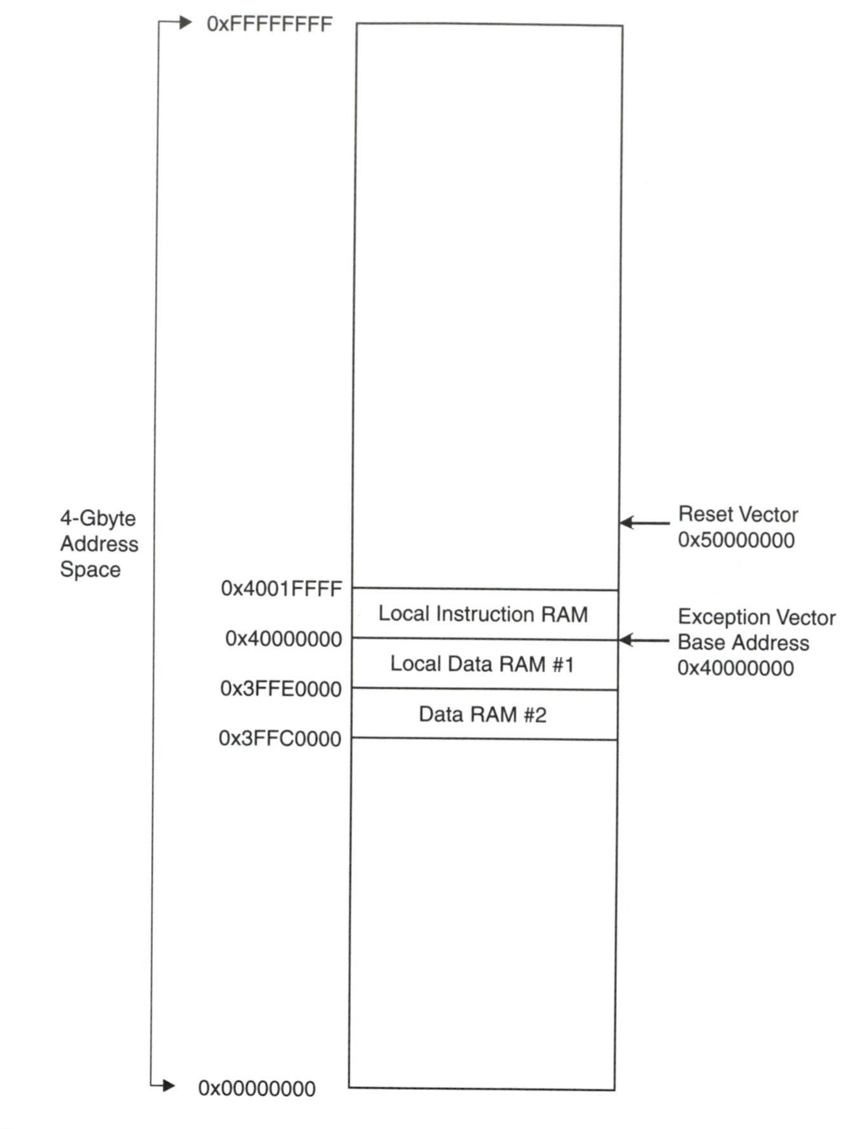

■ **FIGURE 12.4**

The Diamond 545CK DSP core's pre-configured address space maps the exception vectors to local-memory and the reset vector to memory located in the PIF (non-local) memory address space.

8-entry write buffer to accommodate as many as eight simultaneous outstanding write transactions. The 545CK DSP core's inbound-PIF transaction buffer is 16 entries deep.

The 545CK DSP core has three internal 32-bit timers that can generate interrupts at regular intervals. The processor also has nine external, level-triggered interrupt input pins and one edge-triggered, non-maskable interrupt pin.

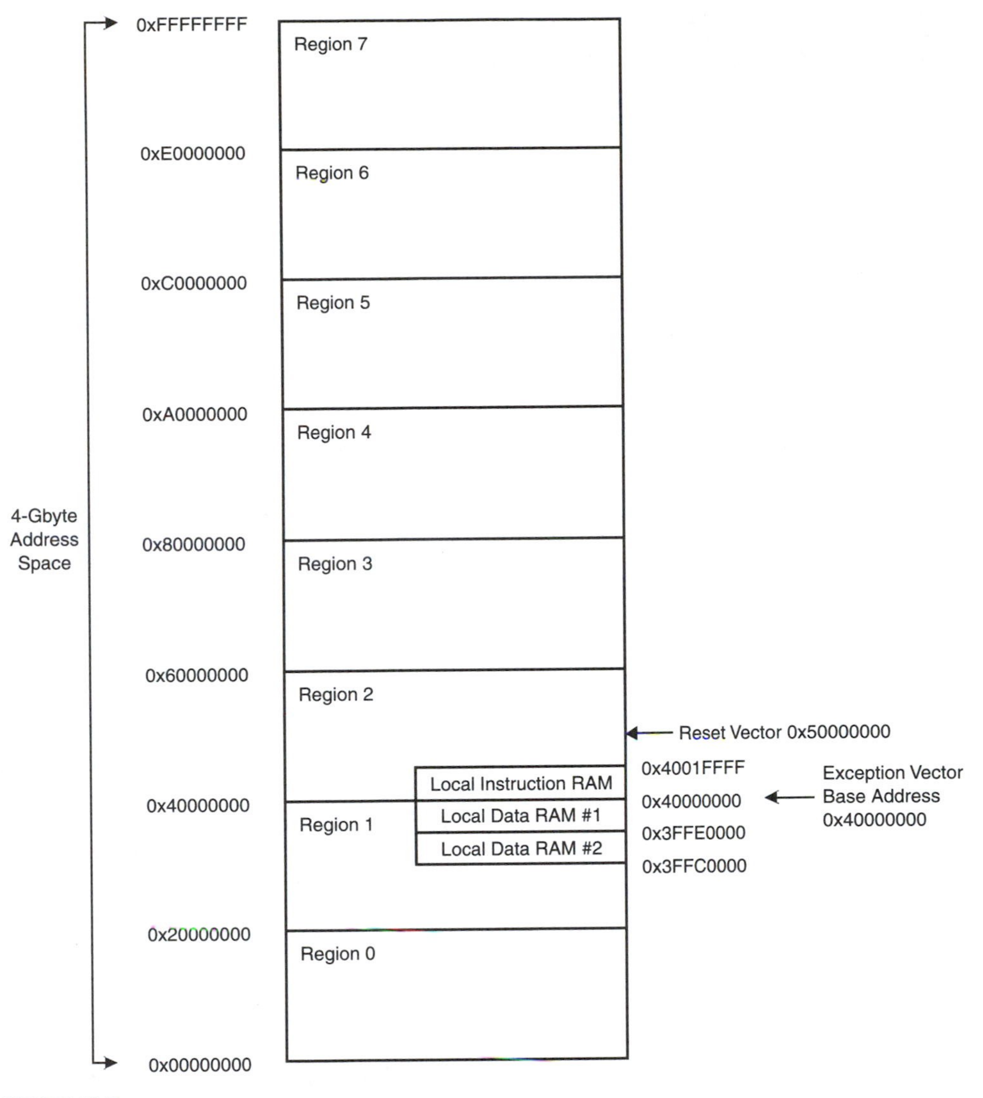

■ **FIGURE 12.5**

The Diamond 545CK DSP core's RPU divides the processor's memory space into eight protected regions. The processor's two local data-memory address blocks map into memory-protection region 1 and its local instruction-memory address space maps into memory-protection region 2 so the RPU can prevent accidental writes to instruction memory through the proper use of its protection mechanisms.

12.5 THE DIAMOND 545CK DSP CORE'S REGION-PROTECTION UNIT

The 545CK DSP core uses the same region-protection unit (RPU) discussed in Chapter 8 on the 212GP processor. Figure 12.5 shows how the Diamond RPU divides the 545CK DSP core's 4-Gbyte memory space into eight equally

TABLE 12.4 ■ Diamond RPU access modes

Access-mode value	Access-mode name	Instruction-fetch behavior	Load behavior	Store behavior
0000	No Allocate	Instruction-fetch exception	No allocate	Write through/No allocate
0001 (see Note)	Write-through/No write allocate	Allocate	Allocate	Write through/No write allocate
0010	Bypass	Bypass	Bypass	Bypass
0011	Not supported	Undefined	Load exception	Store exception
0100	Write-back	Allocate	Allocate	Write back/ Write allocate
0101–1101	Reserved	Instruction-fetch exception	Load exception	Store exception
1110	Isolate	Instruction-fetch exception	Direct processor access to memory cache	Direct processor access to memory cache
1111	Illegal	Instruction-fetch exception	Load exception	Store exception

Note: RPU access-mode 1 forces Diamond core data caches to operate in write-through mode even though all Diamond processor core data caches are pre-configured as write-back caches.

sized, 512-Mbyte regions. The 545CK DSP core's local data-memory address spaces fall into memory-protection region 1. Its local instruction-memory address space falls into memory-protection region 2. Thus the RPU can prevent accidental writes to instruction memory through the proper use of its protection mechanisms. The 545CK DSP core's non-local address space (assigned to the PIF) falls into all eight memory-protection regions, so the RPU is also useful for protecting PIF-attached memory and devices.

The Diamond 545CK DSP core sets the memory-protection attributes for each region independently by setting 4-bit access-mode values in separate, 8-entry instruction and data TLBs. (*Note*: The three-letter abbreviation TLB stands for *translation lookaside buffer* but, for Diamond processor core RPUs, its definition is widened to mean "translation hardware.") Each TLB has an entry for each of the eight memory-protection regions.

The access modes control both the protection level and the cache behavior for each of the eight memory-protection regions. The access modes appear in Table 12.4 and descriptions of the modes appear in Table 12.5. (*Note*: Tables 7.3 and 7.4 from Chapter 7 are repeated in this chapter as Tables 12.4 and 12.5.)

12.6 INPUT- AND OUTPUT-QUEUE INTERFACES

The Diamond 545CK DSP core has a 32-bit input-queue interface and a 32-bit output-queue interface. These queue interfaces respectively connect

TABLE 12.5 ■ Diamond RPU access-mode descriptions

RPU access mode	Access-mode description
No allocate	Do not allocate a cache line for this address. If the address is already cached, fetch or load the cached value. If the address has an allocated cache line but the cache line is not already in the cache, fetch or load the value from main memory and place the value in the cache.
Bypass	Do not use the cache.
Write back	Write the value to the cache and then update main memory when the cache line is evicted or when the processor forces the cache line to be written to main memory.
Isolate	Permits direct read/write access to the cache's data and tag RAM arrays.
Illegal	Any access causes an exception.

TABLE 12.6 ■ Input- and output-queue-interface instructions

Instruction	Definition
CHECK _IPQ	Check ready status of input queue
CHECK_OPQ	Check ready status of output queue
READ_IPQ	Read value on input queue
SETB_EXPSTATE	Set any bit in the output port, parameter is immediate bit index
WRITE_OPQ	Write value to output queue

directly to the head or tail of a FIFO memory. Each queue interface has a pair of hardware handshaking wires that automatically sequence the flow of data between the attached FIFO memory and the processor. The input-queue handshake signals are called *PopReq* (Pop Request) and *Empty*. The output-queue handshake signals are called *PushReq* (Push Request) and *Full*.

Table 12.6 lists the instructions added to the Diamond 545CK DSP core to manage the input- and output-queue interfaces. These instructions can appear in each of the three operation slots in the core's 64-bit instruction word.

The Diamond 545CK DSP core's input- and output-queue interfaces have been designed to mate directly to synchronous FIFO devices. The queue interfaces are synchronous in that all data transfers between the Diamond 545CK DSP core and the external FIFOs occur on the rising edge of the clock signal. There should be no combinational paths from *PopReq* to *Empty* in the input-queue interface logic or from *PushReq* to *Full* in the output-queue interface logic.

If an input queue becomes empty as a result of a pop request from the Diamond 545CK DSP core, the queue should assert its *Empty* signal in the next clock cycle, following the clock edge marking the data transfer. Similarly, if an output queue becomes full as a result of a push request, it

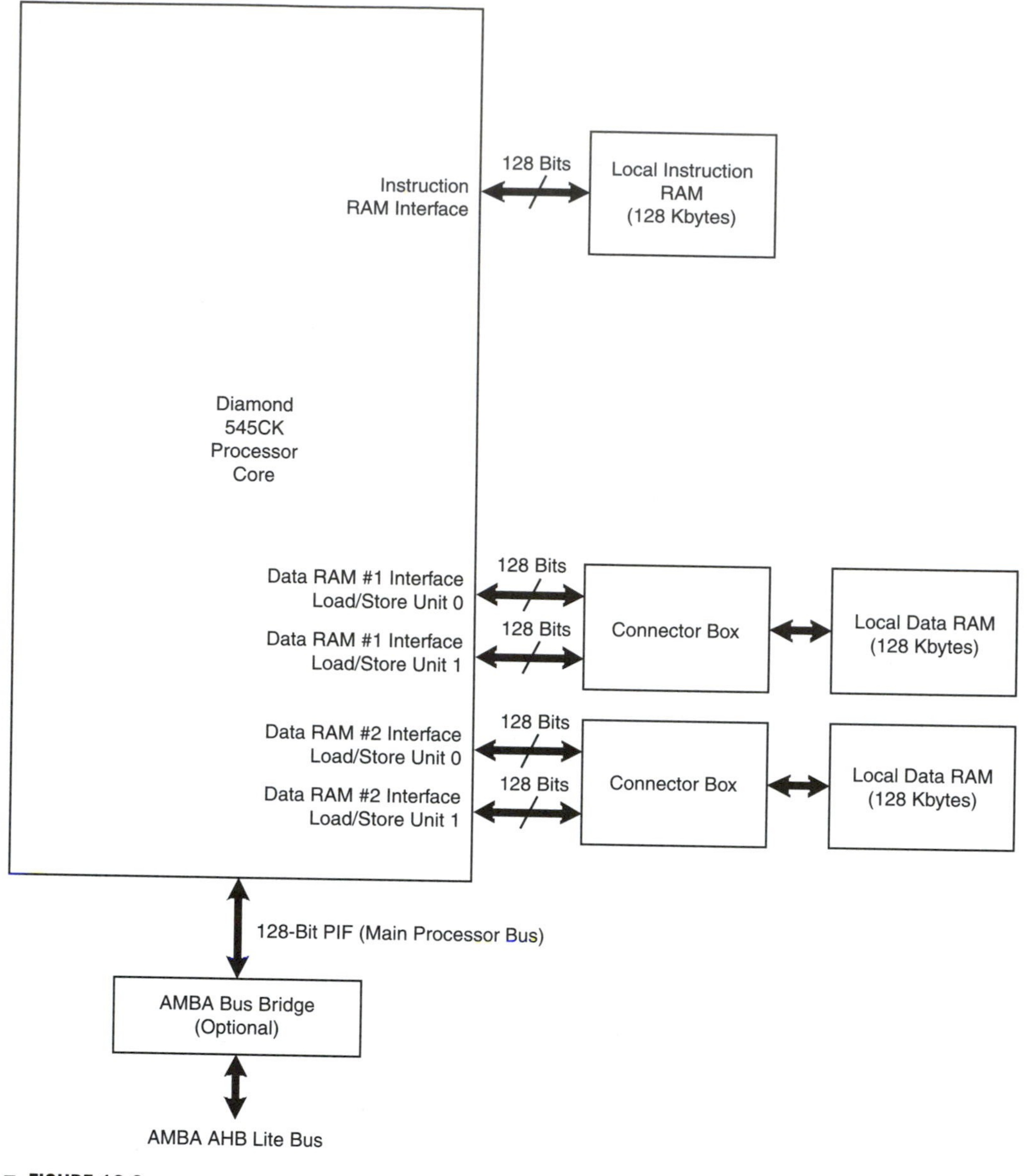

■ **FIGURE 12.6**

Connector boxes link the memory-interface ports from the 545CK DSP's two load/store units to local data-memory blocks.

should assert its *Full* signal in the next clock cycle after the clock edge, when it accepts the data from the processor.

The Diamond 545CK DSP core can assert its *PopReq* signal even when an input queue is empty and it can assert its *PushReq* signal even when an output queue is full. External queue designs must protect the queue FIFOs against input-queue underflows and output-queue overflows by ignoring the processor's requests when an input queue is empty or when an output queue is full.

The Diamond 545CK DSP core core should be the only device connected to the read port of an input FIFO or the write port of an output FIFO because the Diamond core's queue interfaces are not designed to connect to external queues that are shared between multiple clients. During normal queue operation, the status of an input queue must change from "not empty" to "empty" only in response to a pop request from the Diamond 545CK DSP core. Similarly, the status of an output queue must change from "not full" to "full" only in response to a push request from the Diamond 545CK DSP core. The only situation where it may be appropriate to violate this particular restriction (for both input and output queues) is during a system flush. Such a flush might occur, for example, before data processing begins on a new stream of data.

System designers must ensure that the queue interfaces are used in a manner consistent with the recommendations outlined in this section. Not doing so may result in unexpected or incorrect queue behavior.

12.7 SYSTEM DESIGN WITH THE DIAMOND 545CK DSP CORE

The 545CK DSP core is unique among the Diamond cores in having two load/store units. These units allow the DSP to perform two simultaneous memory transactions, often called XY memory access by DSP system designers. Each of the 545CK DSP's load/store units has its own interface to local memory so these two interfaces to each local-memory block must be joined using logic that multiplexes and arbitrates access to the associated memory block. Figure 12.6 illustrates a system with logic blocks called "Connector Boxes" that provide the multiplexing and arbitration functions.

BIBLIOGRAPHY

Diamond Standard Processors Data Book, Tensilica, Inc., February 2006.

Leibson, S. and Kim, J., "Configurable Processors: A New Era in Chip Design," *IEEE Computer*, July 2005, pp. 51–59.

Rowen, C. and Leibson, S., *Engineering the Complex SOC*, Prentice-Hall, 2004.

Xtensa Microprocessor Programmer's Guide, Tensilica, Inc., July 2005.

Staff of Berkeley Design Technology, Inc, *"An Independent Analysis of the Tensilica LX Processor with Vectra LX,"* available at www.bdti.com, 2005.

Staff of Berkeley Design Technology, Inc, *"The BDTImark2000™: A Summary Measure of Signal Processing Speed,"* available at www.bdti.com, 2005.

CHAPTER 13

USING FIXED PROCESSOR CORES IN SOC DESIGNS

The requirement for processing power will be 1000× in the next ten years.
—International Technology Roadmap for Semiconductors: 2005.

Designer-defined Xtensa processor cores and pre-configured Diamond cores are SOC building blocks. It's critically important for the SOC design team to have the most flexible building blocks available to maximize the team's ability to reach project performance goals while staying within budget and on schedule. However, it's also critically important for the design team to employ productive system-design strategies that effectively and efficiently use those building blocks. Many SOC design teams employ system-design strategies rooted in the previous century. These design strategies were developed when ASIC gate counts were in the hundreds of thousands, long before SOCs broke the million-gate threshold. Consequently, these strategies are now obsolete. They cannot make the design-team members sufficiently productive to produce mega-gate SOC designs economically.

13.1 TOWARD A 21st-CENTURY SOC DESIGN STRATEGY

The organization that perhaps pays more attention to chip-level system-design-strategies than any other in the world is the ITRS (International Technology Roadmap for Semiconductors). The ITRS assesses all semiconductor technology requirements (for design, manufacturing, and reliability) to ensure continued global advancements in integrated circuit performance. This assessment, which produces a new industry road map every two years, is a worldwide cooperative effort of the industry's manufacturers and suppliers, government organizations, consortia, and universities. It is the way that the semiconductor industry has chosen to map the route for the continued enforcement of Moore's law.

The ITRS continually assesses the available design technologies, technology trends, and possible future developments that shape the way design teams develop integrated circuits, including SOCs. It produced an updated road map at the end of 2005 with many new additions. Several of these additions to the road map directly apply to the way SOCs are designed. In fact, the ITRS is extremely concerned with the way ICs are designed. Here's a quote from the Design section of the 2005 ITRS Report on IC design:

> *The main message in 2005 remains—Cost (of design) is the greatest threat to continuation of the semiconductor roadmap. Cost determines whether differentiating value is best achieved in software or in hardware, on a programmable commodity platform or on a new IC. Manufacturing non-recurring engineering (NRE) costs are on the order of millions of dollars (mask set + probe card); design NRE costs routinely reach tens of millions of dollars, with design shortfalls being responsible for silicon re-spins that multiply manufacturing NRE. Rapid technology change shortens product life cycles and makes time-to-market a critical issue for semiconductor customers.*
>
> *Manufacturing cycle times are measured in weeks, with low uncertainty. Design and verification cycle times are measured in months or years, with high uncertainty. Without foundry amortization and return-on-investment (ROI) for supplier industries, the semiconductor investment cycle stalls. ITRS editions prior to 2003 have documented a design productivity gap—the number of available transistors grows faster than the ability to meaningfully design them. Yet, investment in process technology has by far dominated investment in design technology.*
>
> *The good news is that enabling progress in DT [design technology] continues. The estimated design cost of the power-efficient system-on-chip (SOC-PE) defined in the System Drivers chapter is near $20 M in 2005, versus around $900 M had DT innovations between 1993 and 2005 not occurred. . . . The bad news is that software can account for 80% of embedded-systems development cost; test cost has grown exponentially relative to manufacturing cost; verification engineers outnumber design engineers on microprocessor project teams; etc. Today, many design technology gaps are crises.*

The ITRS assessment divides the many aspects of the design-technology crisis into two broad groups:

1. silicon complexity
2. system complexity

The issues surrounding silicon complexity relate to the physical design of a chip. Some of the physical-design challenges identified by the 2005 ITRS report include:

- Non-ideal scaling of device parasitics and supply/threshold voltages (leakage, power management, circuit/device innovation, current delivery).

- Coupled high-frequency devices and interconnects (noise/interference, signal integrity analysis and management, substrate coupling, delay variation due to cross-coupling).
- Manufacturing variability (statistical process modeling and characterization, yield, leakage power).
- Complexity of manufacturing handoff (reticle enhancement and mask writing/inspection flow, NRE cost).
- Process variability (library characterization, analog and digital circuit performance, error-tolerant design, layout reuse, reliable and predictable implementation platforms).
- Scaling of global interconnect performance relative to device performance (communication, synchronization).
- Decreased reliability (gate insulator tunneling and breakdown integrity, joule heating and electromigration, single-event upset, general fault-tolerance).

The issues connected with silicon complexity threaten some long-standing IC-design paradigms:

1. Systemwide clock synchronization becomes infeasible due to power-dissipation limits and the rising cost of system robustness (statistical timing slack) in the face of increasing manufacturing variability.
2. CMOS transistors exhibit ever-larger behavioral variability.
3. Fabrication of perfect chips with 100% functional transistors and interconnects is becoming prohibitively expensive.

These issues connected with escalating silicon complexity result in other issues related to system complexity. The ITRS report ties on-chip system complexity directly to the exponentially increasing transistor counts produced by the semiconductor industry's adherence to Moore's law, which is spurred by consumer demand for increased product capabilities, lower cost, the perpetual craving—spurred by marketing—for "the new, new thing," and increasingly global competition among system and silicon vendors.

Note that consumer demand did not always serve as the semiconductor industry's main driver. In the 1960s and 1970s, relatively low-volume military system requirements drove much of the industry's development. In the 1980s and 1990s, the demands of computer systems became the semiconductor industry's leading driver. Computer systems, particularly PCs, sold in higher volumes than did the military systems and used massive quantities of certain key ICs, especially DRAMs. Only in the late 1990s did consumer electronics ascend to the leading position as the semiconductor industry's main volume driver. The voracious demands of high-volume, consumer-product categories for low cost, good performance, and rapid improvement (to promote frequent replacement or upgrade purchases) now largely define the semiconductor industry.

These changes have severely challenged the existing, tried-and-true methods used to architect and develop electronic systems (and therefore SOCs). System complexity has exploded due to the development of increasingly advanced forms of digital media and broad consumer demand yet the system-design techniques and design styles used to develop the latest systems are often the same ones used during earlier, simpler eras.

The ITRS road map identifies many 21st-century system-design challenges including:

- Block reuse and support for hierarchical design
- Verification and test
- Embedded software design and hardware codesign
- Design process management (design team size and geographic distribution).

Any updated SOC design style must address all of these issues.

13.2 THE ITRS PROPOSAL FOR SOC DESIGN

In addressing the design needs of the 21st century, the consensus opinion represented in the ITRS 2005 report clearly identifies several possible approaches to resolving the rise in design complexity. Two of the key elements the ITRS report identifies are reuse productivity and allowing designers to work at a higher level of abstraction. The most reusable logic-block element the electronics industry has produced to date, by far, is the programmable microprocessor and its brethren (DSPs, microcontrollers, network processors, graphics processors, etc.). By its very nature, the microprocessor can serve in a very wide number of roles through its programmability. The broad use of microprocessors to implement tasks dovetails with the use of high-level programming languages, particularly C and C++, to initially describe the function of both systems and subsystems.

Therefore, it's no surprise that the ITRS 2005 report plots a rise in "processing elements" for future SOC designs, as shown in Figure 13.1. The numbers shown for processing engines are absolute and the values shown for logic gates and memory bits are normalized to the year 2005. In 2005, the ITRS 2005 report assumes the "typical" SOC in 2005 had 6.5 million gates, which included 16 processing engines. Again, the ITRS estimates are an industry consensus based on input from the semiconductor industry, system vendors, university researchers, and government agencies.

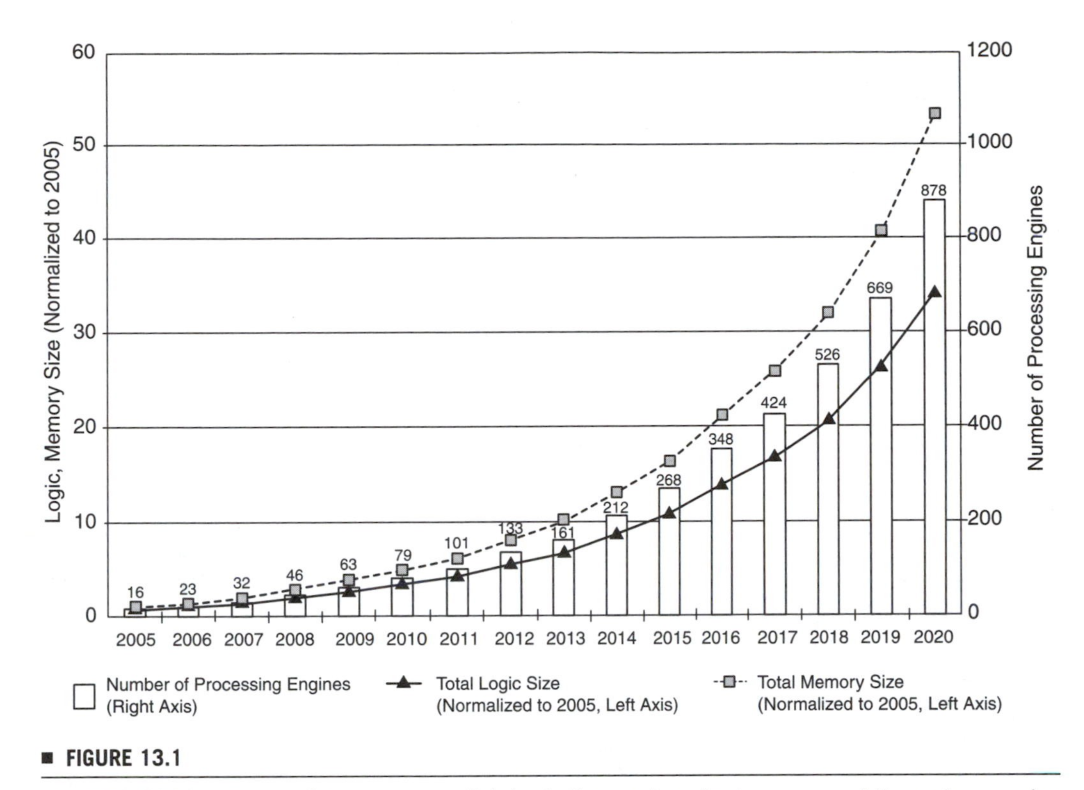

■ **FIGURE 13.1**

The ITRS 2005 report predicts an exponential rise in the number of gates, memory bits, and processing engines used to build an SOC. The numbers shown for processing engines are absolute values. The values shown for logic gates and memory bits are normalized to the year 2005.

Perhaps more important than the ITRS estimate of the number of processors on future SOCs is the organization's ideas for how SOCs using those processors will be architected. Figure 13.2 shows one such possible architecture template, taken directly from the ITRS 2005 report. This architecture template replicates the classic, now-obsolete, 20th-century "one-bus-fits-all" system-design approach that should be anathema to anyone who has read the preceding 12 chapters of this book. The single bus appearing in Figure 13.2 that ties all of the system's processing engines together represents a heavily shared resource—one that's all but certain to become overloaded as tens and then hundreds of processors are connected. A single-bus architecture also fails catastrophically if the bus fails, making built-in fault tolerance impossible. This single-bus system architecture is simply not a good design for multiple-processor SOCs (MPSOCs).

A far more useful picture of an SOC floor plan appears in the ITRS 2005 report alongside the architectural template shown in Figure 13.2. It appears in Figure 13.3. This floor plan shows the grouping of multiple processing engines into function-specific domains. This sort of illustration is more useful because it shows closely related processing engines

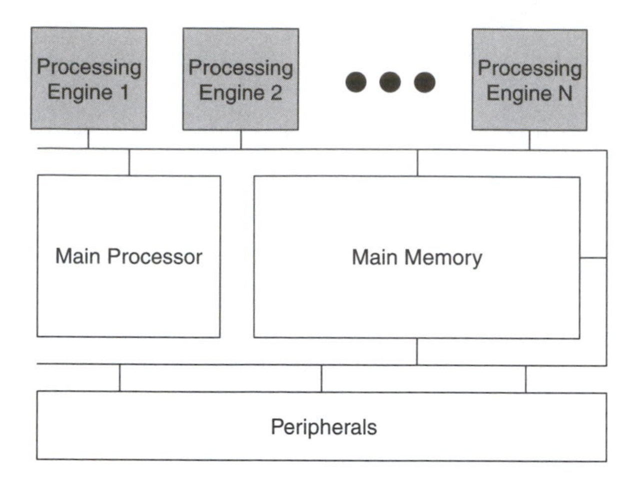

▪ **FIGURE 13.2**

All processing elements share the bus in this multiple-processor system. Bandwidth overload is certain as tens and then hundreds of processors are connected to the bus and the system will fail catastrophically if the bus fails.

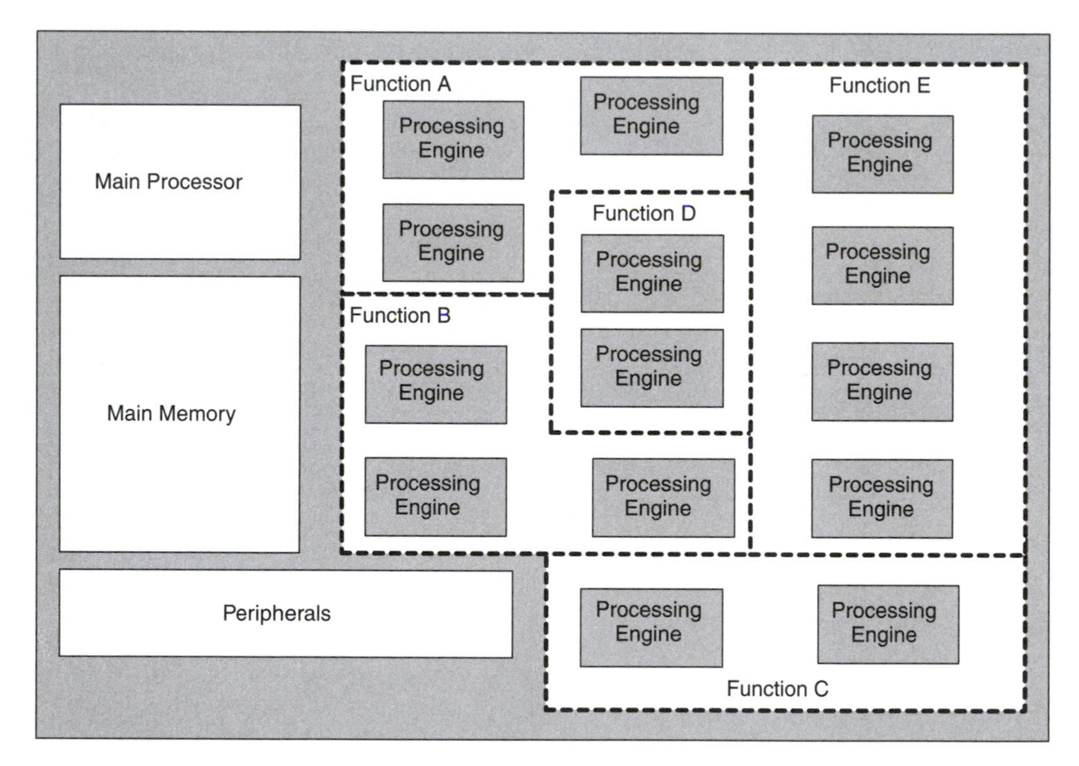

▪ **FIGURE 13.3**

Grouping related processing engines gives a better picture of the inter-processor communications required to implement a complex SOC.

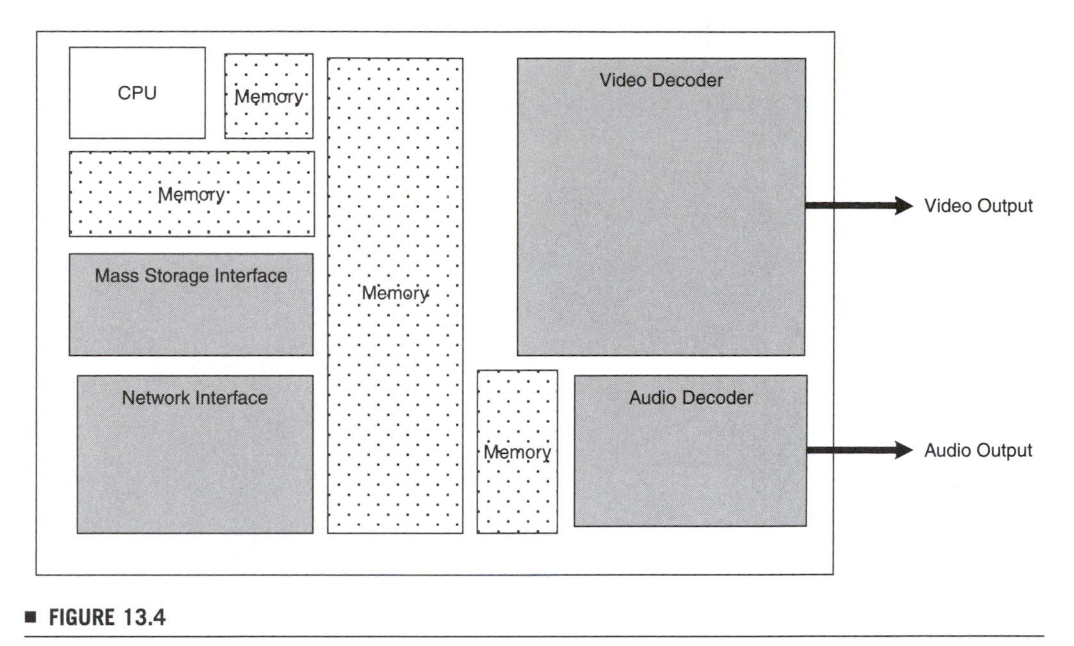

■ **FIGURE 13.4**

Floor plan for a multimedia SOC.

physically grouped together. Graphically grouping related processing engines produces a better picture of the inter-processor communications required to implement a complex SOC. The processing engines within a group will likely need high-bandwidth inter-processor communications and the communications channels within each processor group should probably be independent of other communications paths on the SOC to decouple bandwidth requirements. Otherwise, managing the complexity of bandwidth allocation across the entire chip will quickly become impossible.

Figure 13.3 shows a vague, general-purpose SOC floor plan. Figure 13.4 shows a more specific diagram, that of a multimedia system, that represents many 21st-century SOCs. The multimedia system has function-specific blocks for processing audio and video and it has blocks for interfacing to a network and to mass-storage devices. This SOC floor plan could belong to a chip designed to be used in a mobile telephone handset, a personal multimedia player, a set-top box, or a number of other end products.

Each of the function-specific blocks in this multimedia SOC could contain one or more processors while some of the blocks may contain no processors at all, as shown in Figure 13.5. The SOC designers will determine how to design each block based on a number of factors including the availability of existing block designs and the amount of work to be performed in each block. In Figure 13.5, for example, one processor has been assigned the task of producing standard-resolution video and another processor has been assigned the task of generating stereo audio.

Figure 13.6 shows a similar multimedia SOC designed for multimedia systems with higher performance than the SOC shown in Figure 13.5.

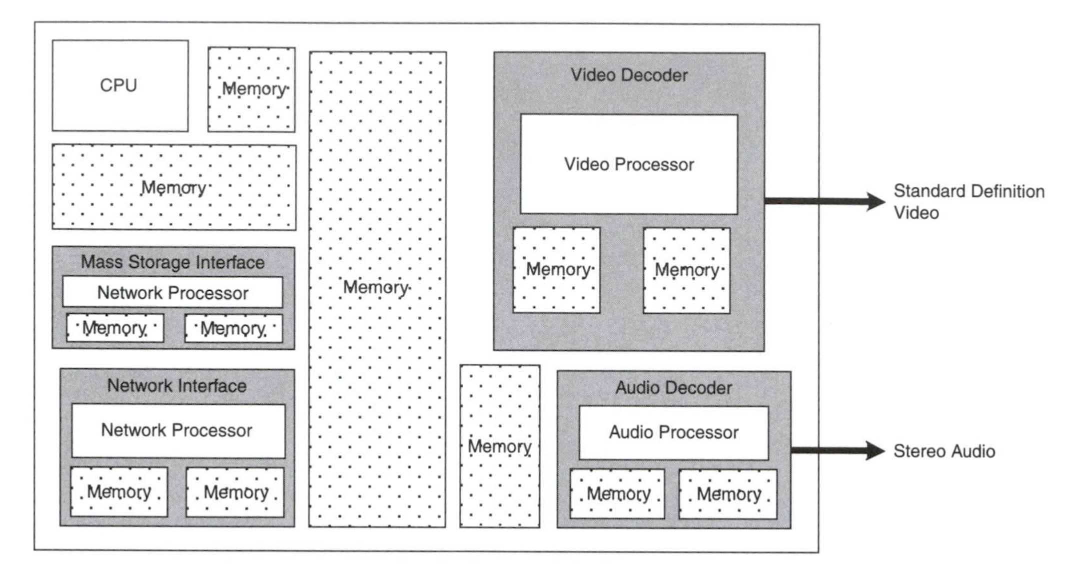

■ **FIGURE 13.5**

This floor plan for a multimedia SOC shows processor assignments for each function block.

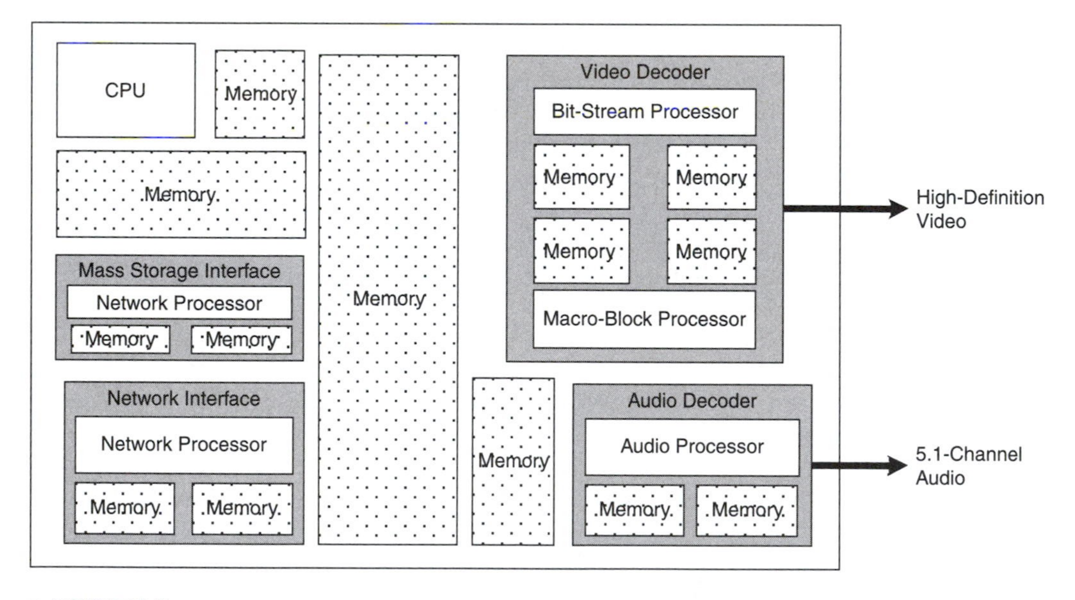

■ **FIGURE 13.6**

This floor plan for a high-performance multimedia SOC shows processor assignments for each function block.

The more powerful multimedia SOC produces high-definition video so the video block consists of two processors. This SOC also produces 5.1-channel audio, but one processor is still sufficient to generate all of the audio channels.

13.3 ON-CHIP COMMUNICATIONS FOR SOCs

Previous chapters in this book have discussed a number of ways to connect multiple processors together to achieve high bandwidths. Some of these include the use of bridged buses (shown in Figure 13.7) and pipelined data-flow architectures (or systolic-processing systems) using FIFO-based inter-processor communications (Figure 13.8). Many such possible architectures exist.

All of the SOC architectures shown so far are ad hoc designs. The inter-processor connections in these designs all follow the intended function of the chip. The ad hoc approach to processor connectivity can produce the

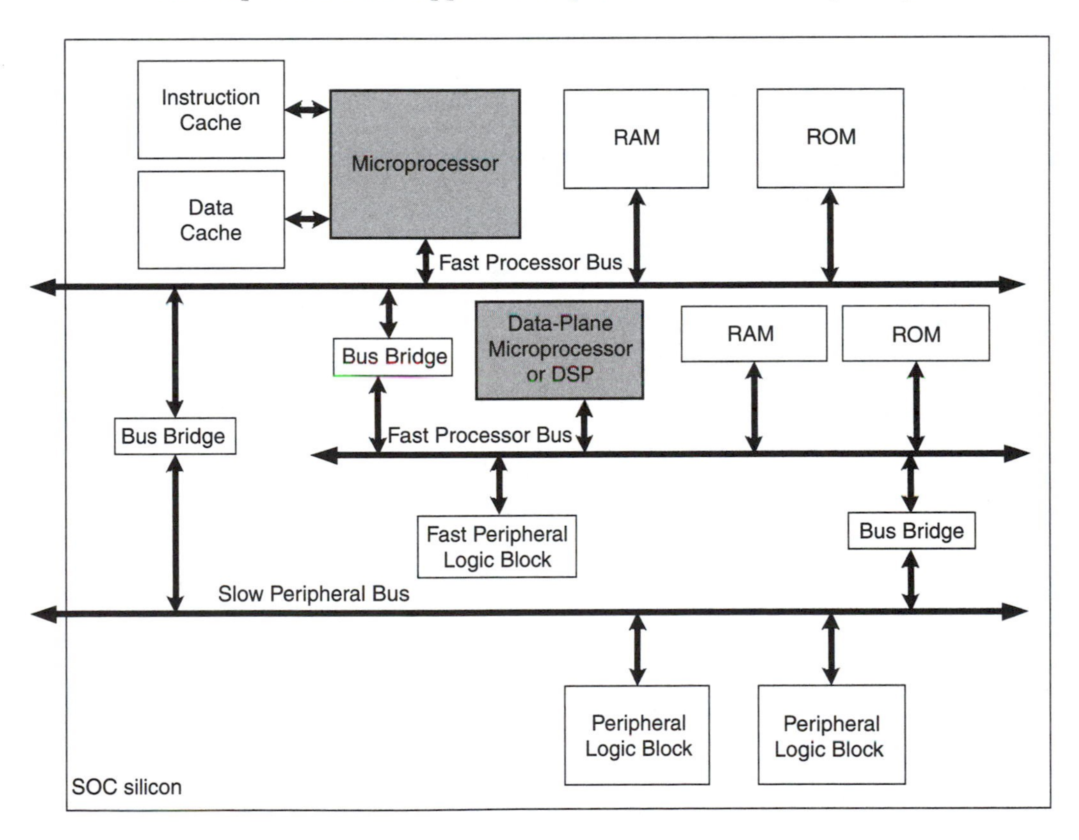

■ **FIGURE 13.7**

Bridged-bus, MPSOC architecture.

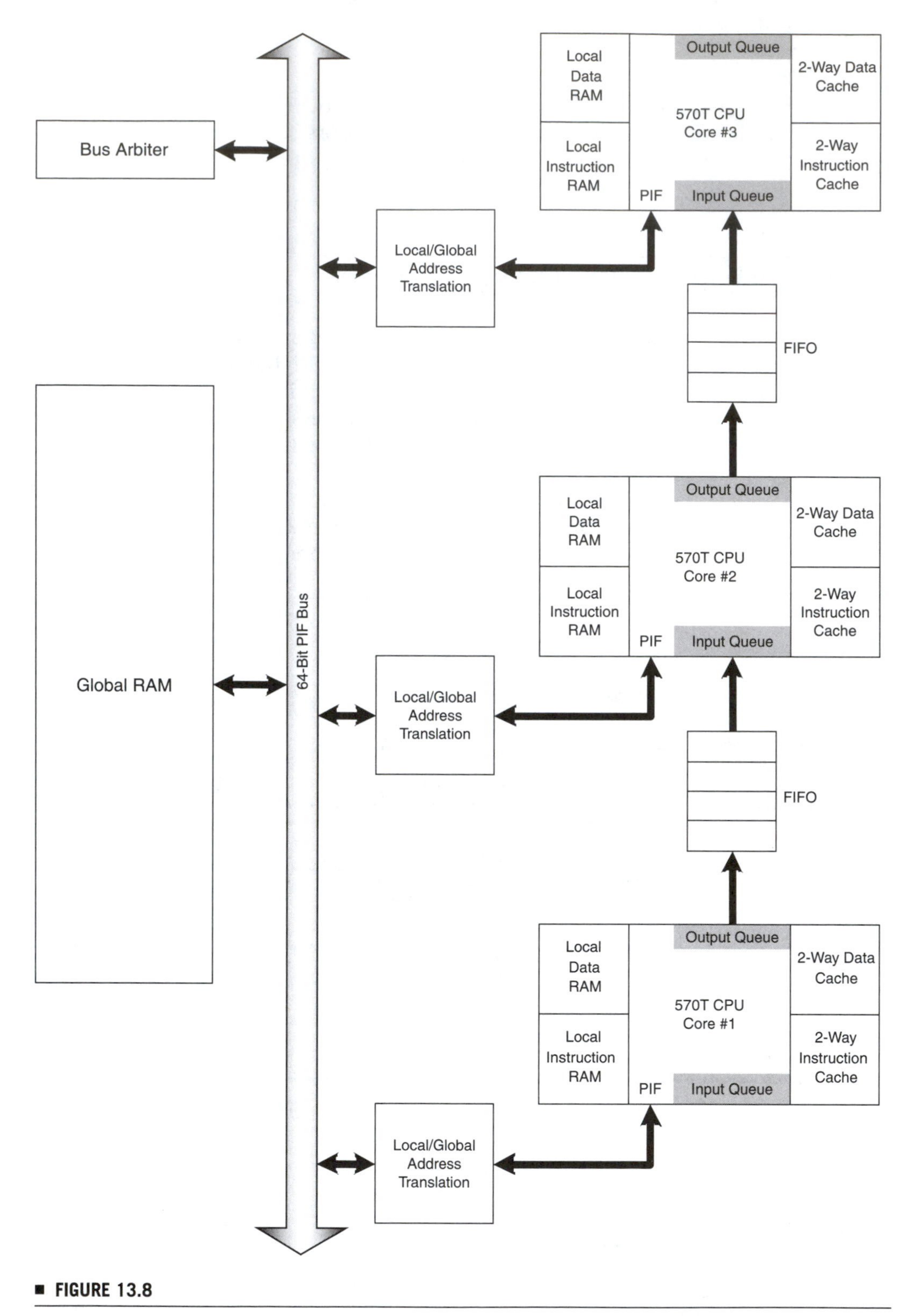

■ **FIGURE 13.8**

Systolic, MPSOC architecture.

most efficient interconnection amongst processors, if the data-movement requirements of the system are well understood. However, these needs are often not so well understood or they may change during the SOC's operation. In such cases, the SOC design may need a more flexible approach to moving data among processors. One way of building such flexible connections is through on-chip networks or "networks on chip" (NoCs).

13.4 NoC

Once you accept that complex, 21st-century SOC designs must incorporate large numbers of processors, the next logical step is to think about efficiently connecting those on-chip processors together. NoCs do not attempt to replicate the deep, multilayer, protocol-heavy schemes used for macro-level, Ethernet-style networking amongst computer systems. Instead, NoCs attempt to create lightweight, high-bandwidth networks that are agile enough to serve as links amongst on-chip processor cores. NoC designs and concepts vary widely, but there are some common NoC characteristics:

- They are more than a single, shared medium (like a bus). They are truly networks.
- They provide point-to-point connections between any two hosts attached to the network either by employing true, point-to-point crossbar switches or through virtual point-to-point connections.
- They provide high aggregate bandwidth through parallelism.
- They clearly separate communication from computation.
- They take a layered approach to communications, such as used in other macro networking schemes although they may not employ as many network layers because such complexity is prohibitively expensive in terms of on-chip area.
- They have implicit pipelining and provide intermediate storage points for the data as it moves from sender to receiver.

When evaluating NoC performance, there are two key figures of merit for performance: throughput and latency. Throughput is simply the maximum amount of data that senders can pump through the network at any time. Complex network topologies can make it difficult to measure the true throughput of a network. One approach to measuring this parameter is to measure aggregate throughput. Another method is to measure the bisection throughput: the amount of data crossing an imaginary line drawn to bisect the network.

Latency measures the time needed for data to traverse the NoC from sender to receiver. The latency depends on the number of network switches the message must travel through, the amount of buffering or storage there is in each network switch, and the amount of NoC traffic. Throughput and latency are closely intertwined and both depend on network loading. Research indicates that network congestion becomes a concern as the NoC messaging load increases above about 35%.

There are two fundamental approaches to building NoCs: circuit switching and packet switching. Circuit-switched NoCs establish a connection between sender and receiver before the data transfer starts. Once established, the circuit connection is reserved for the entire duration of the message transfer and then closed when the transmission ends or when there's no longer a need for that connection. Packet-switched NoCs chop the data into packets. Each packet contains a header that directs it to a destination address. Separate packets (also called sub-packets, flits, or phits) can take different routes through the network and must then be reassembled, in the proper order, by the receiver. Circuit- and packet-switched networking schemes are hardly new. The original telephone network was a

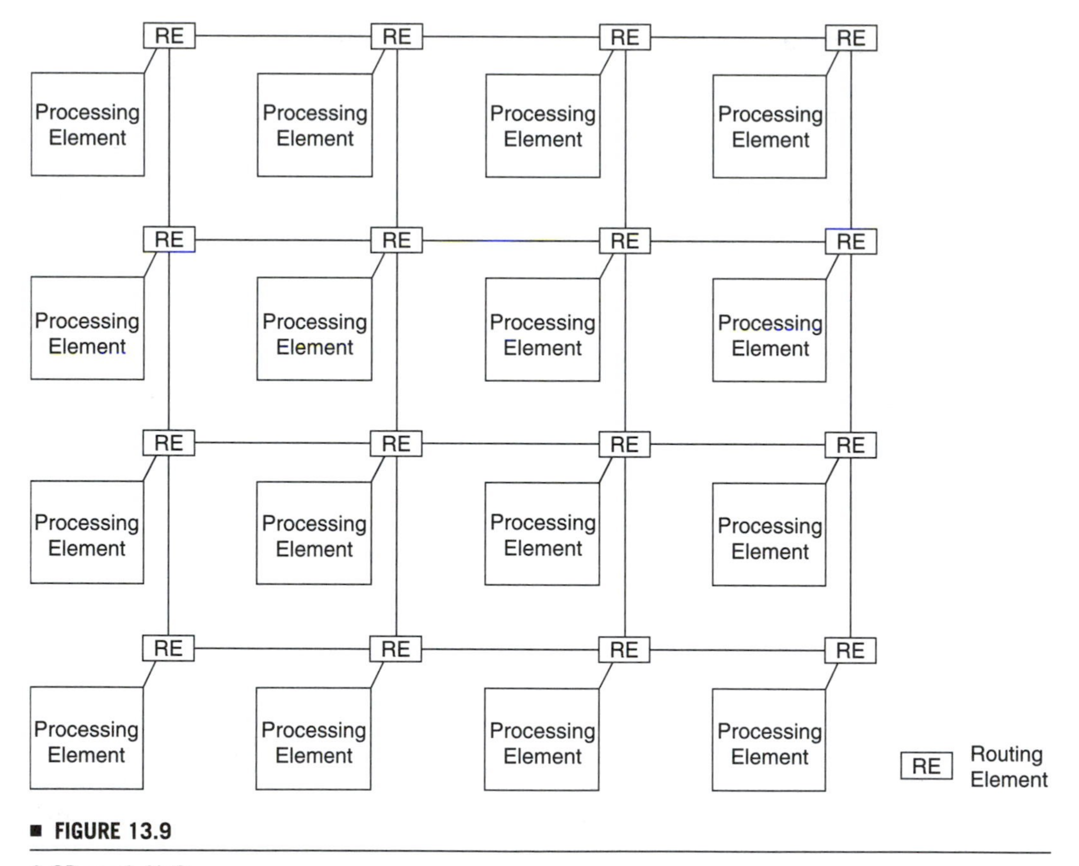

■ **FIGURE 13.9**

A 2D mesh NoC.

circuit-switched network and today's Internet is the world's largest and most familiar packet-switched network.

NoCs are also classified by network topology. Mesh networks arrange on-chip sender and receiver blocks in a regular grid. Each sender or receiver within the mesh is associated with a network switch and each network switch is connected to its nearest four neighbors (North, South, East, and West). Figure 13.9 shows a 2D mesh NoC.

The 2D mesh network is an intuitive network topology. It allows any sender to talk to any receiver and it seems to have won the popularity contest among academic NoC researchers. Variations of the mesh network are the torus and folded torus, which respectively link network switches at two or four of the mesh edges to the corresponding switch at the opposite side of the mesh. This topology reduces the number of switches a message must traverse (the number of "hops"). Figure 13.10 shows a torus version of the 2D mesh network shown in Figure 13.9.

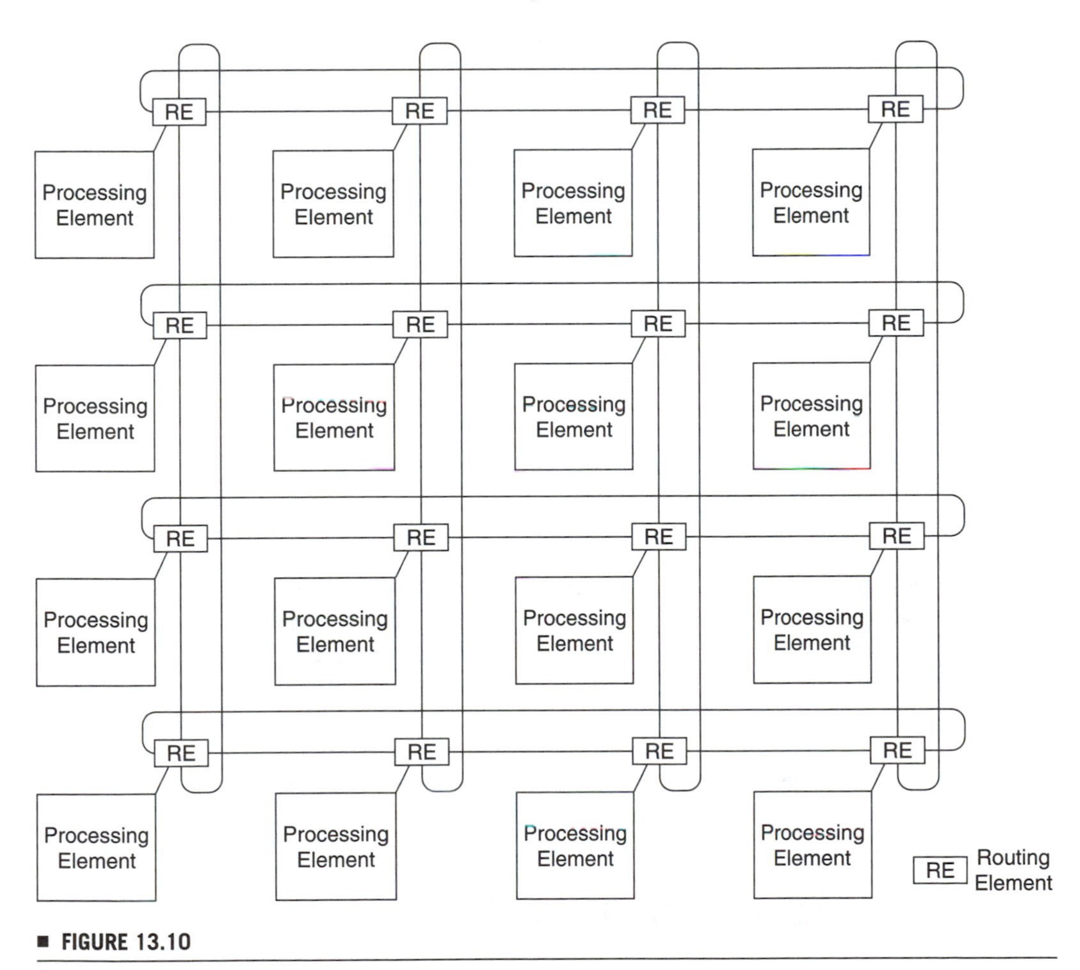

■ **FIGURE 13.10**

A torus mesh NoC.

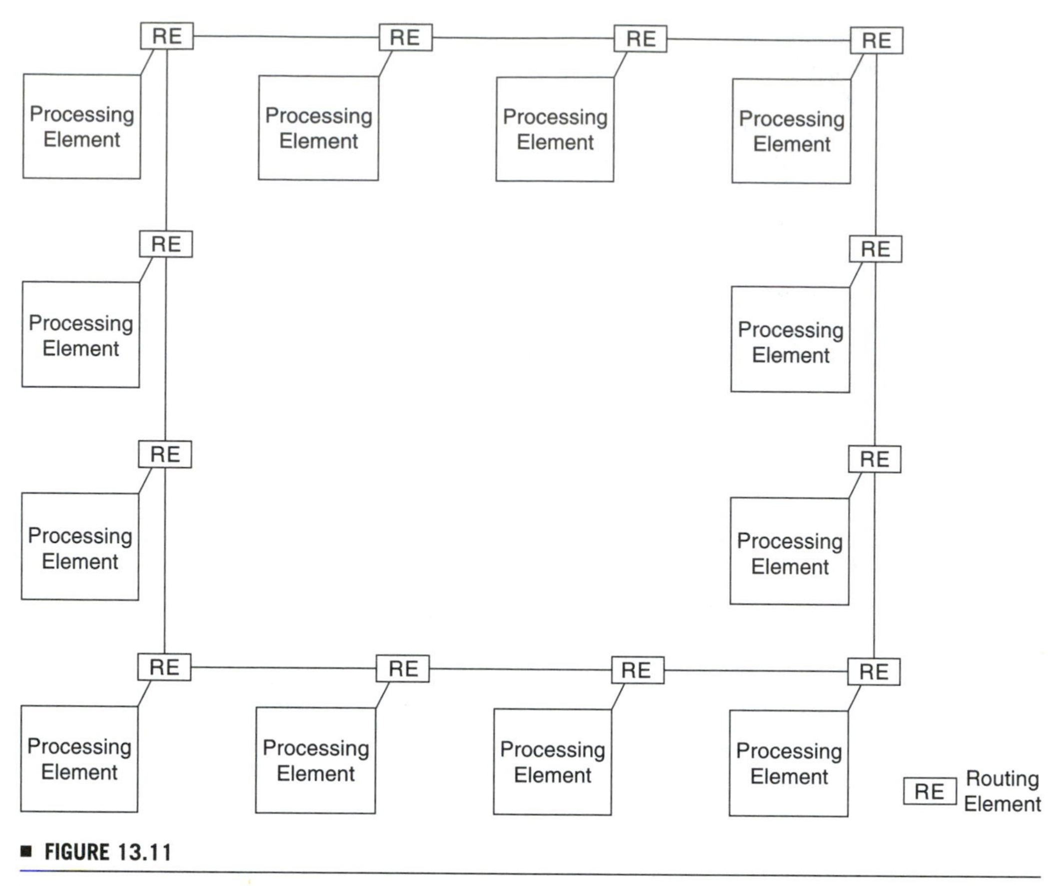

■ **FIGURE 13.11**

A ring NoC.

The ring network is another familiar networking topology and appears in Figure 13.11. Rings can be unidirectional or bidirectional. However, published research suggests that a ring is the least efficient NoC topology.

Although the NoCs shown above make excellent university research vehicles, the amount of logic needed to implement some of these NoCs can easily cause many commercial SOC design teams to discard any ideas for NoC use. There is a different line of NoC research that has a goal of developing very lightweight on-chip networks. Advocates of such networks believe that the processing units or clusters on the chip that perform the actual work of the SOC should get the majority of the silicon and that the on-chip network should only be sufficiently complex to serve the needs of these processing units.

13.5 THE THREE NoC TEMPTATIONS

Dr. Se-Joong Lee of KAIST (Korea Advanced Institute of Science and Technology) discusses what he calls the three NoC temptations. The first

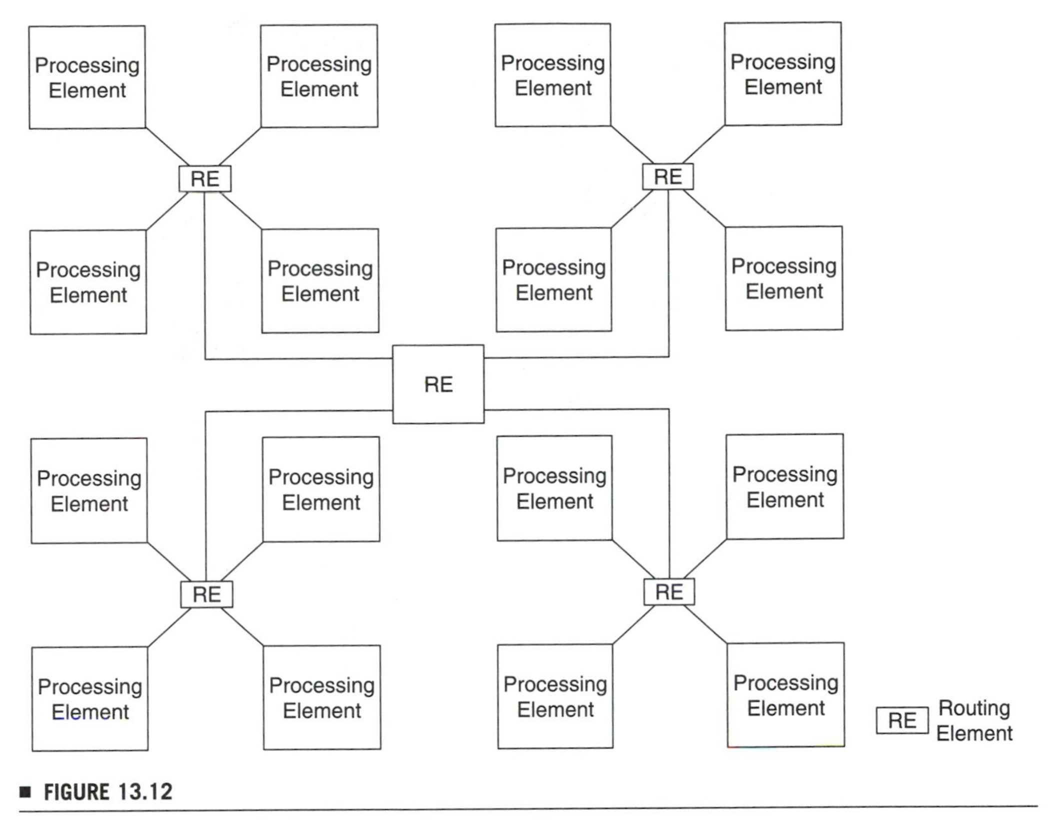

■ **FIGURE 13.12**

A 2-level, hierarchical-tree NoC.

such temptation is the siren song of the mesh network. With their elegant, regular architectures; good scalability; and opportunities for exploring adaptive routing algorithms, mesh networks have become the darlings of academic NoC research. However, Lee argues, mesh networks aren't that attractive under close scrutiny. They are inefficient for global (broadcast) transactions and they are not nearly so regular and orderly as they appear in published academic articles. When actually laid out on silicon, mesh networks complicate the physical layout of the SOC and they require many network switches, each of which can often be as large as the 32-bit RISC processor it serves. Instead, says Lee, 2-level star or tree networks provide equivalent bandwidth and better latency while consuming far less silicon and much less energy. Figure 13.12 shows a 2-level, hierarchical-tree NoC and Figure 13.13 shows the ITRS MPSOC floor plan from Figure 13.3 with a tree NoC overlay added. Note that every processor within a function cluster in Figure 13.13 is on the same leaf level (connected to the same routing element, RE) as the other processors in its cluster. This design ensures that local communications stay local to the cluster.

Mesh networks provide redundant paths, which leads to the second NoC temptation: adaptive routing. At first blush, adaptive routing looks

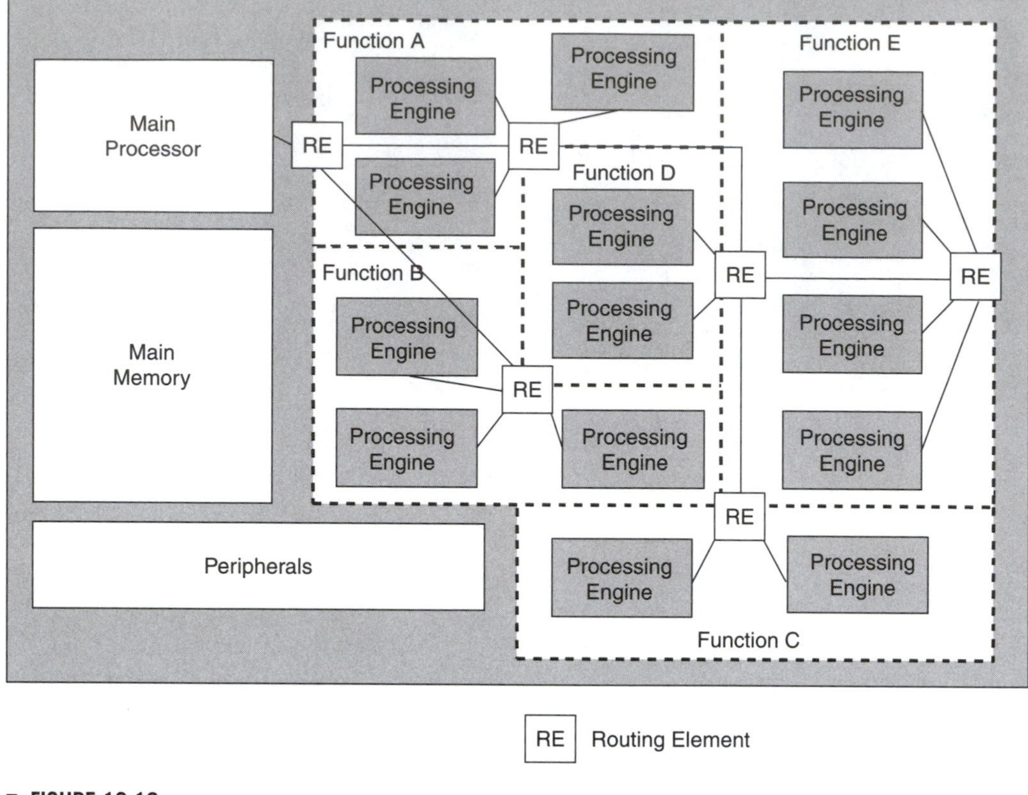

■ **FIGURE 13.13**

A MPSOC with a star-topology NoC.

promising for NoCs. It allows for intelligent message routing that can bypass congested communication hot spots and defective switch nodes or broken links. However, adaptive routing incurs the problem of packet ordering, as shown in Figure 13.14.

In Figure 13.14a, three packets arrive at RE 1 and need to pass on through the network to the right of RE 4. Packet 1 is routed from RE 1 to RE 2 (shown in Figure 13.14b) because RE 4 is busy. However, packet 2 is sent directly to RE 4 (shown in Figure 13.14c) because the busy condition clears. Meanwhile, packet 1 passes to RE 3 and packet 3 passes to RE 2 because packet 2 is already occupying RE 4. Eventually packets 1, 2, and 3 pass through this section of the NoC's mesh (shown in Figure 13.14d), but the operation of the adaptive routing algorithm has reordered the packets. With adaptive routing, each network packet can take a different path, which results in different latencies for each packet. Consequently, packets can arrive out of order. Each receiver must therefore have a

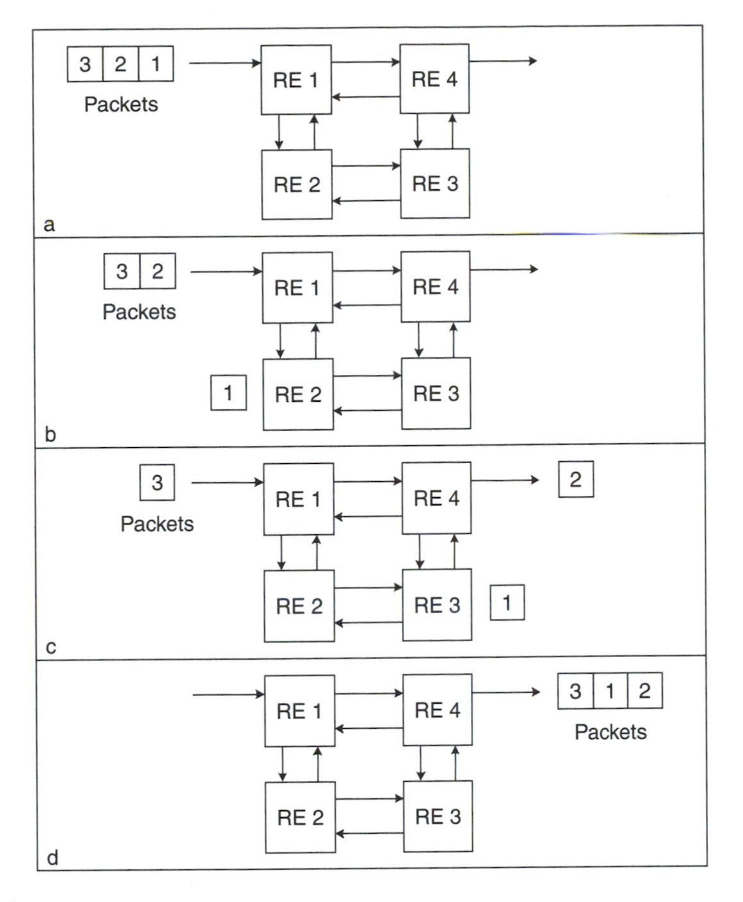

■ **FIGURE 13.14**

Adaptive routing can cause packet reordering when passing through a network's routing elements.

packet-reordering buffer, which increases the silicon overhead required by the network and also increases network latency.

NoCs with a star or tree topology cannot support adaptive routing because there is only one path through the network between each sender and receiver. All routes are pre-determined so no time is required to search for an "optimal" route for each packet. However, with only one route possible, traffic blockages or NoC "hot spots" that appear cannot be bypassed. Hot spots determine the maximum available network bandwidth, so each segment of the NoC must be designed accordingly.

The third NoC temptation is packet switching itself, which permits multiplexing of multiple messages onto one segmented communications channel. Like adaptive routing algorithms, channel segmentation adds packet buffers to the network and therefore consumes additional silicon that star and tree networks don't need.

13.6 GALS ON-CHIP NETWORKS

GALS (globally asynchronous, locally synchronous) NoCs address another growing problem in SOC design—one noted by the ITRS. There's no escaping the conclusion laid down by the ITRS:

> *. . . as it becomes impossible to move signals across a large die within one clock cycle or in a power-effective manner, or to run control and dataflow processes at the same rate [best effort versus guaranteed service], the likely result is a shift to [an] asynchronous (or, globally asynchronous and locally synchronous (GALS)) design style.*

This statement melds two SOC design problems: maintaining a constant clock skew across a large chip and efficiently conveying best-effort and guaranteed-service traffic among many blocks in a complex SOC.

The GALS approach to NoC design represents one way to attack these two problems. It recognizes the difficulty of maintaining near-constant clock skews across a complex SOC by discarding the effort entirely. Complex SOCs are already partitioned into many self-contained blocks (as demonstrated in the many SOC block diagrams shown in this chapter) and the GALS approach allows each of those blocks to be internally synchronous—they can even run at different clock rates—but inter-block communications is asynchronous, which eliminates the need for a global, low-skew reference clock. Commercial GALS design tools, such as the CHAINworks tool suite offered by Silistix, are just starting to appear on the market.

13.7 SOFTWARE CONSIDERATIONS FOR MPSOCs

Although this book is largely about hardware design of SOCs, any book that advocates the broad use of processors for implementing on-chip tasks must address the software issues. The rest of this chapter discusses this topic.

At a superficial level, MPSOCs look like large multiprocessor systems that present many problems to software developers. The electronics industry has yet to develop effective, automated methods for partitioning large programs and distributing the load across large numbers of processors. However, the heterogeneous collections of task-specific processing nodes described throughout this book are not at all like large homogeneous processing arrays and they are not programmed in the same manner. Task specificity is the key to dividing and conquering large problems.

Figure 13.15 again presents the floor plan of a multimedia SOC discussed earlier in this chapter. In addition to the SOC's main CPU, the figure shows four main processing blocks: audio, video, network, and mass

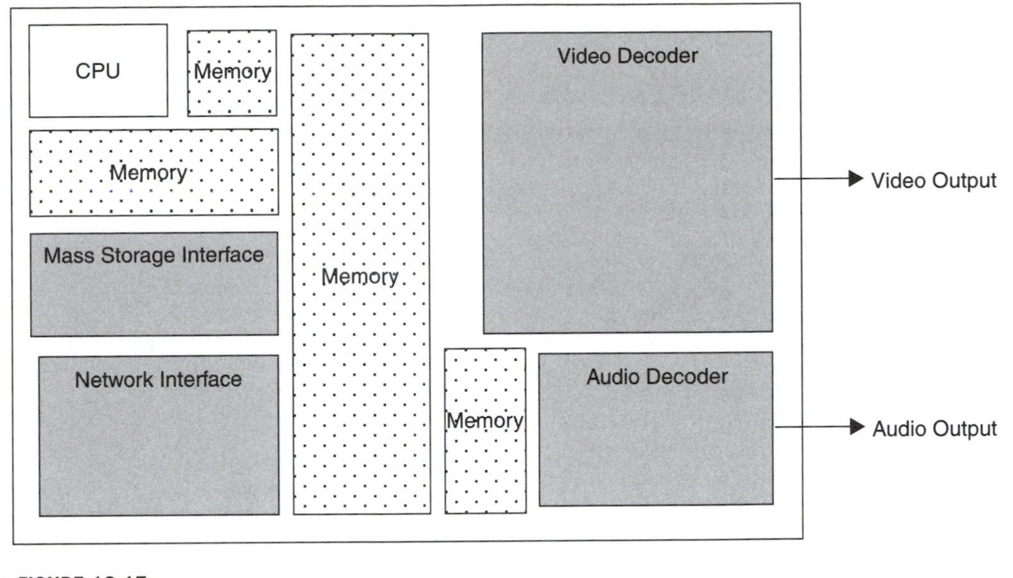

■ **FIGURE 13.15**

A multimedia SOC.

storage. Each of these processing blocks has a clearly defined job on the SOC and there's very little question about the duties of each block at this high level. Consequently, the code that needs to run on each block is also clearly evident. In fact, the code for this SOC will not have been written in one monolithic block. The audio and video decoders will likely run audio and video codecs written as standalone C programs. Similarly, the network block will run software network stacks developed as a separate networking package. Each of these programs will likely communicate with the others using messaging protocols.

This "divide-and-conquer" approach to the programming of large MPSOCs is essential to handling the exploding complexity of 21st-century systems. For one very practical reason, code for these systems cannot be written as a monolithic block to be later partitioned. Quite simply, if the code for a large system is written as a single, large program then code development will most assuredly miss the project schedule, probably by a wide margin. Figure 13.16, taken from Jack Ganssle's article "Subtract software costs by adding CPUs" that's noted in the chapter references, tells the entire, dismal story.

Figure 13.16 plots programmer productivity versus program size, where program size is measured in thousands of lines of code (KLOC). The plot is based on Barry Boehm's COCOMO (Constructive Cost Model), which considers 15 cost drivers related to software development. As software complexity rises, programming productivity plummets. Larger programs require more programmers (if completed in the same amount of

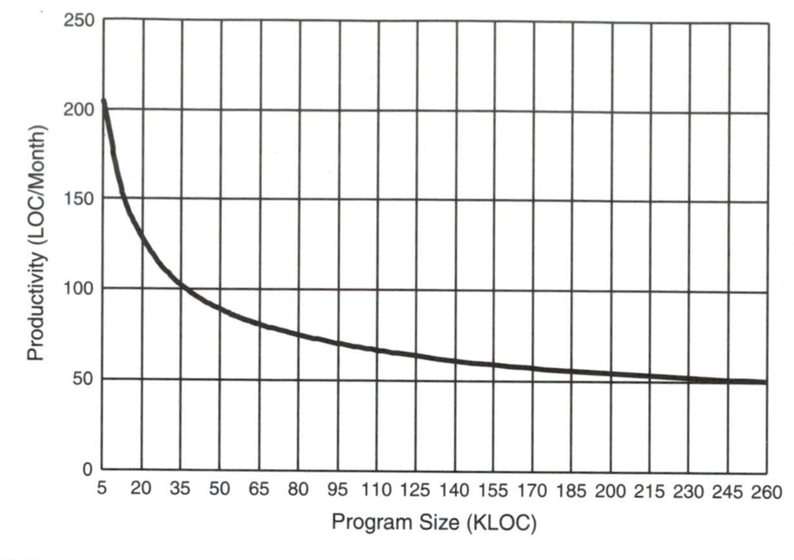

■ **FIGURE 13.16**

Programmer productivity plummets as a function of increasing program size, as predicted by the COCOMO model.

time) and the presence of more programmers creates many more communications channels between programmers, which in turn creates many more opportunities for miscommunication—and therefore opportunities for the insertion of program bugs that further degrade productivity. Figure 13.17 shows the effect of applying a "divide-and-conquer" software-development strategy. By slicing large programs into 20-KLOC chunks, the programming schedule stays linear with the number of LOC instead of rising exponentially.

There are at least two ways to run the partitioned software. One way is to run all of the smaller programs on one large, fast processor. This is the PC's software model. It requires multi-GHz processors that dissipate many tens of Watts. This is not an effective model for SOC development. As industry observer and commentator Jack Ganssle writes:

> *A single CPU manages a disparate array of sensors, switches, communications links, PWMs, and more. Dozens of tasks handle many sorts of mostly unrelated activities. A hundred thousand lines of code all linked into a single executable enslaves dozens of programmers all making changes throughout a Byzantine structure no one completely comprehends. Of course development slows to a crawl.*

The other way to run partitioned software is to distribute task-specific code across several heterogeneous processing blocks. This is the approach

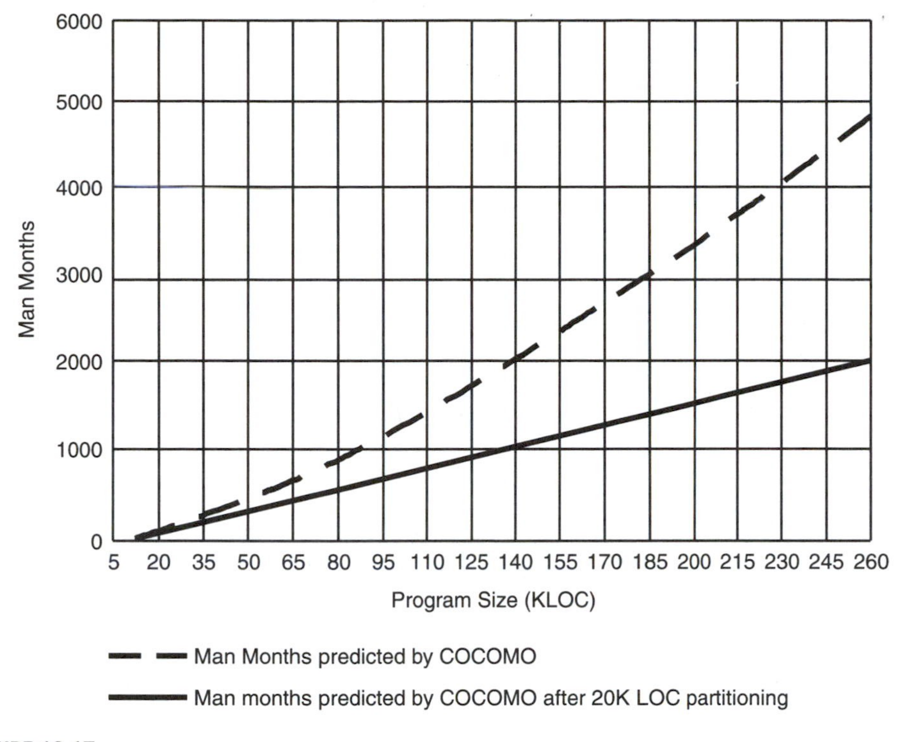

FIGURE 13.17

Programmer productivity stays linear with increasing program size if the programming task is sliced into 20-KLOC chunks.

advocated by this book and by the ITRS road map. It's also the approach advocated by Jack Ganssle, who gets the last words in this chapter:

> *Suppose the monolithic, single-CPU version of the product requires 100 K lines of code. The COCOMO calculation gives a 1,403 man-month development schedule.*
>
> *Segment the same project into four processors, assuming one has 50 KLOC and the others 20 KLOC each. Communication overhead requires a bit more code so we've added 10% to the 100-KLOC base figure. The schedule collapses to 909 man-months, or 65% of that required by the monolithic version.*
>
> *Maybe the problem is quite orthogonal and divides neatly into many small chunks, none being particularly large. Five processors running 22 KLOC each will take 1,030 man-months, or 73% of the original, not-so-clever design.*
>
> *Transistors are cheap—so why not get crazy and toss in lots of processors? One processor runs 20 KLOC and the other nine each run 10-KLOC programs. The resulting 724 man-month schedule is just half of the single-CPU approach. The product reaches consumers'*

hands twice as fast and development costs tumble. You're promoted and get one of those hot foreign company cars plus a slew of appreciating stock options. Being an engineer was never so good.

BIBLIOGRAPHY

Furber, S., *"Future Trends in SOC Interconnect,"* SOC 2005, Tampere, Finland, November 15–17, 2005.

Ganssle, J., *"Subtract Software Costs by Adding CPUs," Embedded Systems Design*, May, 2005, pp. 16–25, http://www.embedded.com/showArticle.jhtml?articleID=161600589.

Greiner, A., *"Distributed Micronetwork for GALS,"* SOC 2005, Tampere, Finland, November 15–17, 2005.

International Technology Roadmap for Semiconductors: 2005, http://public.itrs.net/.

Lee, S.-J., *"Realization of an On-chip Network for Practical Application to System-on-Chip—A Chip Designer's View,"* SOC 2005, Tampere, Finland, November 15–17, 2005.

Leibson, S., "NOC, NOC, NOCing on heaven's door: Beyond MPSOCs: *A report from the seventh-annual International Symposium on System-on-Chip design,"* EDN, December 8, 2005, http://www.edn.com/article/CA6289284.html.

Nurmi, J., *"Network-on Chip: A New Paradigm for System-on-Chip Design,"* SOC 2005, Tampere, Finland, November 15–17, 2005.

Smit, G., *"An Energy-Efficient Reconfigurable Circuit-Switched Network-on-Chip,"* SOC 2005, Tampere, Finland, November 15–17, 2005.

CHAPTER 14

BEYOND FIXED CORES

Anything you can do, I can do better.
I can do any thing better than you!
—Annie Oakley in
"Annie Get Your Gun," 1946

Because many applications just don't run fast enough on standard embedded microprocessor cores even with an auxiliary DSP core, engineering teams have hand-coded parts of many SOC designs in Verilog or VHDL to achieve system-level performance goals. However, custom, manually-coded RTL logic takes a long time to design and longer to verify. In addition, RTL blocks can't be easily changed once they're designed because of verification issues, yet changes are often needed to accommodate new standards or product features.

14.1 A VIABLE ALTERNATIVE TO MANUAL RTL DESIGN AND VERIFICATION

Configurable processors like Tensilica's Xtensa cores can be used as alternatives to manually-coded RTL blocks. Application-tailored Xtensa cores use the same data-path structures as traditional RTL blocks: deep pipelines, parallel execution units, task-specific state registers, and wide data buses to local and global memories. Tailored, task-specific processors can sustain the same high-computation throughput and support the same data interfaces as RTL hardware designs.

Migrating an SOC design team's design style from heavy use of RTL data paths and finite state machines (FSMs) to application-tailored processors with firmware control has many important implications:

1. *Flexibility*: changing the firmware is all that's needed to change a block's function.
2. *Software-based development*: Fast and low-cost software tools can be used to implement or modify most chip features.
3. *Faster, more complete system modeling*: For a 10-megagate design, even the fastest RTL logic simulator may not exceed a few simulation

cycles per second. By contrast, firmware simulations for extended processors will run on instruction-set simulators at hundreds of thousands or millions of cycles per second.

4. *Unification of control and data*: No SOC consists solely of hardwired logic. There's always at least one processor on the chip running software. Moving more RTL-based functions into processors removes artificial partitioning between control and data-processing functions.

5. *Time-to-market*: Using configurable processors simplifies SOC design, accelerates system modeling, and speeds hardware finalization. State machines implemented with firmware on an application-tailored processor can easily accommodate changes to standards or to the SOC's functional definition.

6. *Designer productivity*: The engineering manpower needed to manually develop and verify hardware RTL is greatly reduced. A processor-centric SOC design approach permits graceful recovery when (not if) a bug is discovered through firmware changes instead of chip respins.

The benefits of being able to make changes in firmware rather than hardware with a processor-centric design approach cannot be understated. Application-tailored processor cores reduce the risks associated with state-machine design by replacing hard-to-design, hard-to-verify, hardware FSMs with pre-designed, pre-verified processor cores running FSM code.

14.2 THE CONVENTIONAL, EMBEDDED SOFTWARE-DEVELOPMENT FLOW

Figure 14.1 illustrates a conventional development flow for creating embedded application software. Design work starts with the algorithm. Application developers generally use high-level design tools and languages such as C or C++ and they may also purchase pre-developed algorithms already written in those high-level languages.

Next, code developers translate the main algorithm and sub-algorithms into C to create a portable, processor-independent, application code base. After simulation and integration of the sub-algorithms and other application software modules into a coherent whole, the entire program is compiled for a target processor and the resulting application code is tested and profiled.

Often, to meet performance goals, software teams must convert critical sections of code into hand-tuned assembly code once a processor is

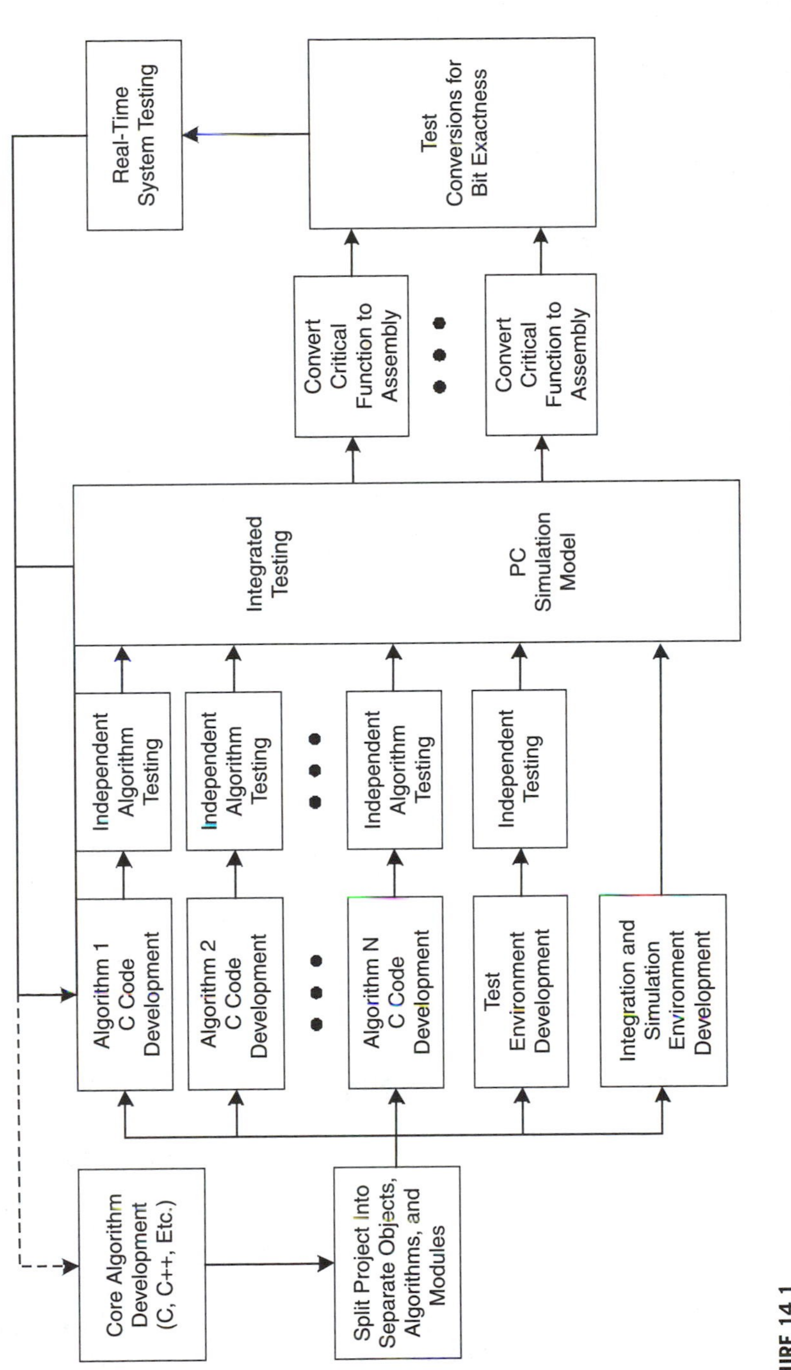

■ **FIGURE 14.1**

The conventional embedded software-development process.

selected. Assembly-code developers must carefully dovetail their variables into the processor's available registers because there's no way to add more registers to a fixed-ISA processor if the processor's existing register set proves inadequate.

14.3 FIT THE PROCESSOR TO THE ALGORITHM

Configurable, extensible processors allow developers to tailor the processor to the target algorithms. Designers can add special-purpose, variable-width registers; specialized execution units; and wide data buses to reach an optimum processor configuration for specific algorithms. They perform this work using a surprisingly small variation of the conventional development flow shown above in Figure 14.1. In addition to recoding critical routines in assembly language, a configurable processor gives the SOC development team the option of adding instructions, registers, and register files to the processor that speed critical inner loops.

This processor tailoring can occur at the same point in the software-development process as the assembly-coding step, as shown in Figure 14.2. One box (with a drop shadow for emphasis) has been added. This box represents the step where the SOC designers add processor extensions to an Xtensa processor core using the Tensilica Instruction Extension (TIE) language.

As with hand-tuned assembly language, code-optimization points for an application-tailored processor become apparent through code profiling. Optimization targets typically reside within the innermost software loops that execute many thousands or millions of times per second. Reducing the instruction count of the object code inside of these critical inner loops greatly improves system performance without raising clock rate.

The examples in the following sections show the performance improvements possible when using the TIE language to tailor Tensilica's Xtensa processors.

14.4 ACCELERATING THE FAST FOURIER TRANSFORM

The heart of the decimation-in-frequency FFT algorithm is the "butterfly" operation, which resides at the algorithm's innermost loop. Each butterfly operation requires six additions and four multiplications to compute the real and imaginary components of a radix-2 butterfly result. Using the TIE language, it's possible for a design team to augment the Xtensa processor's pipeline with four adders and two multipliers so that the processor can compute half of an FFT butterfly in one cycle.

The Xtensa processor's configurable memory interfaces can be configured to be as wide as 128 bits so that all four real and imaginary integer

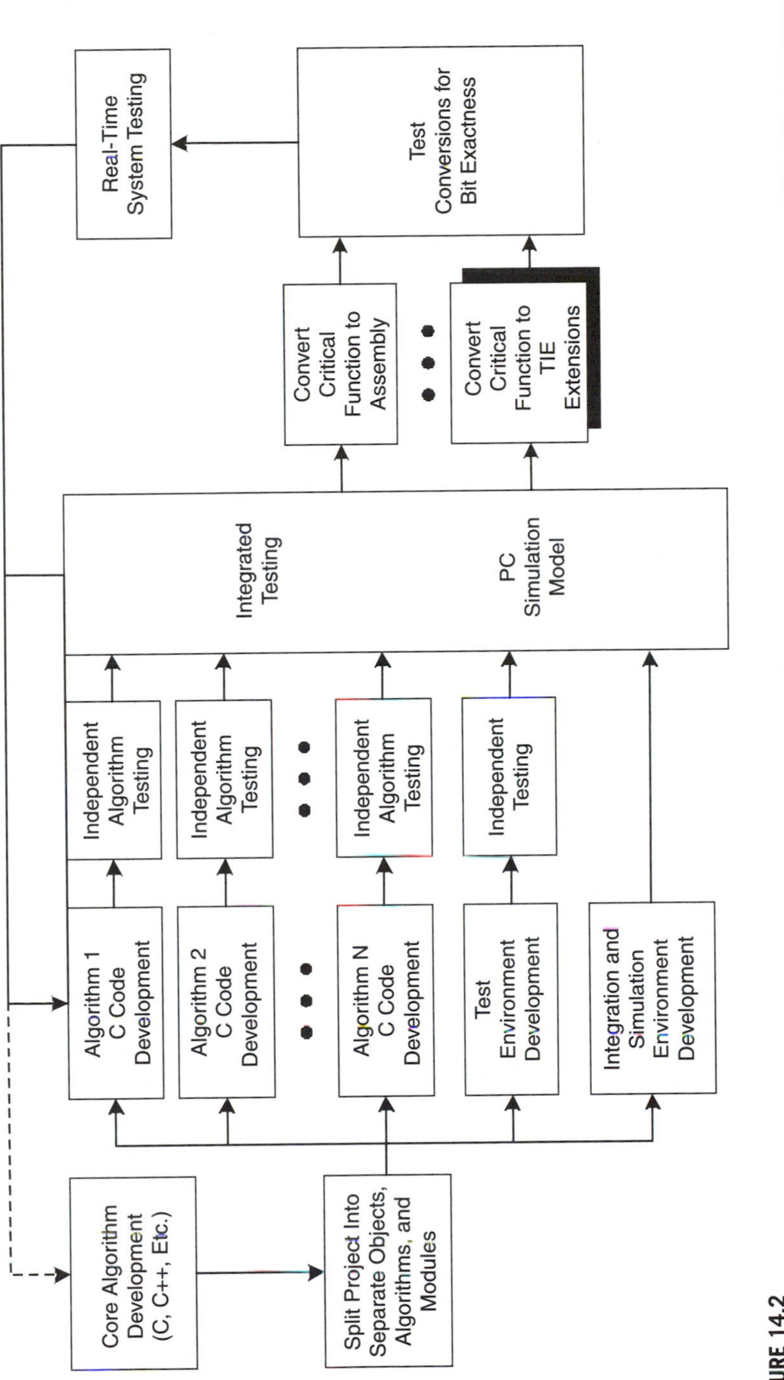

■ **FIGURE 14.2**

The embedded software-development process with configurable processors.

TABLE 14.1 ■ Acceleration results from processor augmentation with FFT instructions

		C code with software multiplication	C code with hardware multiplication	C code with TIE-based butterfly instruction extension	Performance improvement
Code size		430 bytes + libraries	430 bytes	158 bytes	
	FFT length				
Performance (cycles)	128-point	763548	169739	2269	337x
	256-point	1787645	386498	4711	379x
	512-point	3975245	867133	9841	404x
	1024-point	9241893	1922644	20603	449x

input terms for each butterfly can be loaded into special-purpose FFT input registers in one cycle. A processor with 128-bit memory interfaces can store all four computed output components into memory in one cycle as well.

Practically speaking, it's very hard to create single-cycle, synthesizable multipliers for SOCs that operate at clock rates of several hundred MHz. It's better to stretch the multiplication across two cycles so that the multiplier doesn't become the critical timing element on the SOC. The additional multiplier latency does not affect the throughput of the butterfly computations in this example and, if necessary, even longer latencies can be accommodated through additional state storage in the butterfly execution unit.

This approach adds a SIMD (single-instruction, multiple data) butterfly computation unit to the processor (using fewer than 35,000 gates including the two 24×24-bit multipliers). The performance improvements appear in Table 14.1. The table also shows the code size of the FFT programs with and without the TIE extensions.

14.5 ACCELERATING AN MPEG-4 DECODER

One of the most difficult parts of encoding MPEG-4 video data is motion estimation, which searches adjacent video frames for similar pixel blocks to detect inter-frame movement in the picture. The motion-estimation search algorithm's inner loop contains an SAD (sum of absolute differences) operation consisting of a subtraction, an absolute value, and the addition of the resulting value with the previously computed value.

For a QCIF (quarter common image format, 176×144 pixels) image frame, a 15-Hz frame rate, and an exhaustive-search motion-estimation scheme, SAD operations require about 641 million operations/sec. As shown in Figure 14.3, it's possible to add SIMD SAD hardware

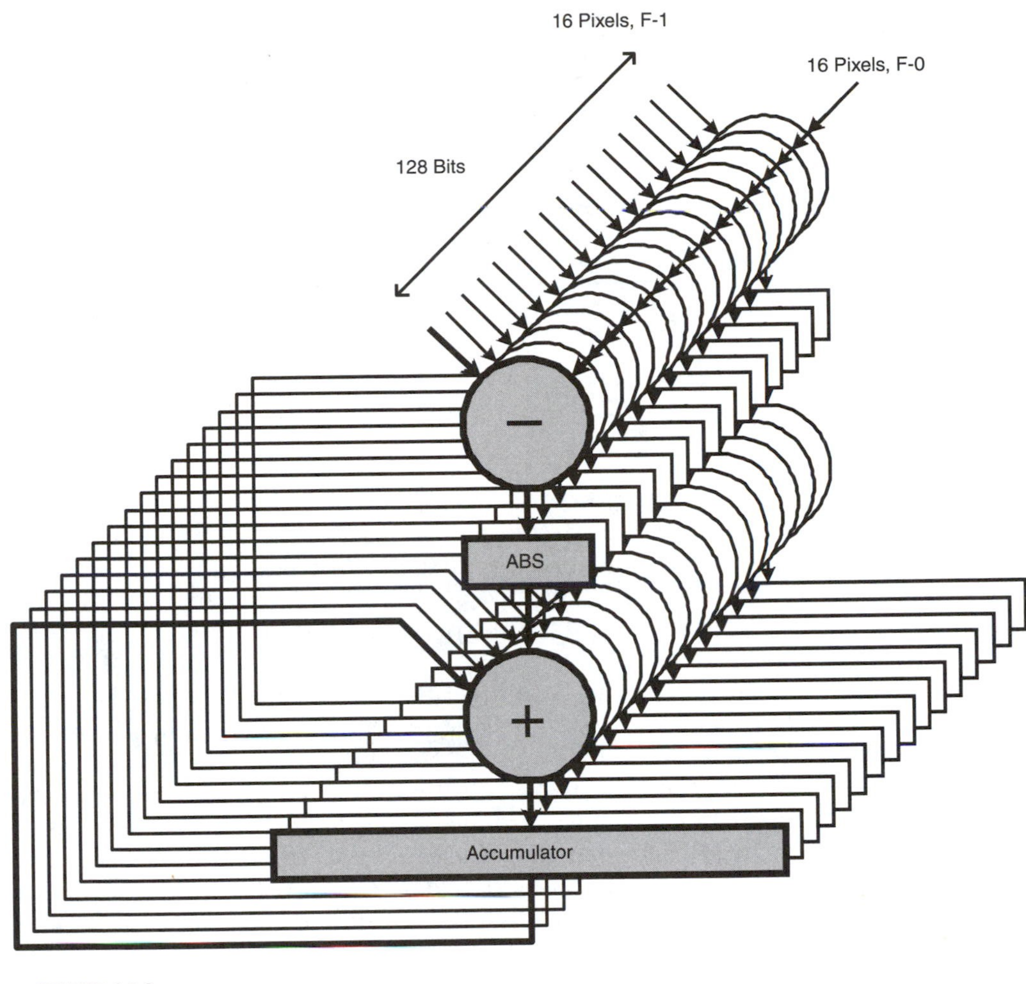

■ **FIGURE 14.3**

MPEG-4 SIMD SAD instruction execution hardware.

capable of executing 16 pixel-wide SAD instructions per cycle using TIE. (*Note*: Configuring the Xtensa processor's memory bus to be 128 bits wide makes it possible to load 16 pixels worth of data using one load instruction.)

Executing all three SAD component operations (subtraction, absolute value, addition) at once for 16 pixel values simultaneously reduces the 641 million operations/sec requirement to 14 million instructions/sec, a substantial reduction in cycle count, which should result in a reduced clock rate. This MPEG-4 motion-estimation accelerator is part of a MPEG-4 decoder reference design developed by Tensilica. The MPEG-4 decoder adds approximately 100,000 gates to the base Xtensa processor

TABLE 14.2 ■ MPEG-4 decoder acceleration results from processor augmentation with FFT instructions

Video clip	Original MPEG-4 decoder performance (# of execution cycles)	Optimized MPEG-4 decoder performance (# of execution cycles)	Clock frequency (15 frames/sec)	TIE speedup
Miss America	3.126 G cycles	76.81 M cycles	7.7 MHz	40.1x
Suzie	3.389 G cycles	102.19 M cycles	10.3 MHz	33.2x
Foreman	10.045 G cycles	359.5 M cycles	13.5 MHz	27.9x
Car phone	9.222 G cycles	308.7 M cycles	12.2 MHz	29.9x
Monsters Inc.	29.327 G cycles	822.8 M cycles	8.6 MHz	35.6x

and implements a 2-way QCIF video codec operating at 15 frames/sec or a QCIF MPEG-4 decoder that operates at 30 frames/sec using approximately 30 MIPS for either operational mode.

Other MPEG-4 algorithms also can be accelerated including variable-length decoding, iDCT, bitstream processing, dequantization, AC/DC prediction, color conversion, and post filtering. When instructions are added to accelerate all of these MPEG-4 decoding tasks, creating an MPEG-4 SIMD engine within the tailored processor, the results can be quite surprising.

As Table 14.2 shows, the resulting SIMD engine acceleration drops the number of cycles required to decode the MPEG-4 video clips from billions to millions and the required processor operating frequency by roughly 30x to around 10 MHz. Without the additional, application-tailored instructions, the processor would need to run at roughly 300 MHz to perform the MPEG-4 decoding. Clearly, there is a substantial difference in power dissipation and process-technology cost between a 10 MHz and a 300 MHz processor. It's unlikely that any amount of assembly language coding could produce similarly large drops in the clock rate.

As shown in the examples above, it's possible to accelerate the performance of embedded algorithms using configurable and extensible microprocessor cores. Designers can add precisely the resources (special-purpose registers, execution units, and wide data buses) required to achieve the desired algorithmic performance instead of attempting to shoehorn algorithms into the computational assets of a fixed-ISA processor.

This design approach only requires that the design team be able to profile existing algorithm code and to find the critical inner loops in that profiled code (two tasks they already perform). From these profiles, the design team can then define new processor instructions and registers that accelerate these critical loops. The result of this new approach is to greatly accelerate algorithm performance. In most cases, designers can replace entire RTL blocks with configurable processors tuned for the exact application, saving valuable design and verification time and adding an extra level of flexibility because of the inherent programmability of this approach.

14.6 BOOST THROUGHPUT WITH MULTIPLE OPERATIONS PER CYCLE

The Xtensa LX processor is not limited to performing one operation per cycle. TIE provides two ways to perform two or more operations at a time. SOC designers can create single-cycle TIE instructions that draw operands from several sources, perform multiple operations on these operands concurrently, and then output the results to several destinations. Alternatively, SOC designers can use the Xtensa LX processor's FLIX (flexible length instruction extension) technology to create wide-word instructions that perform multiple independent operations during the same instruction cycle. The Diamond 570T CPU, Diamond 330HiFi DSP, and Diamond 545CK DSP all incorporate FLIX-format instructions, which were previously described in the appropriate chapters. In some applications, both methods (compound-operation instructions, and FLIX instructions) will work equally well. In other applications, the flexibility of FLIX instructions will make function-block development much easier.

The example code below shows a short but complete example of a very simple long-instruction word processor described in TIE with FLIX technology. It relies entirely on built-in definitions of 32-bit integer operations, and defines no new operations. It creates a processor with a high degree of potential parallelism even for applications written purely in terms of standard C integer operations and data-types. The first of three slots supports all the commonly used integer operations, including ALU operations, loads, stores, jumps, and branches. The second slot offers loads and stores, plus the most common ALU operations. The third slot offers a full complement of ALU operations, but no loads and stores.

```
1: length ml64 64 {InstBuf[3:0] == 15}
2: format format1 ml64 {base_slot, ldst_slot, alu_slot}
3: slot_opcodes base_slot {ADD.N, ADDX2, ADDX4, SUB, SUBX2, SUBX4,
    ADDI.N, AND, OR, XOR, BEQZ.N, BNEZ.N, BGEZ, BEQI, BNEI, BGEI, BNEI,
    BLTI, BEQ, BNE, BGE, BLT, BGEU, BLTU, L32I.N, L32R, L16UI, L16SI,
    L8UI, S32I.N, S16I, S8I, SLLI, SRLI, SRAI, J, JX, MOVI.N}
4: slot_opcodes ldst_slot {ADD.N, SUB, ADDI.N, L32I.N, L32R, L16UI,
    L16SI, L8UI, S32I.N, S16I, S8I, MOVI.N}
5: slot_opcodes alu_slot {ADD.N, ADDX2, ADDX4, SUB, SUBX2, SUBX4, ADDI.N,
    AND, OR, XOR, SLLI, SRLI, SRAI, MOVI.N}
```

The first line of this example code declares a new instruction length (64 bits) and specifies the encoding of the first 4 bits of the instruction that determine the length. The second line declares a format for that instruction length, `format1`, containing three slots: `base_slot`, `ldst_slot`, and `alu_slot` and names the three slots within the new format. The fourth line lists all the TIE instructions that can be packed into the first of those slots: `base_slot`. In this case, all the instructions happen to be pre-defined

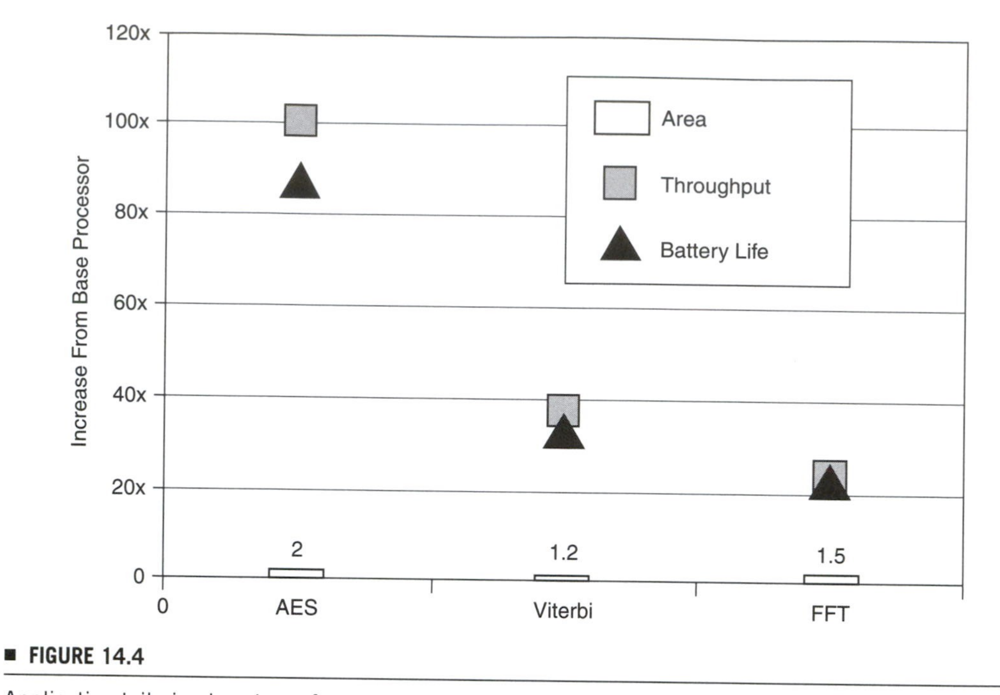

■ **FIGURE 14.4**

Application tailoring boosts performance and energy efficiency.

Xtensa LX instructions but new instructions could also be included in this slot. The processor generator also creates a NOP (no operation) for each slot, so the software tools can always create complete instruction bundles, even when no other operations for that slot are available for packing into a long instruction. Lines 4 and 5 designate the subset of instructions that can go into the other two slots. In fact, the code in Figure 14.4 goes a long way towards creating a Diamond 570T CPU.

The following block of example code defines a long-instruction-word architecture with a mix of built-in 32-bit operations and new 128-bit operations. It defines one 64-bit instruction format with three sub-instruction slots (`base_slot`, `ldst_slot`, and `alu_slot`). The description takes advantage of the Xtensa processor's pre-defined RISC instructions, but also defines a large new register file and three new ALU operations on the new register file:

```
1: length ml64 64 {InstBuf[3:0] == 15}
2: format format1 ml64 {base_slot, ldst_slot, alu_slot}
3: slot_opcodes base_slot {ADD.N, ADDX2, ADDX4, SUB, SUBX2, SUBX4,
     ADDI.N, AND, OR, XOR, BEQZ.N, BNEZ.N, BGEZ, BEQI, BNEI, BGEI, BNEI,
     BLTI, BEQ, BNE, BGE, BLT, BGEU, BLTU, L32I.N, L32R, L16UI, L16SI,
     L8UI, S32I.N, S16I, S8I, SLLI, SRLI, SRAI, J, JX, MOVI.N}
4: regfile × 128 32 ×
```

```
5: slot_opcodes ldst_slot {loadx, storex}  /* slot does 128b load/store*/
6: immediate_range sim8 -128 127 1 /*8 bit signed offset field */
7: operation loadx {in x   *a, in sim8 off, out x d} {out VAddr, in
   MemDataIn128}{
8: assign VAddr = a + off; assign d = MemDataIn128;}
9: operation storex {in x   *a, in sim8 off, in x s} {out VAddr,out
   MemDataOut128}{
10: assign VAddr = a + off; assign MemDataOut128 = s;}
11: slot_opcodes alu_slot {addx, andx, orx} /* two new ALU operations on
    x regs */
12: operation addx {in x a, in x b, out x c} {} {assign c = a + b;}
13: operation andx {in x a, in x b, out x c} {} {assign c = a & b;}
14: operation orx {in x a, in x b, out x c} {} {assign c = a | b;}
```

The first three lines of this code block are identical to those of the previous example. The fourth line declares a new register file 128 bits wide and 32 entries deep. The fifth line lists the two load and store instructions for the new wide register file, which can be found in the second slot of the long instruction word. The sixth line defines a new immediate range, an 8-bit signed value, to be used as the offset range for the new 128-bit load and store instructions.

Lines 7–10 fully define the new load and store instructions, in terms of basic interface signals Vaddr (the address used to access local data memory), MemDataIn128 (the data being returned from local data memory), and MemDataOut128 (the data to be sent to the local data memory). The use of 128-bit memory data signals also guarantees that the local data memory will be at least 128 bits wide. Line 11 lists the three new ALU operations that can be put in the third slot of the long instruction word. Lines 12–14 fully define those operations on the 128-bit wide register file: add, bit-wise AND, and bit-wise OR.

With this example, any combination of the 39 instructions (including NOP) in the first slot, three instructions in the second slot (loadx, storex, and NOP), and four instructions in the third slot can be combined to form legal instructions—a total of 468 combinations. This example shows the potential to independently specify operations to enable instruction-level parallelism. Moreover, all of the techniques for improving the performance of individual instructions—especially fusion and SIMD—are readily applied to the operations encoded in each sub-instruction.

14.7 HIGH-SPEED I/O FOR PROCESSOR-BASED FUNCTION BLOCKS

The two bottlenecks in high-speed SOC block design are I/O performance and computational performance. The previous sections in this chapter have discussed improvements in a processor's computational performance through TIE-based extensions. This section and following sections discuss the ways that TIE can be used to improve a processor's I/O performance.

Every microprocessor core's main bus represents a major I/O bottleneck. All data into and out of the processor must pass over this main bus. Consequently, two factors constrain I/O traffic in and out of the processor. First, a bus can only perform one transfer at a time so other pending transfers must wait for the current transfer to clear. Second, because processor main buses are designed to accommodate many system configurations, they tend to require multiple cycles to effect bus transactions. As a result of these limitations, processor cores have lacked the I/O bandwidth required by many tasks performed in SOCs.

Tensilica's Xtensa LX processor incorporates several features that allow an SOC design team to improves its I/O bandwidth through application-specific tailoring. In fact, these features allow the Xtensa LX processor to deliver I/O transfer rates that can match those achieved by manually designed hardware RTL blocks. However, the Xtensa LX processor achieves high I/O data rates with automatically generated, pre-verified hardware that greatly reduces the time required to design and verify the SOC. In addition, the resulting function block is firmware-programmable, which means that it can be changed at a later date to accommodate a new or revised industry standard, to add a feature, or to fix a bug in the system design without changing the silicon.

The key features that allow the Xtensa LX processor to achieve these high data-transfer rates are TIE ports and queue interfaces, which allow designers to add many new input and output ports that lead directly into and out of the processor's execution unit. Most of the Diamond Standard series processor cores have ports or queue interfaces, which have been discussed in previous chapters. Each Diamond core (except for the Diamond 232L CPU core) has a pair of 32-bit ports, a pair of 32-bit queue interfaces, or both. The Xtensa LX processor can have as many as 1024 ports and each port can be as wide as 1024 bits. Consequently, it's possible to add as many as one million I/O pins to an Xtensa LX processor.

The Xtensa LX ports and queues can be directly invoked by instruction extensions written in TIE, so that input and output operations become implicit in the execution of a computation. This approach maximizes I/O bandwidth in a manner quite similar to the high I/O bandwidth achieved by manually designing function blocks using RTL. However, the processor-based approach requires much less effort from the SOC development team to design and verify the hardware because the Xtensa LX processor is generated automatically by Tensilica's Xtensa Processor Generator.

14.8 THE SINGLE-BUS BOTTLENECK

The sole data highway into and out of most processor cores is the main processor bus. Because processors often interact with other types of bus masters including other processors and DMA controllers, their main buses have sophisticated transaction protocols and arbitration mechanisms for sharing the bus among masters. These extra mechanisms result in bus transactions that occur over several clock cycles.

Read transactions on the Xtensa LX processor's main bus (PIF) take a minimum of six cycles and a write transaction takes at least one cycle, depending on the speed of the target device connected to the PIF. From these transaction timings, we can calculate the minimum number of cycles needed to perform a simple flow-through computation, where two numbers are loaded from memory, added, and stored back into memory. The assembly code to perform this computation might look like this:

```
L32I reg_A, Addr_A          ; Load the first operand
L32I reg_B, Addr_B          ; Load the second operand
ADD reg_C, reg_A, reg_B     ; Add the two operands
S32I reg_C, Addr_C          ; Store the result
```

To simplify this code, we assume that pointers to memory locations storing values A, B, and C are already initialized in registers `Addr_A`, `Addr_B`, and `Addr_C`. If not, then more time will be needed for this computation.

The minimum cycle count required to perform this computation is:

```
L32I reg_A, Addr_A          ; 6 cycles
L32I reg_B, Addr_B          ; 6 cycles
ADD reg_C, reg_A, reg_B     ; 1 cycle
S32I reg_C, Addr_C          ; 1 cycle
```

Total: 14 cycles
(*Note*: This cycle count is a minimum number. Because the Xtensa LX processor is pipelined, the total number of cycles will be slightly greater than 14 but the additional cycles will overlap the execution of other instructions. If this code sequence sits within a zero-overhead loop, the cycle count for each loop iteration is 14 cycles.)

Note that the two load (`L32I`) instructions consume 6 cycles each. This cycle count is the minimum required to return the requested information over the processor's main bus (the PIF). Loads must complete before the next instruction that uses the resulting data executes. Also note that the store (`S32I`) instruction consumes only one cycle because the stored value is immediately placed in a store buffer. Once the value enters the processor's store buffer, the store instruction completes. The processor's bus-control logic subsequently moves the stored value to the target location.

For high-speed data that must flow through this function block, the large number of required cycles for this flow-through operation is often a major factor in deciding to design a purpose-built block of RTL hardware to perform the task because a conventional processor would be too slow.

One way to solve this problem is to conduct the load and store transactions over a faster bus to improve the overall I/O bandwidth. Some Xtensa processor configurations have a local-memory bus interface called the XLMI that implements a simpler transaction protocol than the processor's main PIF bus. XLMI transaction protocols are simpler than PIF protocols, so load and store operations can occur in as little as one cycle. By conducting loads and stores over the XLMI bus instead of the PIF, the above computation timing becomes:

```
L32I reg_A, Addr_A        ; 1 cycle
L32I reg_B, Addr_B        ; 1 cycle
ADD reg_C, reg_A, reg_B   ; 1 cycle
S32I reg_C, Addr_C        ; 1 cycle
```

Total: 4 cycles (with the same caveat regarding the processor pipeline)

This code sequence represents a 3.5x improvement in the function's cycle count compared to the equivalent transaction conducted over the PIF, which may mean the difference between acceptable and unacceptable performance for a particular task. However, even with the performance improvement gained from faster bus transactions, the XLMI bus still conducts only one transaction at a time, so loads and stores must occur sequentially. Even the 4-cycle count achieved with single-cycle loads and stores over a fast local bus can be too slow for certain SOC tasks. TIE ports and queue interfaces significantly boost the I/O bandwidth of the Xtensa LX processor.

Ports and queues are very simple, direct communication structures. Like XLMI transactions, transactions over ports and queues occur in one cycle. However, transactions conducted over ports and queue interfaces are not activated by addresses supplied by the processor. These simpler structures are activated by specially written processor instructions that implicitly initiate port and queue transactions. Consequently, one designer-defined instruction can initiate transactions on several ports and queues at the same time, which boosts I/O bandwidth.

Using TIE, it's possible to create queue interfaces especially for the example problem discussed in this section. Three queue interfaces are needed: two input queues for the input operands and one output queue for the result. With these three queue interfaces defined, it's possible to define an addition instruction that:

1. Implicitly draws input operands A and B from their respective input queues.
2. Adds A and B together.
3. Outputs the result of the addition (C) on the output queue.

The TIE code needed to create the three queue interfaces is:

```
queue InQ_A 32 in
queue InQ_B 32 in
queue OutQ_C 32 out
```

The first two statements declare input queues named InQ_A and InQ_B that are 32 bits wide. The third statement declares an output queue named OutQ_C that is also 32 bits wide. Using the TIE queue statement causes the Xtensa Processor Generator to create an additional I/O port with the handshake lines needed to connect to an external FIFO memory.

TIE instructions can implicitly read from several input queues and write to several output queues during one clock cycle. Input queues serve as input operands to TIE instructions and output queues can be assigned values just like any other output operand. (Note that input and output ports and queues need not be 32 bits wide. They can be as narrow as 1 bit or as wide as 1024 bits.)

The following example describes a TIE instruction called ADD_XFER that reads operands from each of the input queues defined above, adds the values together, and writes the result to an output queue.

```
operation ADD_XFER {} {in InQ_A, in InQ_B, out OutQ_C} {
        assign OutQ_C = InQ_A + InQ_B;
}
```

With this new instruction, the example problem reduces to one instruction:

```
ADD_XFER
```

This instruction takes five cycles to run through the Xtensa LX processor's 5-stage pipeline but the instruction has a latency of only one clock cycle. By placing this instruction within a zero-overhead loop, the processor can deliver an effective throughput of one ADD_XFER instruction per clock cycle. Thus the computation and data movement occur in the absolute minimum number of clock cycles, namely one. Even hand-coded, hardwired RTL cannot perform this function any faster than the application-tailored Xtensa LX processor.

14.9 ALONE, FASTER IS NOT NECESSARILY BETTER

The above examples demonstrate that application-tailored processors can run specific code blocks one or two orders of magnitude faster than fixed-ISA processors without the need for faster clock rates. However, most SOCs also require that power dissipation and energy consumption (power dissipated over time) not rise commensurately with the performance improvement. Here too, configurable processors excel. Figure 14.4 demonstrates the performance and power benefits of application tailoring for three applications: AES (advanced encryption standard) cryptographic coding and decoding, Viterbi decoding, and the FFT. In each case, ISA extensions have been added to an Xtensa processor to improve its performance. As a result of the extensions, throughput improves by factors of more than 20x–100x, depending on the application.

A second result shown by Figure 14.4 is the reduction in energy required to execute the application over the same time period (as measured in terms of battery life). Energy consumption improves by nearly as much as performance. Although the size of each tailored processor is larger than the untailored version, the drop in the required clock rate more than compensates for the additional capacitance incurred by the extra silicon needed for the enhancement logic.

The Diamond processor cores, which are based on Xtensa configurable cores, deliver more performance and low-power operation relative to other fixed-ISA processors, as shown in Figure 14.5, which plots performance

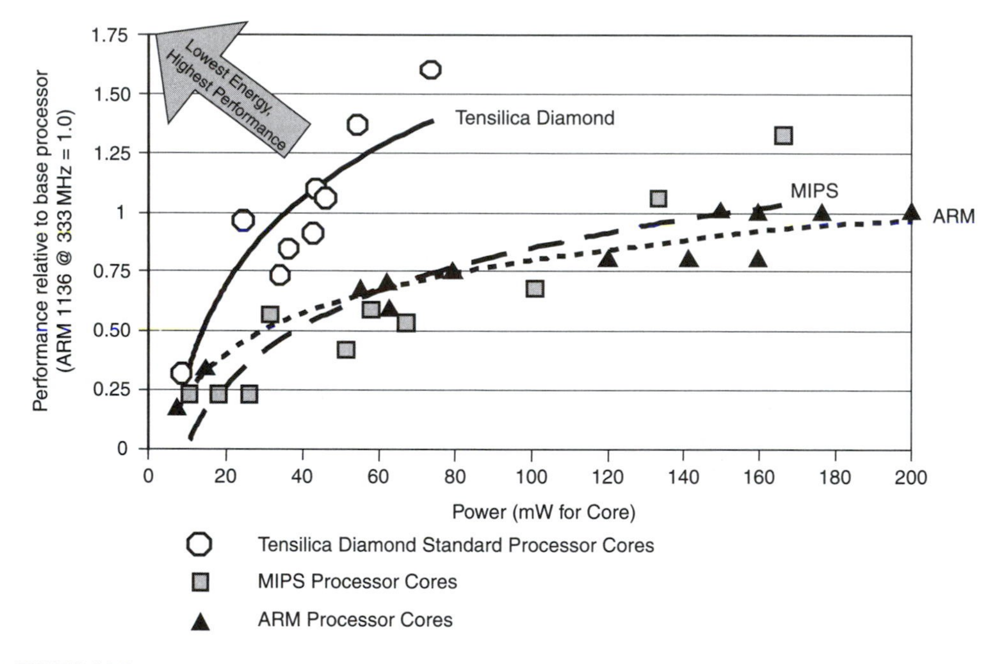

■ **FIGURE 14.5**

Performance/power numbers for three fixed-ISA processor families.
Performance on EEMBC benchmarks aggregate for Consumer, Telecom, Office, Network, based on ARM 1136J-S (Freescale i.MX31), ARM 1026EJ-S, Tensilica Diamond 570T, Xtensa V, and Xtensa 3, MIPS 20K, (NECVR5000). MIPS M4K, MIPS 4Ke, MIPS 4Ks, MIPS 24K, ARM 968E-S, ARM 966E-S, ARM 926EJ-S, and ARM7TDMI-S scaled by ratio of Dhrystone MIPS within architecture family.
All power figures from vendor websites, 2/23/2006.

and power dissipation for several ARM, MIPS, and Tensilica Diamond processors as reported for the EEMBC suite of benchmarks. Note that each processor family demonstrates a characteristic logarithmic curve linking performance and power dissipation. As performance improves (mostly due to clock-rate increases), power dissipation rises. As clock frequencies increase, performance increases as well, but not as quickly as power dissipation. Hence the logarithmic curves.

As shown in Figure 14.6, application tailoring Xtensa processors for each of the benchmarks produces performance/power numbers in an entirely different league. Figure 14.6 must be drawn to an entirely different scale to accommodate the performance range of the configurable Xtensa processor.

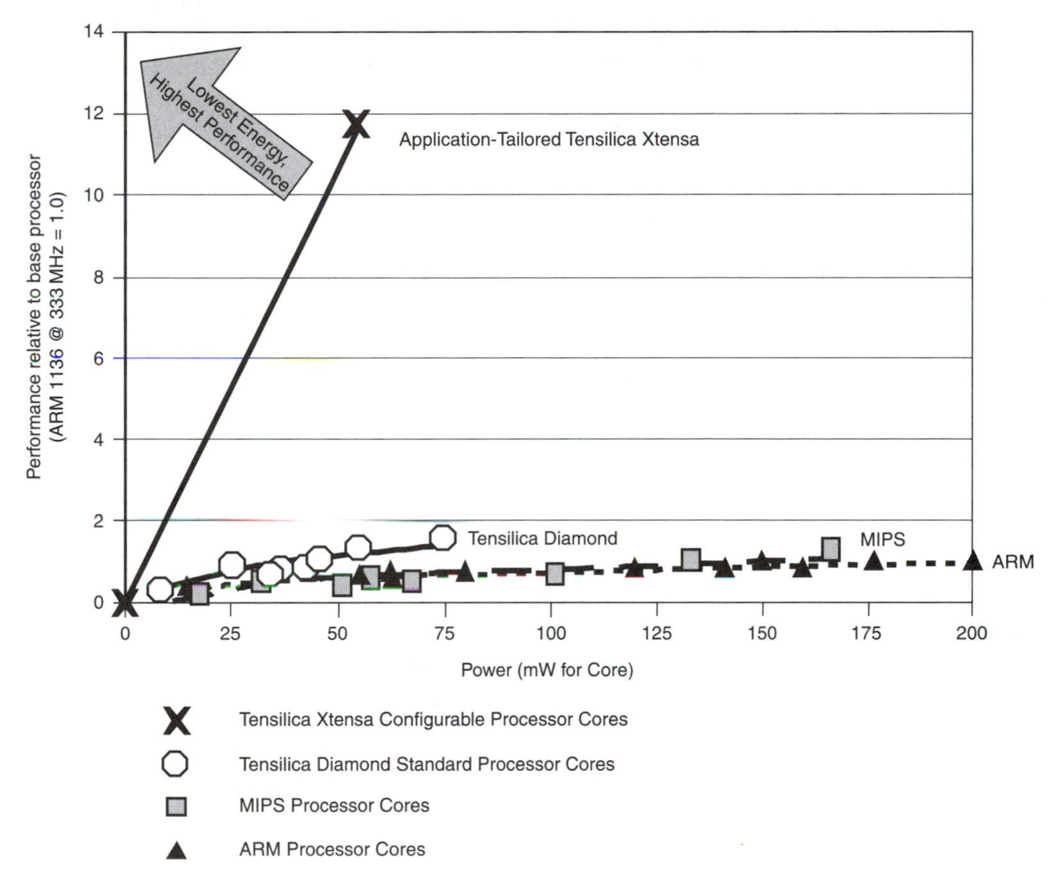

FIGURE 14.6

Performance/power numbers for three fixed-ISA processor families and application-tailored Xtensa processor cores.
Performance on EEMBC benchmarks aggregate for Consumer, Telecom, Office, Network, based on ARM 1136J-S (Freescale i.MX31), ARM 1026EJ-S, Tensilica Diamond 570T, Xtensa V, and Xtensa 3, MIPS 20K, (NECVR5000). MIPS M4K, MIPS 4Ke, MIPS 4Ks, MIPS 24K, ARM 968E-S, ARM 966E-S, ARM 926EJ-S, and ARM7TDMI-S scaled by ratio of Dhrystone MIPS within architecture family.
All power figures from vendor websites, 2/23/2006.

The performance-versus-power-dissipation equation looks even better for configurable processor cores when looking at multiple cores. Figure 14.7 shows the silicon footprints of three ways to get high processor throughput using 90 nm SOC process technology. Figure 14.7a shows an ARM Cortex A8 processor core. The core runs at 800 MHz, consumes about 5 mm^2 of silicon, dissipates 800 mW, and delivers about 1600 peak instructions/sec. Figure 14.7b shows the 3-way superscalar Diamond 570T CPU core, which runs at 525 MHz, consumes about 0.6 mm^2 of silicon, dissipates about 90 mW, and delivers 1575 peak instructions/sec. Figure 14.7c shows three Xtensa 6 processor cores. Each Xtensa 6 core runs at 525 MHz, the three cores together dissipate about 50 mW and consume about 0.36 mm^2 of silicon, and the three processors together deliver 1575 peak instructions/sec.

The approach shown in Figure 14.7a represents a conventional method for achieving performance: use advanced process technology and a deep pipeline to boost the processor's clock frequency. The second approach

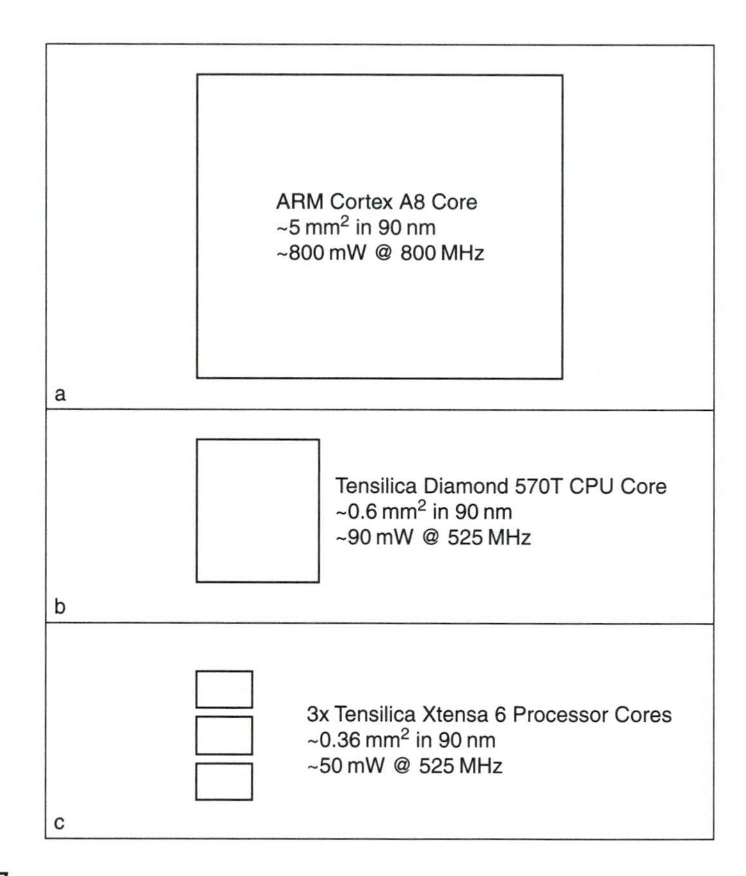

■ **FIGURE 14.7**

Three ways to get 1600 peak instructions/second from 90 nm SOC process technology.

applies more modern processor architecture and compiler technologies to achieve equivalent performance at a lower clock rate, with less power dissipation, and a smaller silicon footprint. The approach shown in Figure 14.7c has all of the advantages, however. The three processor cores are independent. Together, they consume the least amount of power of the three alternatives, they use the least amount of silicon, and they can be used independently. When not needed, they can be put to sleep, which greatly reduces dynamic power dissipation.

Figure 14.7 encapsulates many significant reasons why using multiple on-chip processors to increase an SOC's processing capability is a good idea:

- Lower clock rate
- Lower power dissipation
- Smaller silicon footprint
- More options for effectively using the available processing power.

BIBLIOGRAPHY

A Quick Guide to High-Speed I/O for SOC Function Blocks, Tensilica white paper, www.tensilica.com.

Configurable Processors: What, Why, How?, Tensilica white paper, www.tensilica.com.

FLIX: Fast Relief for Performance-Hungry Applications, Tensilica white paper, www.tensilica.com.

Increasing Computational Performance Through FLIX (Flexible Length Instruction Extensions), Tensilica white paper, www.tensilica.com.

Queue- and Wire-based Input and Output Ports Allow Processors to Achieve RTL-Like I/O Speeds on SOCs, Tensilica white paper, www.tensilica.com.

Rowen, C., *Low Power is the New "Killer App"*, Globalpress 2006, March 28, 2006.

Rowen, C. and Leibson, S., *Engineering the Complex SOC*, Prentice-Hall, 2004.

Think Outside the Bus: High-Speed I/O Alternatives for Inter-Processor Communications on SOCs, Tensilica white paper, www.tensilica.com.

Xtensa LX Microprocessor Data Book, Tensilica, Inc, August 2004, www.tensilica.com.

Xtensa 6 Microprocessor Data Book, Tensilica, Inc, October 2005, www.tensilica.com.

CHAPTER 15

The Future of SOC Design

Cost [of design] is the greatest threat
to continuation of the semiconductor roadmap…
Today, many design technology gaps are crises.
—ITRS:2005

We have to have major technology shifts every
10 to 12 years to keep up with silicon
—Gary Smith, Gartner Dataquest

In March, 2006, researchers at the National Nano Fab Center located at the Korea Advanced Institute of Science and Technology (KAIST) announced successful fabrication of an experimental FinFET with a gate length of 3 nm. (A finFET is a field-effect transistor with a 3D gate structure that surrounds the device's two vertical channels on three sides.) According to the KAIST development team, achieving this milestone means that we can expect Moore's law to be enforceable until at least the year 2026 without resorting to new and exotic non-silicon structures such as carbon nanotubes.

The expanding ability to imprint many billions of tiny FETs on a chip drives system designers to find efficient, cost-effective ways to use all those FETs in their system-on-chip (SOC) designs. According to the ITRS:2005 report, many significant challenges must be overcome if the industry doesn't want to continue losing its ability to fully exploit the transistor bounty it enjoys under Moore's law. This final chapter explores the identified pitfalls that await the SOC-design community and then discusses how ideas discussed in previous chapters can help avert those pitfalls.

15.1 GRAND CHALLENGES AND DISASTER SCENARIOS

SOC designers have become increasingly unable to fully exploit all of the transistors available to them starting with the 90-nm process node in 2003–2004, as shown in Figure 15.1. The figure incorporates data presented by Gary Smith of Gartner Dataquest at DAC 2003 that is based on ITRS survey data and shows that silicon-manufacturing capacity and

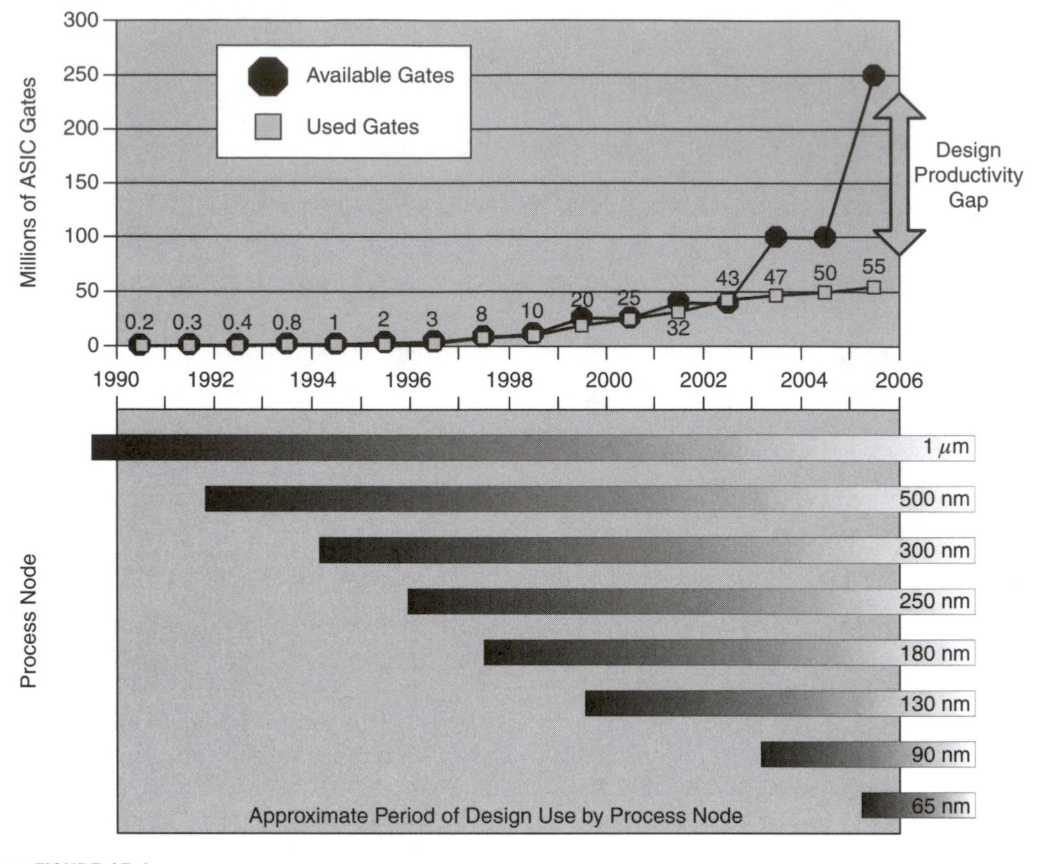

▪ **FIGURE 15.1**

Per-chip silicon capacity and the SOC-design team's ability to exploit that capacity tracked well, until the 90 nm node.

SOC-design capacity have tracked well since the widespread adoption of hardware-description languages (HDLs) in the early 1990s during the 1-micron era. HDLs first appeared years before, in the early 1980s, but did not become popular until gate- and transistor-level schematic drafting—the well-established IC-design method of the day—could no longer handle chip complexities of several hundred thousand gates. As Figure 15.1 shows, IC gate counts reached this level in the early 1990s.

As IC gate counts grew, thanks to continuous improvements and revolutionary developments in IC manufacturing, IC-design teams got bigger to keep pace with the rising design load. However, Figure 15.1 shows that the current HDL-based design style seems to be reaching a plateau at around 50 million gates, just as the schematic-drafting design style reached its descriptive limits at a few hundred thousand gates. The HDL-based design style alone apparently cannot efficiently manage the complexity of larger chips. Figure 15.1 clearly shows that the gap between what can be designed and what can be manufactured is now increasing at a rapid rate.

The ITRS:2005 report minces no words about the state of SOC design:

Today, many design technology gaps are crises.

The ITRS:2005 report identifies five "grand challenges" related to SOC design:

1. Productivity
2. Power
3. Manufacturability
4. Interference
5. Reliability

Failure to adequately address all of these challenges would be disastrous for the electronics industry because it would curtail the rapid advancements that the commercial, consumer, and financial markets have come to expect from semiconductor advances and semiconductor vendors. Thus it's possible to interpret failure to address the five grand challenges as five disaster scenarios for the electronics industry.

15.1.1 SOC Disaster Scenario 1: Insufficient Productivity

Each new IC process node doubles the number of available transistors. To keep up, SOC-design teams must also become at least twice as productive with each new process generation. A major part of the required productivity improvement must come from design reuse. Originally, "reuse" meant using a hardware block that had been designed earlier by someone working for the same company. The market for purchased IP blocks didn't exist at that time because chips were too small to exploit large digital blocks and the required small blocks were relatively easy to design. As chips have gotten much larger, with millions of available gates, the market for large blocks in the form of purchased IP has grown.

In fact, the IP market for SOCs looks very much like the board-level market for LSI chips in the late 1970s and 1980s. Back then, systems engineers had become proficient in developing complex (for the era) digital circuits from basic 7400-series TTL building-block chips. However, some standard functions such as microprocessors, interrupt and DMA controllers, and UARTs became available as LSI chips and were the more cost-effective choice.

Similarly, early SOC designers tended to code everything in HDL (the SOC equivalent of TTL chips). Even then, some large-block IP—especially the microprocessor—was almost always purchased. However, large portions of many SOC designs are still hand built and hand verified due to inertia and engineers' natural tendencies to stick with traditional design styles (whatever's traditional at the time).

Some of the major design-productivity obstacles noted in the ITRS:2005 report are:

1. *Verification.* No single factor looms larger in the design-productivity disaster scenario. Already, more than half of an SOC-design team's resources are devoted to verification and the problem is getting worse, not better.

2. *Reliable and predictable silicon implementation fabrics.* Today's ad hoc approach to system design makes increasingly complex systems hard to analyze. As a result, design teams can miss performance goals by factors of 2x, 4x, or more.

3. *Embedded software design.* Ad hoc system-design techniques cause design teams to sprinkle dissimilar processor architectures (such as general-purpose processors and DSPs (digital signal processors) across the SOC fabric. Software teams familiar with programming one processor architecture become inefficient when trying to program a completely different architecture.

4. *Organizational hypergrowth.* The temptation is to grow the number of people on the development team as the SOC-design task becomes larger and more complex. This management style was thoroughly debunked for large software projects way back in 1975 when Frederick Brooks published *The Mythical Man Month* (Brooks law: "Adding manpower to a late software project makes it later.") and Figure 15.1 clearly shows that this project-management strategy is no longer effective for SOC design as well.

15.1.2 SOC Disaster Scenario 2: Excessive System Power Dissipation

Since the introduction and widespread adoption of the 130-nm process node, classical device scaling no longer delivers the same large drops in power and energy to compensate for the increased number of available transistors. At the same time, increasingly ambitious SOC performance goals have emphasized the use of faster and faster clock rates at the expense of power dissipation. The net result is ever-higher SOC energy consumption in an era that increasingly values end-product performance characteristics such as battery-life, talk time, and compactness. All of these valued end-product characteristics are at odds with increased SOC energy consumption.

The trends are clear. Disaster looms. SOCs that dissipate more power and energy require larger, more expensive, and often noisier cooling systems and bigger power supplies or batteries. Contemporary system-design styles waste far too much energy:

- Large, global buses force bus masters to dissipate more energy than needed for data transfers.

- Piling tasks onto a multitasked processor core drives processor power dissipation in the wrong direction.
- Large sections of an SOC that could be quiescent during long periods are often clocked anyway because the design team didn't have time to determine when each SOC block could be safely put into a sleep mode.

In the name of reigning in device cost, system-design styles save transistors (which are cheap and only get cheaper with each new process node) while wasting design-team hours, system power, and overall energy (which are not cheap). Labor rates and battery development are not on a Moore's-law curve. Efforts to adhere to Moore's law grow on-chip transistor counts by approximately 40–50% per year while engineering productivity has historically grown at about half that rate (according to ITRS reports) and numerous estimates place constant-volume battery-capacity improvement at only 3–9% per year. In other words, semiconductor device counts double every two years or less (Moore's law), battery capacity doubles about every decade (a rule of thumb that's glibly been dubbed "Eveready's law"), and engineering productivity improves at a rate that falls somewhere in between.

15.1.3 SOC Disaster Scenario 3: Loss of Manufacturability

The current SOC-design style assumes a perfect chip. Flawed chips are simply thrown away, a decades-old practice. Memory and FPGA vendors have long designed redundant circuits in the form of extra rows and columns into their products to boost manufacturing yields. Nanometer silicon manufacturing can no longer provide the luxury of perfect chips to SOC designers.

One of the wonders of the semiconductor industry is that it has pushed manufacturing yields high enough to shield application-specific integrated circuit (ASIC) and SOC designers from the need to develop manufacturable systems for their chip designs. As system designers move to higher design-abstraction levels and become further removed from the "bare metal" circuitry of their systems, they need to devote energy toward developing system design styles that produce manufacturable designs. The classic approach to developing fault-tolerant systems is through the addition of redundant components, at the circuit level and at higher levels.

Yield is not the only manufacturability issue. Device-test time and test costs are another. Testability has always been the system designer's unloved stepchild. In general, designers wish to design new systems, not develop ways to test them. As a result, designers of semiconductor-test equipment have had to scramble to develop increasingly innovative ways of testing larger and more complex SOCs without letting test times become completely unreasonable.

Future step functions in potential SOC logic complexity brought on by each new semiconductor process node will be accompanied by a similar, disastrous increase in SOC test times unless the system designer's basic approach to design for testability (DFT) changes. This warning cry is now decades old, so it's unlikely that system designers will adopt DFT methods en masse unless the DFT methods dovetail nicely with system designers' main interest: designing shiny new and ever more powerful systems.

15.1.4 SOC Disaster Scenario 4: Excessive Signal Interference

Long on-chip interconnects, high-frequency signals, sharp clock edges, global clocking, and global buses are all part of the system-design status quo. The result of this design style is growing electrical and magnetic interference between on-chip signals. Physical design rules for wire spacing and inductance management have previously sufficed to keep this problem at bay. Nanometer silicon substantially increases this challenge.

Long buses interconnecting everything on a chip must go. Bus hierarchies are also in peril. Increased wire density makes victim signal wires even more susceptible to inductively and capacitively coupled noise. Repeaters and slew-rate control methods that have controlled signal-interference problems in the past are increasingly ineffective and must be augmented with more systemic design solutions. Even without inter-signal interference, fast signals experience signal-integrity problems—including delay, attenuation, reflection, and edge deterioration—while traversing long wires on nanometer SOCs, which are additional reasons why long wires must go.

15.1.5 SOC Disaster Scenario 5: Deteriorating Chip Reliability

All SOC designs can fail due to manufacturing defects. Nanometer ICs are more susceptible to manufacturing errors due to their finer geometries. Flaws are best caught on the manufacturing floor during testing because the repair cost of a chip failure is much higher after the chips are in the field. If properly designed, the same redundancy added to ICs to counter-manufacturing defects can serve double duty as a way to patch failures in fielded systems.

15.2 AVOIDING THE SOC DISASTER SCENARIOS

Progress in IC development will be seriously curtailed if any of the disaster scenarios discussed in the previous section come to pass. Consequently, much current research is being devoted to creating new design techniques and manufacturing processes that will forestall these scenarios. To varying extents, new SOC-design techniques, such as the ones advocated in this book, can help to avoid all of these disaster scenarios.

15.2.1 Avoiding Disaster Scenario 1: Insufficient Productivity

Certainly, new SOC-design techniques can effectively bridge the productivity gap between the number of transistors available on nanometer silicon and the number of transistors that designers can effectively use (as shown in Figure 15.1). Often, discussions of system-design styles involve the phrase "moving to a higher level of abstraction" without fully defining that new abstraction level. To some, that new abstraction level is a new, system-level design language such as UML (the Unified Modeling Language). System Verilog and SystemC are also contenders.

System design is closely linked with high-level modeling. After design, SOCs must be modeled through simulations because the actual hardware cannot be prototyped economically. Consequently, any new system-design style must dovetail with a system-modeling language. Although the idea of a more rigorous system-level design methodology has been given lip service for more than a decade, SOC designers continue to jump into the RTL (register transfer level) implementation level using a textual system specification as a starting point. In the software domain, many design activities start with hacking some C code to rough out a system despite abundant research that points to a better way.

As Grant Martin observed at the *Electronic Design Process Workshop* held in 2002, system-level design as practiced today has many of the attributes of a cult. Many advocates of system design at the "cutting-edge" veer toward arcane notations and obscure mathematical jargon. The field has many "gurus and acolytes, robes and sandals, weird incantations and self-flagellating rituals ...," which system-level design appears to share with cult religions. (When he gave this presentation, Martin was in the office of the CTO at Cadence Design Systems, working at Cadence Berkeley Labs. He subsequently left Cadence and joined Tensilica as Chief Scientist.)

Martin, using some of the same religious trappings he attributed to system-level design, predicted six future possibilities for high-level modeling and system-level design styles. These possibilities are:

1. *The cult:* High-level modeling and system-level design continue to be the preserve of a few gurus and acolytes. Ritual replaces results.
2. *The niche:* High-level modeling and system-level design succeed in niche design domains, as they have in the past.
3. *Platform-based design:* Almost all electronics product design will be based on developing derivatives of complex fixed and reconfigurable, application-oriented platform chips.
4. *Return to hardware:* System-design styles move away from software and reconfigurable logic to more efficient hardware implementations.

5. *Growth of software:* System design becomes largely a matter of developing software variations for a few fixed hardware platforms.

6. *True system-level design:* Designers really do move up in abstraction to system-level descriptions. Design involves the building, modification, and reuse of high-level models in an overall system-design methodology with appropriate implementation design flows.

All of these design styles will likely be used in the future, to a greater or lesser extent. The design style that best addresses the design-productivity problem while keeping other design factors in line will be the most successful. In addition to designer productivity and the closely associated design costs, other design drivers include project schedule, system power dissipation, and SOC manufacturing cost. Each SOC project weights each of these drivers differently so no single design style is likely to completely prevail over the others.

This book advocates a system-design style that uses elements of styles 3, 5, and 6. The emphasis on the use of processors to implement SOC blocks whenever possible recognizes that many algorithms used in 21st-century SOCs are first developed in HLLs (high-level languages, especially C or C++) and debugged on inexpensive PCs. The structures of these algorithms reflect this development style; these software algorithms are defined by the associated C or C++ code and their design is heavily influenced by the HLL-based design style. Embedded processors can directly execute this algorithmic code. Recasting the algorithms in hardware requires a far more complex translation. Thus the processor-centric system-design style maximizes designer productivity when software algorithms already exist.

There is a cost for the processor-centric design style. Processor-based blocks will usually require more transistors than non-programmable block designs, when used to implement only one algorithm. That is the engineering tradeoff. Processor-centric system design emphasizes designer productivity (because the design team's available hours are few and expensive) over transistor count (because transistors are plentiful and cheap).

15.2.2 Avoiding Disaster Scenario 2: Excessive System Power Dissipation

A processor-centric SOC-design style directly addresses issues related to system power dissipation. Current ad hoc system-design styles waste power in many ways. Design efforts directed at saving transistors through processor multitasking and similar schemes drive on-chip clock rates ever higher, which drastically increases power dissipation. This design style, which evolved when transistors were expensive and chip power levels were low, is no longer valid in the era of nanometer ICs. The use of one or two multitasked, high-speed processors in an SOC also emphasizes the use of large, highly capacitive, global buses which also dissipate excessive power.

The processor-centric design style advocated in this book emphasizes the extensive exploitation of a task's inherent parallelism and concurrency both inside of each processor—to keep processor clock rates low—and through the use of multiple processor cores—to reduce or eliminate multitasking, which drives processor clock rates up.

However, simply running each on-chip processor at the lowest possible clock rate doesn't define the full extent of the processor-centric design style's ability to reduce system power. Properly designed processor cores use extensive, fine-grained clock gating to minimize operating power on a cycle-by-cycle basis. Extensive clock gating is made possible by the processor's fundamental operation. Processors execute instructions. Each instruction can be modeled and its functional needs mapped for each stage in the processor's pipeline. The result is detailed knowledge of clocking requirements throughout the processor for every possible machine state.

Such knowledge is very difficult or impossible to obtain for manually coded RTL hardware and many design teams do not bother. Automated tools to insert clock gating into manually coded RTL hardware depend on exhaustive test benches to exercise the hardware in every conceivable situation. Creating such test benches is manually intensive work. Again, many design teams do not bother to perform this manual work because the schedule will not permit it.

In addition, it is easy to put a properly designed processor into a low-power sleep state and it's similarly easy to awaken the processor when needed. This feature is especially handy when algorithms need to run intermittently. When the algorithm needs to run, the processor is brought out of the sleep state to run it. When the task is complete, the processor can go back to sleep.

Use of many processors to implement on-chip tasks also reduces the need for large, global buses because on-chip communications can more closely follow the actual communications needs of the collected on-chip tasks. Reducing the size and number of large, global buses also reduces SOC power dissipation.

15.2.3 Avoiding Disaster Scenario 3: Loss of Manufacturability

The semiconductor industry has a tried and true approach to dealing with manufacturability issues: make the chip as perfect as possible and bridge any remaining gap between real and desired device yields with redundant structures. Memory and FPGA vendors have successfully used this strategy for many years. They have long put redundant, fuse- and laser-programmable rows and columns into their device designs to maximize yields for a controlled increase in silicon cost.

SOC designers have been shielded from the need to consider redundancy in their designs. No longer. Fabrication defects will be a fact of life for all SOC designers at nanoscale-processing nodes. A processor-centric system-design style based on application-tailored processor cores that retain general-purpose utility can provide high-level redundancy because many tasks can run on any number of similar processors and need not be tied to a particular processor.

The easiest way to provide such system-level redundancy is through a processor array with redundant interconnections. This is essentially the definition of SOCs designed using network-on-chip (NoC) interconnection schemes. The Cisco SPP (silicon packet processor), discussed in Chapter 1, is a good example of such a system-design approach. Cisco's SPP incorporates 192 packet-processing elements, each based on a general-purpose, 32-bit processor core. Four of these packet-processing elements (2%) are treated as spare processors and are used for yield enhancement.

Manufacturability also includes testability, which is also in crisis. For years, semiconductor test engineers and device testers have been tasked with testing increasingly complex devices. Testers can only inject test vectors and extract result vectors from available I/O pins. This situation severely limits test I/O bandwidth. Skyrocketing SOC complexity is leaving tester and test engineer capabilities in the dust.

One way to sidestep the testability crisis is to let the device itself take over much of the on-chip testing. Again, a processor-centric system-level design style can help. The many on-chip processors that have been placed on the chip to perform operational tasks can also serve as BIST (built-in self-test) blocks that can test many portions of the SOC in parallel. Processors can serve as very flexible BIST controllers yet they require no incremental silicon. Thus the on-chip processors bring the same high level of operational concurrency to testing that they bring to normal device operation with essentially no hardware cost. The processor-centric system-design style reverses the long-term trend toward ever increasing SOC test times.

15.2.4 Avoiding Disaster Scenario 4: Excessive Signal Interference

The three largest causes of on-chip signal interference are:

1. Close wire proximity caused by nanometer design rules.
2. High coupling factors caused by long adjacent wire runs.
3. Increased coupling caused by high clock frequencies and sharp signal edges.

A processor-centric system-level design style attacks causes 2 and 3. Assigning tasks to processors creates task islands and reduces the need for large global buses that convey large numbers of signals across long stretches of the SOC. Task-specific processor tailoring reduces the need for high-operating clock frequencies.

15.2.5 Avoiding Disaster Scenario 5: Deteriorating Chip Reliability

Adaptive SOC designs can repair themselves. On-chip processors are good candidates for running adaptive-configuration algorithms that can reroute signals and functions if an SOC suffers a partial failure in the field. The same redundant processors placed on an SOC to boost manufacturing yield can be used to fix field failures with the right sort of system design.

15.3 SYSTEM-LEVEL DESIGN CHALLENGES

A processor-centric, system-level design approach has the potential to address the five disaster scenarios, but that doesn't mean that all of the associated problems are solved and the solutions are well in hand. Many challenges remain.

Of the many remaining challenges, perhaps the most important is transforming the art of SOC design into a more rigorous, engineering discipline while retaining the freedom to innovate architecturally. Some academic research and corporate effort has gone into an attempt to reduce SOC architectural design to a systematic, almost push-button approach. This book does not advocate such a reductionist approach to system-level design. Instead, this book advocates something more akin to the architectural design used in civil engineering, which permits broad freedom in conceiving initial architectures but then supplies rigor in developing the engineering to implement the architectural concepts.

To illustrate this idea, consider a small subset of civil engineering: bridge design. If civil engineers, who have been practicing for thousands of years, had developed a push-button approach to bridge design, there would likely be only one basic type of bridge—a conceptual bridge FPGA if you like. However, there isn't just one basic bridge type; there are several:

1. Girder (beam)
2. Truss
3. Rigid-frame
4. Arch
5. Cable-stayed
6. Suspension

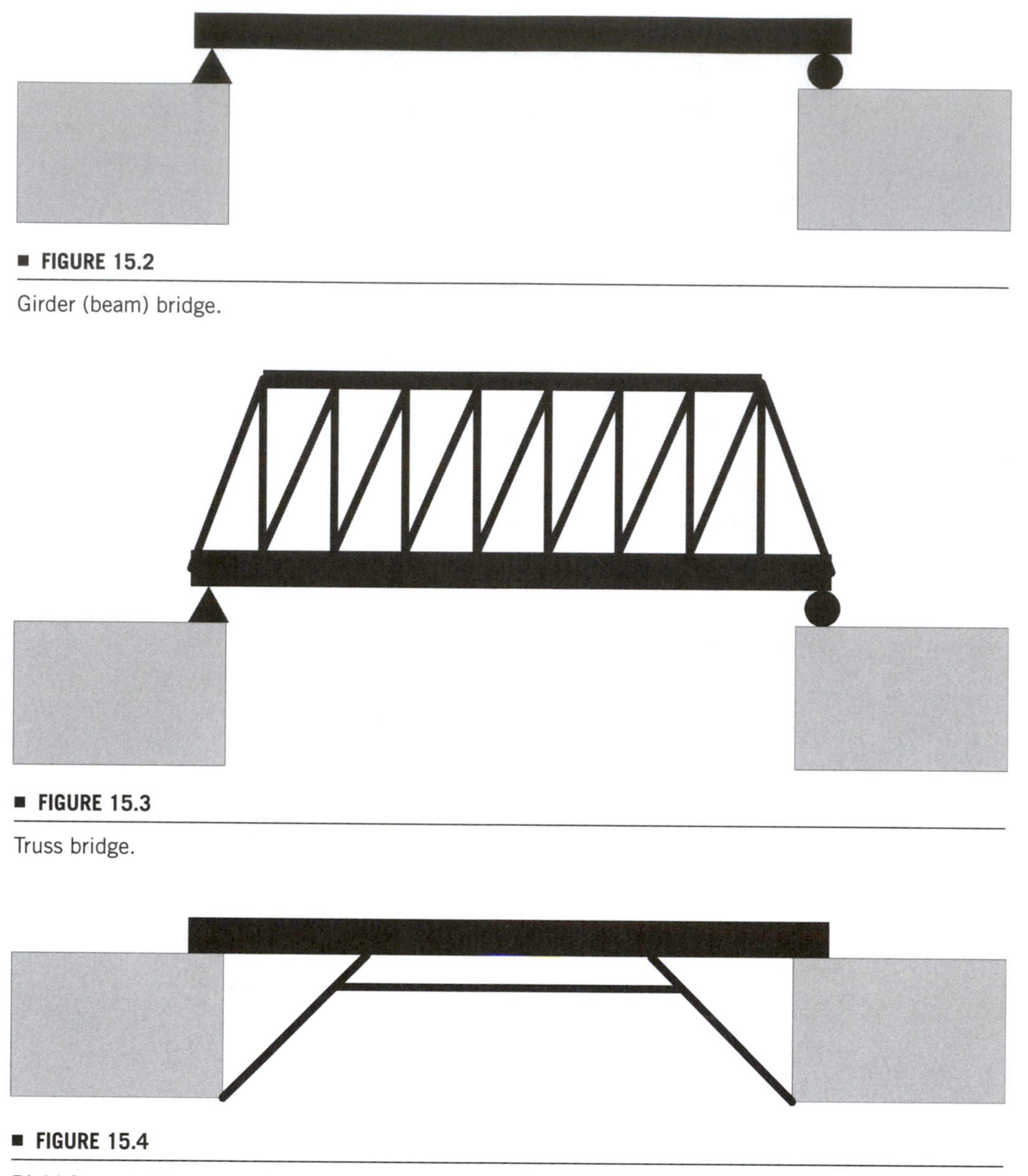

■ **FIGURE 15.2**

Girder (beam) bridge.

■ **FIGURE 15.3**

Truss bridge.

■ **FIGURE 15.4**

Rigid-frame bridge.

Figures 15.2–15.7 illustrate these six bridge types and Table 15.1 lists some of the bridge types and the largest example of each listed type.

Bridge architects also combine the basic bridge types to create compound bridges. For example, the main span of San Francisco's Golden Gate Bridge is a suspension bridge but the side spans consist of arch and truss bridge sections.

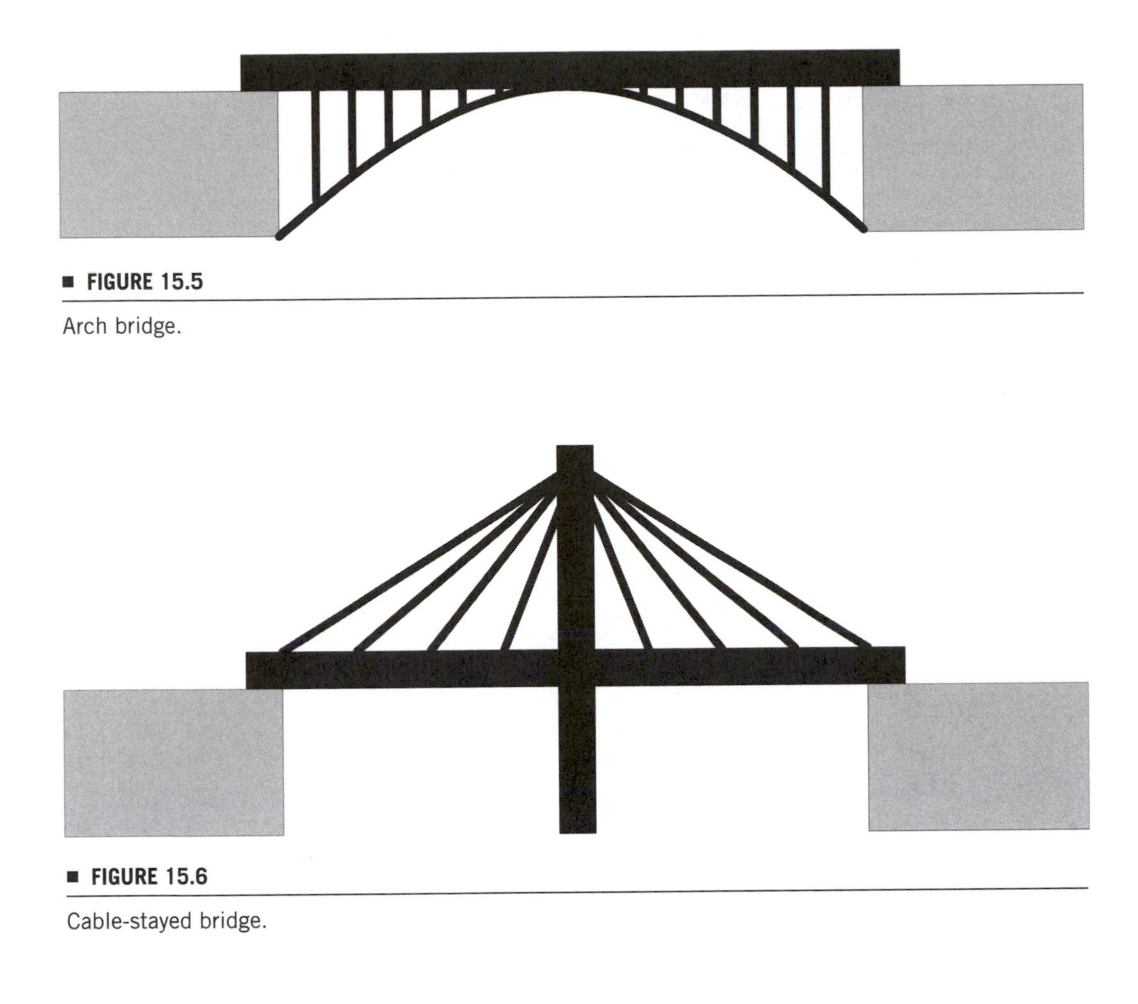

■ **FIGURE 15.5**

Arch bridge.

■ **FIGURE 15.6**

Cable-stayed bridge.

■ **FIGURE 15.7**

Suspension bridge.

Table 15.1 ■ Some basic bridge types and span lengths

Bridge type	Typical span length (m)	World's longest	Length of world's longest (m)
Girder (beam)	10–200	Ponte Costa e Silva, Brazil	700
Truss	40–500	Pont de Quebec, Canada	863
Arch	40–150	New River Gorge Bridge, USA	924
Cable-stayed	110–480	Tatara Bridge, Japan	1480
Suspension	40–1000+	Akashi Kaikyo Bridge, Japan	3911

15.4 THE FUTURE DIRECTION OF SOC DESIGN

Engineers have been designing and building bridges for thousands of years. During those centuries, they have developed these several basic bridge types and many combinations of the basic types. Several different bridge types can often solve a specific bridging problem, so there's no push-button method for selecting one bridge type over another except in situations where there's one obvious solution to a bridging problem. Instead, bridge engineers have developed ways of generating different alternatives by combining the basic bridge types and then comparing the alternatives to pick the "best" one, where "best" includes a wide range of criteria including quantitative (span length, carrying capacity, cost, construction time, etc.) and qualitative (bridge aesthetics, era-specific and location-specific materials limitations, building-site construction restrictions, etc.).

This book advocates a similar approach to SOC architecture and design. SOC designers need rigorous ways of evaluating alternative architectures and rigorous methods for directly implementing the selected architecture without the need to manually translate and implement complex, human-readable specifications.

If you accept the ITRS:2005 prediction that future SOC designs will incorporate tens and then hundreds of processing engines, then you must ponder the following questions:

- How will all of those processing engines be designed?
- Will the processing engines be programmable or fixed-function?
- How will the processing engines be controlled?
- How will all of those processing engines communicate with each other?

The current state of the art in system design cannot answer these questions because only a relatively few people have designed and worked with such complex heterogeneous processing systems. Automated design tools

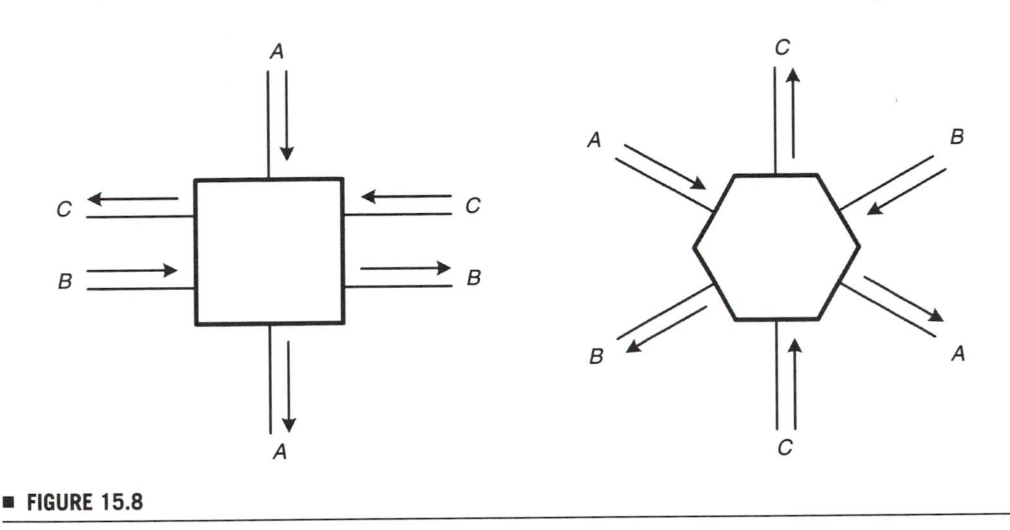

■ **FIGURE 15.8**

Inner-product-step processors for systolic processing.

do not yet exist to deal with such complexity. However, people have thought about such systems and have developed some highly parallel-processing systems. The art of developing such systems still needs to be reduced to an engineering regimen. The following sections briefly discuss some of the most interesting experiments in highly parallel computing that might benefit from SOC implementations.

15.5 SYSTOLIC PROCESSING

Some of the earliest work in highly parallel systems was done by HT Kung and CE Leiserson at Carnegie Mellon University. As early as 1979, Kung had become an early advocate for a parallel-processing style he dubbed "systolic processing." Kung and Leiserson were attempting to develop a general methodology for mapping high-level algorithmic computations onto processor arrays. They envisioned the processor array as resembling an automotive assembly line, with processors in the array working on the same data at different times and the entire array working on many data blocks simultaneously. A key element of the systolic-processing concept was the idea that data—in the form of partial results—could flow among the processors in a systolic array at multiple speeds and in multiple directions.

The systolic-processing array was modular, reconfigurable, and scalable. Kung and Leiserson suggested several types of systolic-processing arrays based on two fundamental types of processors, the rectangular and hexagonal inner-product-step processors. Figure 15.8 shows the inputs and outputs of the inner-product-step processors.

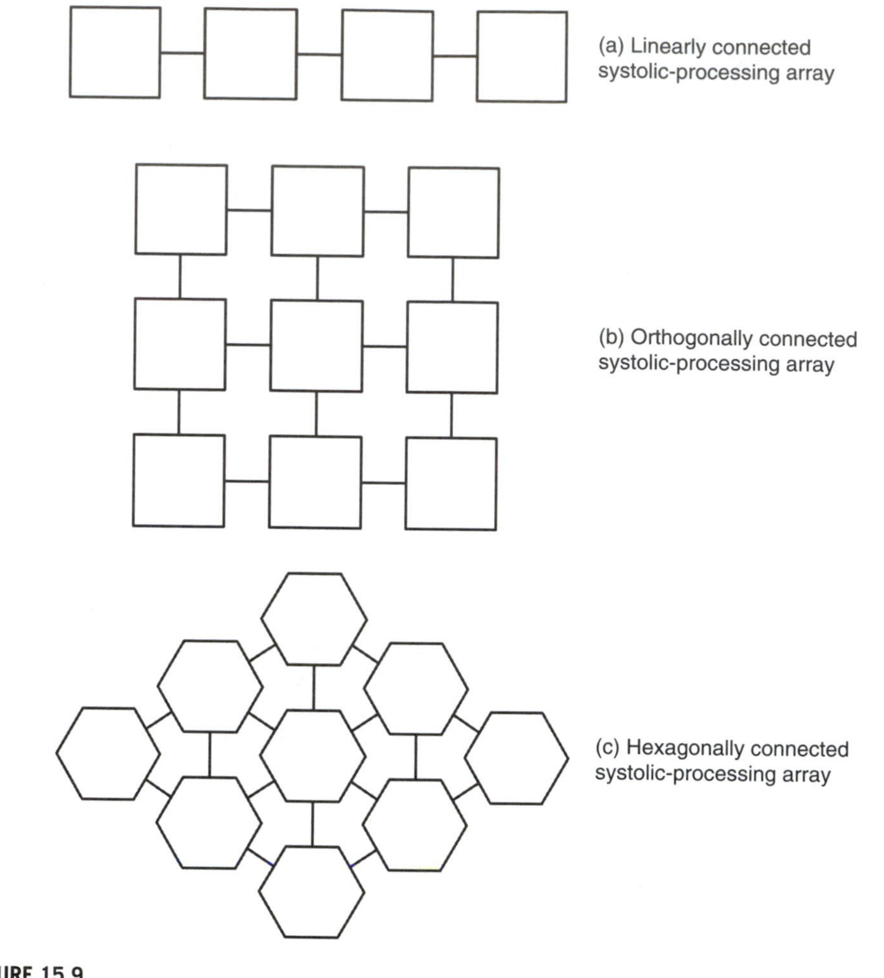

▪ **FIGURE 15.9**

Linear, orthogonal, and hexagonal systolic-processing arrays.

Kung and Leiserson named these processors "inner-product-step processors" because they envisioned the processors executing portions of algorithms at the complexity level of an inner-product computation for matrix arithmetic. That was the complexity level achievable with VLSI semiconductor lithography in 1979. Nanometer silicon allows much larger, more capable, more general-purpose processors to be used today.

With their expanded I/O abilities, the rectangular and hexagonal inner-product processors could be connected in different types of arrays including linear, orthogonal, and hexagonal arrays (shown in Figure 15.9). Note that the I/O configurations on these processors do not resemble the bus structures of conventional microprocessor cores. However, they closely

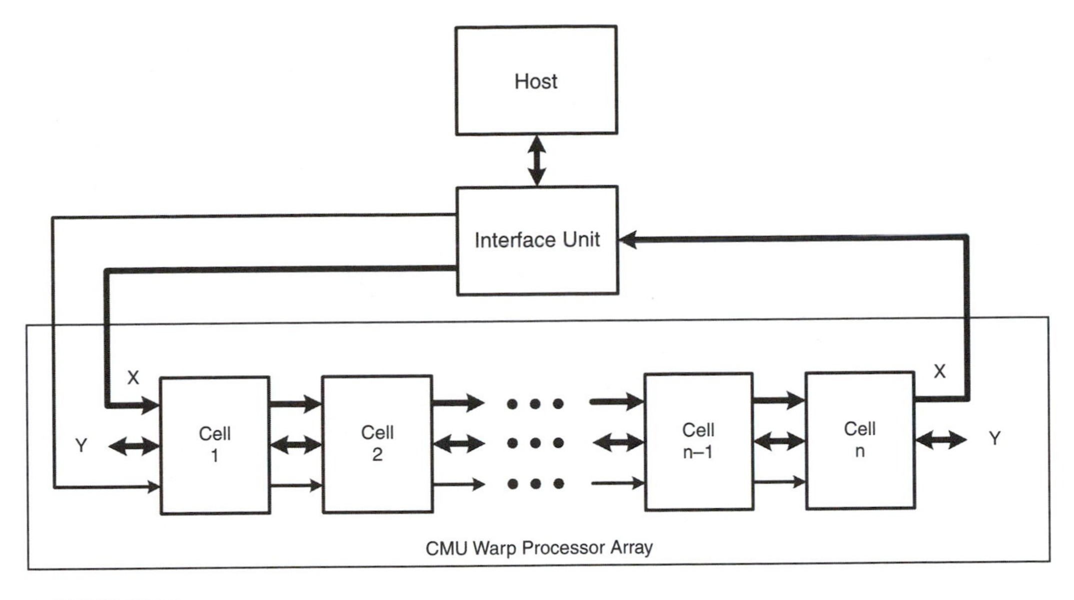

■ **FIGURE 15.10**

The CMU Warp processor developed by HT Kung operated a linear array of floating-point processing cells controlled by a host computer through an interface unit.

resemble the configurations made possible by multiple TIE (Tensilica Instruction Extension) queue interfaces available on some Xtensa and Diamond processor cores.

Kung developed a systolic-processing architecture called WARP at CMU. The CMU Warp processor used a linear array of floating-point computational cells controlled by a host computer through an interface unit, as shown in Figure 15.10.

A 2-cell Warp prototype became operational in 1985 and contractor General Electric delivered a 10-cell pre-production unit to CMU in 1986. One of Kung's graduate students, Monica Lam, developed a parallelizing compiler for the Warp architecture. (Lam eventually became a professor in the Computer Science department at Stanford University. Starting in 1998, she took a sabbatical from Stanford to help found Tensilica. She worked on and supervised work on the HLL compiler and other software tools for the earliest versions of the configurable Xtensa processor.)

The CMU Warp machine was successfully used in several applications including:

- Stereo vision
- Magnetic resonance imaging (MRI)
- Synthetic aperture radar (SAR)
- Radar signal processing

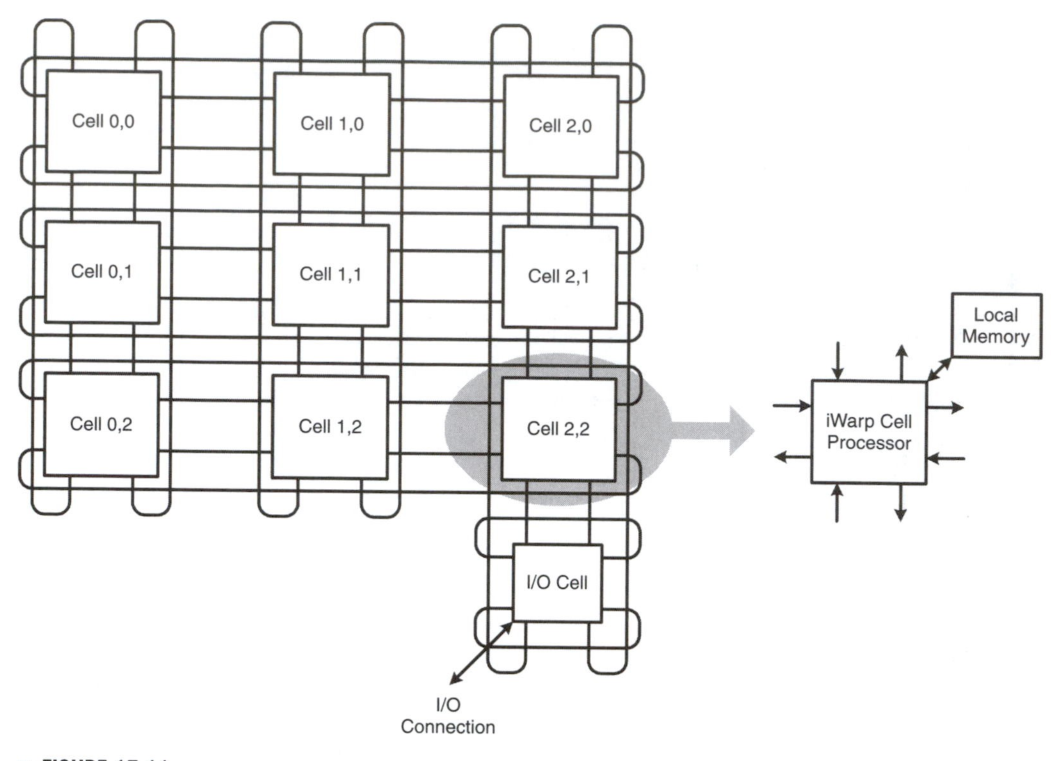

FIGURE 15.11

The Intel/CMU iWarp architecture combined iWarp RISC/VLIW processors into a 2D torus mesh.

- Air pollution monitoring
- Ground motion monitoring.

Intel engaged with CMU to develop a more advanced, integrated version of the Warp architecture dubbed iWarp. The project was cosponsored by DARPA (the US Defense Advanced Research Projects Administration) and the resulting chip, first manufactured in late 1989, implemented a 20-MHz, 32-bit-RISC/96-bit-VLIW (very long instruction word) iWarp cell processor using 650,000 transistors in 1-μm CMOS process technology. Each iWarp cell included eight 40-Mbyte/sec buses for inter-processor communications and a 64-bit, 160-Mbyte/sec interface to the processor's local memory. These iWarp cell processors were packaged into a system using a 2D torus-mesh architecture, as shown in Figure 15.11. The first 64-cell iWarp systems (initially priced at $494,950.00) became operational in 1990 and the architecture was designed to support systems containing as many as 1024 iWarp cells. Intel marketed iWarp systems through its Supercomputing System Division, which was formed in 1992, but the company terminated marketing for the product by 1995. Intel eventually stopped building iWarp systems entirely.

Interest in systolic processing seems to have curtailed with the advent of cluster computing. However, nanometer silicon can easily support large arrays of general-purpose and application-tailored processor cores, so it may well be time to resurrect the idea and see where a new SOC-based renaissance in systolic computation can go.

15.6 CLUSTER COMPUTING AND NoCs

Inexpensive PCs with Ethernet ports have made it possible to build relatively inexpensive cluster computers. A roomful of PCs networked together using the common TCP/IP LAN protocol can form a very powerful, massively parallel computing system. Such clusters of hundreds or thousands of PCs often appear in listings of the fastest computers in the world. The first Google server was just such a PC cluster, built from a collection of bare PC motherboards literally piled one atop the other and jammed into one 19-inch, rack-mount cabinet.

SOCs can also implement computing clusters. However, the relatively microcosmic nature of on-chip computing clusters allows for simpler, lighter networking protocols than TCP/IP. Many such NoC protocols are under development and NoCs are a hot academic research topic in the early 21st century. Much of the learning associated with harnessing PC-based computer clusters can be applied to processor clusters incorporated onto chips.

15.7 THE RESEARCH ACCELERATOR FOR MULTIPLE PROCESSORS PROJECT

During the 32nd International Symposium on Computer Architecture (ISCA) held in Madison, Wisconsin in June 2005, some of the attendees talking together in the hallways between sessions came to several significant conclusions:

1. Most computer architects believe that microprocessor cores will be used in large numbers (hundreds or even thousands) in future chip designs. (This conclusion echoes ITRS:2005.)
2. Straightforward [ad hoc] approaches to the architecture of multiple-core processor systems are adequate for small designs (2 to 4 cores) but little formal research has been conducted on rigorous ways to build, program, or manage large systems containing 64 to 1024 processors.
3. Current compilers, operating systems, and MP hardware architectures are not ready to handle hundreds or thousands of processors.

4. The computer-architecture research community lacks the basic infrastructure tools required to carry out such MP research.

These realizations led to the gestation of the RAMP (research accelerator for multiple processors) project, which has broad university and industry support. The purpose of the RAMP project is to develop a standardized FPGA-based hardware platform that allows researchers to experiment with processor arrays containing as many as 1024 processors. Design and development problems with such arrays that might be invisible for 32- or even 128-processor arrays can become significant for 1024-processor arrays. The RAMP hardware will consist of several boards, each carrying several large Xilinx FPGAs.

Although other methods of simulating parallel-processing systems exist, they suffer from the following problems (which RAMP will attempt to solve):

- *Slowness.* Software simulations of processors are relatively slow. RAMP systems will have high performance by comparison by emulating processors and processor interconnect in FPGA hardware.
- *Target inflexibility.* Simulators based on fixed-core processors emulate processors with different ISAs slowly. The RAMP fabric can more easily model processors with different ISAs (which is common with configurable processors such as the Xtensa processor core).
- *System-level model inaccuracy.* RAMP is designed to emulate large processor arrays "in their full glory." Many other MP simulations trade off simulation speed for model accuracy.
- *Scalability.* RAMP can be expanded with additional FPGAs and FPGA boards using a standardized inter-board communications link.
- *Unbalanced computation and communication.* Clustered computers (as an example) employ layered, bit-serial communication channels between processors. These channels are relatively slow. The RAMP communication channels, contained in the FPGA matrix, can be wide and fast.

A few of the research problems that could be studied using RAMP hardware include:

- Development and optimization of thread-scheduling and data-allocation/migration algorithms for large MP systems.
- Development and evaluation of MP dataflow architectures.
- Evaluation of NoC architectures for large MP systems.

- Development and evaluation of fault-tolerant MP hardware and software.
- Testing the effectiveness of task-specialized processors in MP arrays.

Although PC-based computer clusters can also be used to study some of these problems, clusters cost far more per processor and the inter-processor communications between computers in a cluster are much slower than communications between processors in a closely coupled FPGA matrix.

15.8 21st-CENTURY SOC DESIGN

You can hold one of two opinions about the future of SOC design:

1. Moore's law has run its course due to ceilings on operating frequency and power dissipation. The game is over.
2. The future for SOCs is in massively parallel-processing systems.

Which of these two futures you foresee depends, in part, on how much systems-development work you think there is left to do. If you think the world will continue to be operated by a pointer on a screen that is manipulated by a mouse or by cheap infrared remote controls that have a few dozen rubbery buttons, then there really isn't much left to develop. In actuality, several really tough problems remain to be conquered by electronic systems. An even dozen of these challenges include:

1. Continuous speech recognition with fluent, properly intoned, real-time translation into any chosen language.
2. Google-like searching of photos and video using advanced image and pattern recognition.
3. Natural-sounding text-to-speech synthesis from printed or electronic text.
4. Semi-autonomous and fully autonomous vehicle control—available at consumer-level prices—that protect the unwary traveler from other drivers who are operating their vehicles while chemically impaired, dangerously distracted, or just plain inept.
5. Robotic assistants that can walk or climb stairs and (figuratively) chew gum at the same time without falling over and denting the floor.
6. Rapid, low-cost, diagnostic DNA sequencing performed in same amount of time it currently takes to print a roll of film, fabricate a pair of eyeglasses, or run an MRI (about an hour).

7. Continuous, reliable, high-bandwidth communications via a pervasive global Internet, no matter where you are on or above the planet.
8. Intelligent prosthetics studded with finely controlled actuators and exquisitely sensitive sensors that come far closer to replacing or repairing a person's damaged arm, leg, eyes, ears, etc.
9. Automated security scanners at the airport that can tell the difference between a dangerous hunting knife and a pair of fingernail scissors in someone's carry-on bag and give the appropriate alarm without human assistance.
10. Hands-free medical identification, administration, and billing systems that aren't mistaken 50% of the time, yet still maintain patient privacy.
11. Universal, unobtrusive digital-rights-management systems that allow you to enjoy all of the entertainment and information content you are legally entitled to use in a variety of playback devices without the need to have an engineering degree.
12. Immersive education systems that can engage and teach a population increasingly drugged by entertainment overload.

Meeting all of these challenges will require orders-of-magnitude improvements in computing and processing power.

Time to start working.

BIBLIOGRAPHY

Annaratone, M., Arnould, E., Gross, T., Kung, H.T., Lam, M. and Menzilcioglu, O., "Warp Architecture and Implementation," *Proceedings of the 13th Annual Symposium on Computer Architecture*, Tokyo, Japan, June 1986, pp. 346–356.

Arvind, et al, *RAMP: Research Accelerator for Multiple Processors—A Community Vision for a Shared Experimental Parallel HW/SW Platform*, Technical Report UCB/CSD-05-1412, September 2005.

Brooks, F.P., *The Mythical Man-Month: Essays on Software Engineering, 20th Anniversary Edition*, Addison-Wesley, 1995.

Ganssle, J., "Subtract software costs by adding CPUs," *Embedded Systems Design*, May 2005, pp. 16–25.

International Technology Roadmap for Semiconductors: 2005 (ITRS:2005), http://public.itrs.net/.

Johnson, R.C., "Defects dodged at nanoscale," *Electronic Engineering Times*, March 13, 2006, p. 42.

Kung, H.T., "Why Systolic Architectures?," *IEEE Computer*, January 1982, pp. 37–45.

Lam, M.S., *A Systolic Array Optimizing Compiler*, Kluwer Academic Publishers, 1989.

Martin, G., "The Future of High-Level Modelling and System Level Design: Some Possible Methodology Scenarios," *Electronic Design Process Workshop*, Monterey, April 2002, Session 1.1.

Martin, G., "The Reality of System Design Today: Do Theory and Practice Meet?," *Proceedings of the Third International Conference on Application of Concurrency to System Design (ACSD'03)*, p. 3.

Matsuo Bridge Company, www.matsuo-bridge.co.jp/english/bridges/index.shtm

Peterson, C., Sutton, J., and Wiley, P., "iWarp: A 100-MOPS, LIW Microprocessor for Multicomputers," *IEEE Micro*, June 1991, pp. 26–87.

Rhines, W.C., "Design in the Nanometer Age," keynote speech, DVcon, February 2006.

Smith, G., "The Crisis of Complexity," Gartner Dataquest presentation, DAC 2003.

Young, H.S., "Scientists develop 3 nanometer transistors," Korea Herald, March 15, 2006, www.koreaherald.co.kr

Index

RECEIVED
OCT - 6 2006
ENGINEERING LIBRARY